Advances in Hybrid Supercapacitors: Materials, Devices, Models, Systems, and Applications

Advances in Hybrid Supercapacitors: Materials, Devices, Models, Systems, and Applications

Guest Editors

Xianzhong Sun
Changzhou Yuan
Xiong Zhang

MDPI

Basel • Beijing • Wuhan • Barcelona • Belgrade • Novi Sad • Cluj • Manchester

Guest Editors
Xianzhong Sun
Institute of Electrical Engineering
Chinese Academy of Sciences
Beijing
China

Changzhou Yuan
School of Materials Science & Engineering
University of Jinan
Jinan
China

Xiong Zhang
Institute of Electrical Engineering
Chinese Academy of Sciences
Beijing
China

Editorial Office
MDPI AG
Grosspeteranlage 5
4052 Basel, Switzerland

This is a reprint of the Special Issue, published open access by the journal *Batteries* (ISSN 2313-0105), freely accessible at: https://www.mdpi.com/journal/batteries/special_issues/HW62QKP950.

For citation purposes, cite each article independently as indicated on the article page online and as indicated below:

Lastname, A.A.; Lastname, B.B. Article Title. *Journal Name* **Year**, *Volume Number*, Page Range.

ISBN 978-3-7258-6256-6 (Hbk)
ISBN 978-3-7258-6257-3 (PDF)
https://doi.org/10.3390/books978-3-7258-6257-3

Contents

About the Editors

Xianzhong Sun

Xianzhong Sun earned his B.S. in Mechanical Engineering from Wuhan Institute of Technology (1999), followed by an M.S. in Materials Science and Engineering from Zhejiang University (2006), and a Ph.D. in Materials Science and Engineering from Tsinghua University (2011). He joined the Institute of Electrical Engineering, Chinese Academy of Sciences (IEE CAS) in 2011. His current research interests are in the field of novel electrochemical energy storage devices, including electrical double-layer capacitors (EDLCs), lithium-ion capacitors (LICs), lithium-ion battery-type capacitors (LIBCs), and the supercapacitive swing adsorption (SSA) of carbon dioxide. He has conducted extensive and systematic studies in the field of LICs, covering efficient pre-lithiation techniques for negative electrodes, cell fabrication technology, modeling and simulation, as well as thermal behavior and lifetime prediction of LIC cells. He has published more than 100 peer-reviewed papers in leading journals such as *Advanced Materials, Advanced Functional Materials, Green Energy and Intelligent Transportation, Journal of Power Sources, Journal of Energy Storage, Electrochimica Acta,* and *Batteries,* among others. He has been granted over 30 Chinese invention patents and has made significant contributions to the commercialization of lithium-ion capacitors. Together with his collaborators, he co-authored the book *Lithium-Ion Capacitors* (Science Press, 2021), providing an in-depth and comprehensive overview of LICs, covering the working principles, cathode and anode materials, electrolytes, pre-lithiation technologies for the negative electrode, characterization techniques, cell manufacturing methods, and system integration technologies, as well as application demonstrations of LICs.

Changzhou Yuan

Changzhou Yuan is a distinguished professor of Taishan Scholar in the School of Materials Science and Engineering, University of Jinan, P. R. China. He received his Ph.D. Degree from Nanjing University of Aeronautics and Astronautics in 2009. He then worked as a post-doctoral fellow at Nanyang Technological University, as a visiting scholar at Georgia Institute of Technology, and as a full professor at Anhui University of Technology, before joining the University of Jinan in 2017. He has been listed among the most Highly Cited Researchers (in Materials Science/Cross-Field) by Clarivate Analysis; Most Cited Chinese Researchers (in Materials Science) by Elsevier; Best Materials Science Scientists and the World's Top 2% Scientists by Stanford University/Elsevier; and Highly Ranked Scholars (0.05% or better) by ScholarGPS. His current research focuses on the design and synthesis of micro-/nanomaterials for electrochemical energy-related applications, including supercapacitors, Li/Na/K-ion batteries, Li-S batteries, and energy-storage-type photocatalysts. He has published over 200 SCI-indexed papers in peer-reviewed journals, receiving over 17,000 citations and holding an h-index of 78.

Xiong Zhang

Xiong Zhang received his B.S. degree in Applied Chemistry from Central South University in 2003 and his Ph.D. degree in Applied Chemistry from Beijing University of Chemical Technology in 2008. He then joined the Institute of Electrical Engineering, Chinese Academy of Sciences. His research interest focuses on the synthesis of electrode materials for supercapacitors and the applications of lithium-ion capacitors. He has published more than 200 peer-reviewed papers in leading journals such as *Adv. Mater., Angew. Chem. Int. Ed., Adv. Funct. Mater., Energy Storage Mater., Sci. Bull., J. Energy Chem.,* and *Batteries,* among others.

batteries

MDPI

Article

Analysis of the Energy Efficiency of a Hybrid Energy Storage System for an Electric Vehicle

Florin Mariasiu [1,*] **and Edmond A. Kelemen** [2]

[1] Automotive Engineering and Transport Department, Technical University of Cluj-Napoca, 400114 Cluj-Napoca, Romania
[2] TECOSIM Engineering Srl Cluj-Napoca, 400145 Cluj-Napoca, Romania
* Correspondence: florin.mariasiu@auto.utcluj.ro

Abstract: The large-scale introduction of electric vehicles into traffic has appeared as an immediate necessity to reduce the pollution caused by the transport sector. The major problem of replacing propulsion systems based on internal combustion engines with electric ones is the energy storage capacity of batteries, which defines the autonomy of the electric vehicle. Furthermore, considering the high cost of the battery, it is necessary to consider the implementation of command-and-control systems that extend the life of a battery for as long as possible. The topic covered in this article refers to the analysis by modeling and simulation of the efficiency of a hybrid energy storage system (battery–supercapacitor) adapted for an electric vehicle (e-Golf). Based on the simulations carried out, considering that the operating mode corresponds to the WLTP test cycle, the major conclusion was reached that the use of such a system leads to a decrease in energy consumption by 2.95% per 100 km. Simulations of the model were also carried out to obtain the variation in electricity consumption and vehicle autonomy depending on the number of passengers. Electricity consumption if the vehicle is equipped with a hybrid energy storage system increases by 0.67% on average for each passenger (of 75 kg) added and by 0.73% on average if the vehicle is not equipped with supercapacitors. Moreover, the use of the supercapacitor's properties leads to the reduction in the peaks in energy taken/given by the battery with a direct effect on extending its life.

Keywords: electric vehicle; energy source; Li-ion battery; supercapacitor; efficiency; simulation

1. Introduction

The immediate solution to the global energy crisis and environmental problems caused by the pollutant emissions of internal combustion engine vehicles is considered to be the widespread use of electric vehicles. The main barrier in achieving this is related to the autonomous performance of electric vehicles, performance that is strongly (and directly) affected by the energy storage sources used [1,2]. The immediate requirements of the energy storage systems in electric vehicles are that they must simultaneously offer a pecific energy, a high specific power, and a long lifetime, desires that are difficult to fulfill at the present time with the current sources of energy storage. The usage of a hybrid energy storage system (HESS), which combines an energy storage device with a high power density (such as a supercapacitor), is one of the alternatives suggested and examined for this purpose, together with one with a high density of energy, such as a lithium-ion (Li-ion) battery. Li-ion batteries (among all electrochemical energy storage solutions) are the most used because of their high specific energy, but they have low specific power and a short lifetime, characteristics that limit their application [3]. Compared to Li-ion batteries, supercapacitors have a longer lifetime and higher specific power [4]. As a result, the utilization of an HESS, which combines Li-ion batteries with supercapacitors, benefits from both the Li-ion batteries' high specific energy and the supercapacitors' high specific power. Thus, under the specific conditions of their use in electric vehicles, the overloading of Li-ion batteries can be remarkably reduced with the assistance of supercapacitors, and thus the lifetime of the batteries would be significantly extended.

The existence and technological availability on the market of a much more efficient electrical energy storage system is the main obstacle to the development of electric vehicles, which in the future must offer the performance demanded by customers (especially the travel autonomy) at affordable prices. Energy sources must meet a number of conditions to provide high efficiency and good operational characteristics, including high power and energy density, long lifetime, reliability, and maximum efficiency functioning in a wide range of ambient temperatures. Currently, the most used energy storage systems for electric vehicles are those with electrochemical batteries, hydrogen fuel cells, and supercapacitors, but installing and using only one type of energy storage system is energy inefficient. This has led to the need to carry out studies and research in the field to highlight the effects of the hybridization of energy sources from the point of view of increasing the performance of electric vehicles (autonomy, dynamic, and energetic performances).

There has not been as much research in the field of HESSs as carried out for batteries and, mainly, it has been in related to the identification of domains and areas for application, increasing the energy performance of supercapacitors by developing new materials for components, application in the EV structure and analysis of energy performance for different operating situations, control methods and strategies for energy management of hybrid energy storage sources, etc. [5,6]. The obtained results showed the high potential of their implementation and use [7–9].

From the point of view of identifying the fields of implementation of hybrid storage sources, it was highlighted that HESS systems lend themselves mainly to applications in the field of photovoltaic energy (PV) and electric vehicles (EV) [10,11], but also in the field of portable electronics and industrial energy management [12]. Studies and research are continually being carried out to increase the performance of supercapacitors by optimizing the materials used in their construction. Pseudocapacitive materials, including metal oxides such as TiO_2, MnO_2, Mn_3O_4, V_2O_5, and NiO in combination with conductive polymers (polythiophene—PTP, polyaniline—PANI, polypyrrole—PPY) were considered to improve energy performance [13–15]. Research was also carried out to identify simple and low-cost manufacturing methods of various charge storage materials for high-performance supercapacitors [16,17]. The need to use both the electric energy from the battery and the energy stored in supercapacitors has led to research, development, and optimization of various control methods and strategies integrated into the energy management system [18–21]. Regarding the effect of using an HESS system with supercapacitors as the energy source of an EV, the studies carried out considered different particular or standardized functional cycles (NEDC, China bus driving cycle—CBDC, FTP) to analyze parameters such as autonomy, SoC, and SoH [10,18,22].

A technical parameter of supercapacitors not to be neglected is the one given by the difference in the optimal range of temperatures in which supercapacitors can be used compared to Li-ion batteries. Research carried out on this issue showed that the performance related to energy storage offers an important advantage in the use of supercapacitors, i.e., the wide range of functional temperatures: −20 ... +80 °C [7,23]. This can be an advantage in the implementation of HESS battery–supercapacitor systems in electric vehicles, the use of supercapacitors outside the functional temperature limits of Li-ion batteries (with +15 ... +35 °C being the optimal functional range considered [24]) being beneficial for increasing the lifespan of energy sources. The wider range of operating temperatures also makes it possible to facilitate the fast-charging process, a process demanded more and more by electric vehicle users [25]. A major benefit is that the management of hybrid HESS systems is compatible with (and may be utilized with) both air and liquid thermal management systems used for energy sources with batteries [26,27]. One additional advantage of using HESS battery–supercapacitor systems for electric vehicles is that the supercapacitor does not pose risks of ignition and burning in the event of a traffic accident in certain constructive configurations (depending on the type of electrolyte: aqueous and inorganic solid, for example) [7].

Because supercapacitors are expensive to purchase, despite all the benefits listed above, the HESS battery–supercapacitor systems have not been widely adopted or used (the price of a supercapacitor with the same energy capacity as a Li-ion battery being 275% higher [28]). However, research carried out in this direction has shown that this problem is a false one if the entire lifetime of the EV is considered, due to the large difference between the life cycles of supercapacitors compared to Li-ion batteries. A comprehensive study was carried out regarding the economic efficiency of the use of hybrid HESS systems (battery–supercapacitor) implemented in the propulsion system of different electric buses [28]. The main results obtained showed that adding a supercapacitor unit to the battery module increased the price of the energy storage system by approximately 25%, but by extending the battery life, it reduced its operating costs by 10 to 27% (compared to the use of a battery-based energy storage unit). Similarly, the use of a hybrid HESS system for a micro-EV showed that the operating costs over the entire lifetime were lower by 3.76 ... 15.88% than the use of a single power battery [29]. Furthermore, the application of an HESS battery–supercapacitor in stationary UPS (uninterruptible power supply) systems showed that the battery's capacity faded by up to 60% and the amortized cost of the hybrid system was 17.6% less than that of the battery alone [30].

Energy storage is one of the main challenges not only for EV technology but also for the field of renewable energy, due to its intermittent nature. In this study, the focus is on the analysis of the use of hybrid energy storage systems in EVs with the aim of better understanding the immediate advantages and the possibilities of developing new practical concepts which can contribute to solving the energy problems that appear in electric vehicles. The paper presents a study based on computerized numerical analysis (simulation) on the efficiency of hybrid energy storage systems based on the use of supercapacitors in EVs, with the aim of improving the vehicle's autonomy, protecting the battery from extreme operating demands (peak currents), and thus, extending its life. The novelty of the study is in the fact that it was used to identify the performance of an HESS applied to an EV, the WLTP test cycle (which replaced the old test cycles) and the model were validated based on the data obtained through real testing (and provided by the manufacturers) of an e-Golf vehicle.

2. Current Status of Hybrid Energy Storage Systems (HESSs)

It should be noted that in addition to battery–supercapacitor HESSs (systems that present good functional properties and are used in the construction of several EVs), there are other types of hybrid systems that can be used, such as flywheel, compressed air, and superconducting magnetic energy storage systems. Although these energy storage systems are still in different research phases, they can be further developed and applied in the future to EVs and PHEVs (plug-in hybrid electric vehicles) thanks to the features and characteristics that will be presented in the following sections.

2.1. Hybrid Energy Storage System with Flywheel

Rather than using a battery and supercapacitor HESS, the flywheel system uses a kinetic device. In essence, a flywheel created for energy storage is a rotating disc with a very high moment of inertia that is intended to spin at extremely high rates (20,000–50,000 rpm). Electricity can be created by converting stored kinetic energy into electricity, and vice versa. Typically, the motor (or generator) is connected to the steering wheel, though it can also be the rotor itself. In engine mode, the EV simply increases its speed to increase the kinetic energy stored in the steering wheel, and in generator mode, the flywheel of the electric car drives the shaft mechanically. [31]. The major problem related to this hybrid energy harvesting system is that the high-speed flywheels must be insulated in a vacuum to reduce energy losses due to friction, and the bearings used are of the magnetic type for the same purpose [32]. Another disadvantage in its use is the fact that the flywheels are high-speed devices that require insulating casings of special construction to prevent damage caused in cases of mechanical failure. The advantage of flywheel systems is that they have a long lifetime (typically approx. 20 years [33]) and provide large amounts of energy in a very short time for specific applications for large vehicles (lorries, buses, and locomotives).

Figure 1 shows a hybrid electrical energy storage system that includes a flywheel and a battery. The flywheel can be connected to the DC bus via a bidirectional DC/AC converter and a high-speed electric motor/generator. When the vehicle decelerates, energy is captured by the flywheel by driving the electric car in motor mode and the converter in inverter mode. The electric machine runs in generator mode and the converter is operated in rectifier mode when the energy stored in the flywheel is needed for propulsion [33–35].

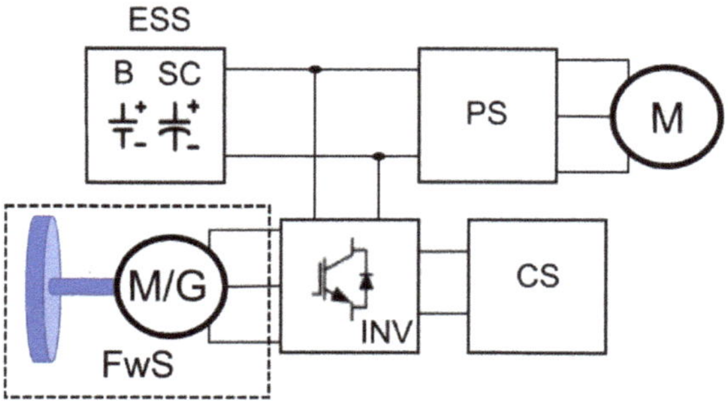

Figure 1. Hybrid energy storage system with flywheel (B—battery, SC—supercapacitor, CS—control system, PS—powertrain system, FwS—flywheel system, M—motor, M/G—motor/generator, INV—inverter).

2.2. Hybrid Energy Storage System with Compressed Air

Hybrid energy storage systems with compressed air may also be future solutions in EV construction. These energy storage options outlast batteries by a significant margin and have an energy efficiency of 75–80% [36]. They feature a reasonably straightforward design and functionality in addition to their high efficiency and long lifespan (Figure 2).

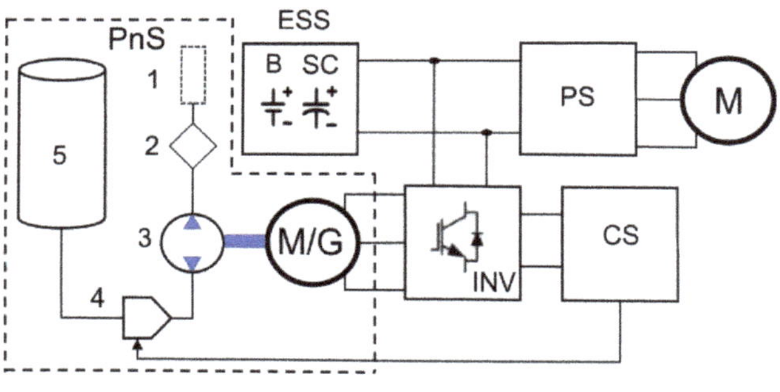

Figure 2. Hybrid energy storage system with compressed air (B—battery, SC—supercapacitor, CS—control system, PS—powertrain system, INV—inverter, PnS—pneumatic system, M—motor, M/G-motor/generator, 1—filter, 2—air dryer, 3—pump, 4—valve, 5—air reservoir).

The DC bus can be used to connect compressed-air-based potential energy storage devices by using a motor/generator and a bidirectional DC/AC–AC/DC converter. The converter is used in inverter mode while the electric machine is used in motor mode during the energy storage process. The compressed air tank's pressure is raised by a pump that the engine powers and functions similarly to a compressor. The converter is operated in rectifier mode and the electric machine is run in generator mode whenever power from the compressed air system is required. In this way, the pneumatic pump is driven by the compressed air that is released from the tank through the valve and drives the electric machine that works in generator mode and provides the necessary energy. The electronic power converter and compressed air tank valve are controlled according to the mode of operation and the amount of power to be delivered to/from the compressed air storage system.

2.3. Superconducting Magnetic Energy Storage Systems

Superconducting magnetic energy storage systems (SMESSs) can store electrical energy in the form of magnetic energy. SMESSs are capable of rapidly transferring large amounts of energy during both charging and discharging with over 95% efficiency [37].

The combination of a battery with SMESS in HESS provides a high power density and high energy density without moving mechanical parts [38–40]. SMESSs can replace both batteries and supercapacitors, but the main disadvantage of these systems is that they operate at very low temperatures (electromagnetic coils are used to act as a superconducting material and reduce ohmic losses). For this, the SMESS must be embedded in a cryogenic system whose complexity and refrigeration power reduce the overall efficiency of the system. Cooling can be ensured by evaporating a cryogenic liquid such as helium, nitrogen, or neon [41], but the application of these strategies to vehicles represents a very large limitation of practical applicability. The most practical and cost-effective immediate option is to set up the conditions for the use of high-temperature superconductors, which would enable SMESSs to be used in the construction of electric vehicles. However, high-temperature superconductor material development is still in its infancy [42–44]. The possibility of implementing SMESSs in EVs is presented in Figure 3.

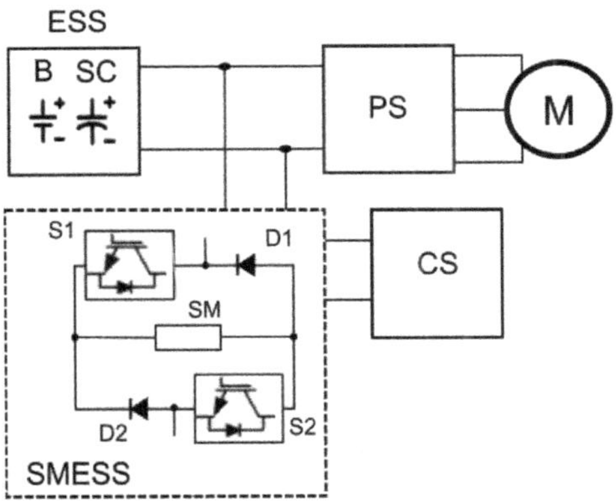

Figure 3. Hybrid energy storage system with SMESS (B—battery, SC—supercapacitor, CS—control system, PS—powertrain system, M—motor, D1 and D2—diodes, S1 and S2—switches).

2.4. Hybrid Energy Storage System with Supercapacitor

Currently, Li-ion batteries are widely used as energy storage systems in EVs due to their high energy density, operational safety, reliability, and increasingly low production cost. However, the peculiarities of their operation and exploitation make frequent charge–discharge cycles directly affect the life of the batteries [3]. By adding more supercapacitors, which will lower the battery's maximum current, battery performance can be increased (Figure 4). Supercapacitors have a long lifespan and a high power density of over 6000 W/kg due to the absence of chemical variations at the electrodes; however, the relatively low energy density of 5 Wh/kg [45] prevents the use of supercapacitors as the primary energy sources in an EV.

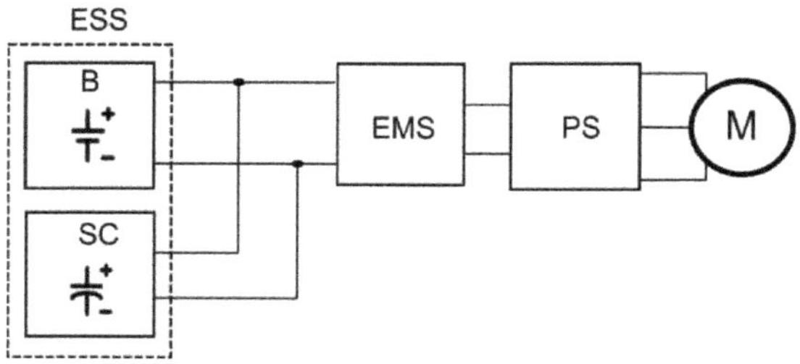

Figure 4. Hybrid energy storage system (battery + supercapacitor) application in EVs (B—battery, SC—supercapacitor, EMS—energy management system, PS—powertrain system, M—motor).

Both electric and hybrid vehicles, as well as fuel cell vehicles, may benefit from hybrid energy storage systems made up of a battery and a supercapacitor because they have high energy and power utilization and long battery and supercapacitor lifespans [46,47]. A battery can store large amounts of energy (of the order of 100 Wh/kg), but it is not suitable for providing a large amount of energy in a very short time, which is due to a low power density [48]. A supercapacitor has a low storage capacity but can deliver a large amount of power (10 kW/kg) [49]. In a hybrid storage system, the battery is used to provide a large (and relatively constant) amount of energy at low loads, while the supercapacitor is used to meet the demands of regenerative braking and acceleration.

Such a scenario also helps to improve battery life, and the combination of battery and supercapacitor also has the advantage of reducing the size and weight of the energy storage system [50,51].

Although significant progress has been made in recent decades in improving the energy performance of batteries, the main problem comes from how they are used. Battery damage can occur even in small electronic devices such as mobile phones and laptops due to sudden power utilization or demand. This condition is permanent in EVs because a variety of circumstances, including traffic, driving habits, the road, unexpected accelerations, etc., produce quick changes in how the battery is used, resulting in a sharp decline in battery performance and life. Due to this, the optimization of energy storage systems in an EV has become vital and the battery–supercapacitor hybrid storage system is one of the systems that can be used in EVs to increase battery life. By using supercapacitors, batteries are protected from high peak currents, which can be harmful to batteries. Additionally, the battery cannot be regularly depleted at a rate fast enough to fulfill the demands of an EV that suddenly consumes power during acceleration. The storage of a high current produced during braking (if a regenerative braking system is present) follows the same rules. The electrolytes may suffer as a result of these significant electrical current variations to and from the battery. The battery life is shortened when this acceleration–braking cycle is repeated (such as when driving in urban agglomerations).

Supercapacitors are electrochemical systems similar to batteries, but the main difference is that they can charge and discharge very quickly. They are good solutions to meet short-term energy needs but cannot be used as the only energy source in electric vehicles (EVs) due to their lower energy density when compared to batteries.

When a vehicle incorporates a hybrid energy storage system, braking energy is stored in both supercapacitors and batteries. The energy from the supercapacitors can be reused to accelerate the vehicle, and the energy stored in the batteries can be used to serve other systems such as the passenger compartment heating, air conditioning, or other electronic devices in the vehicle. Therefore, the total performance of EVs can be significantly impacted by a well-designed hybrid energy storage system with high energy density, extended lifetime, and high power density. Hybrid energy storage systems are mostly used from an economic point of view, for energy recovery from braking in passenger transport vehicles, such as city e-buses [52].

2.5. Classification of Battery–Supercapacitor Hybrid Energy Storage Systems

A battery–supercapacitor hybrid storage system can have a wide range of topologies, and Figure 5 illustrates some fundamental strategies for combining the two energy storage options [53]. In the simplest configuration (Figure 5a), the power distribution between the parallel-connected energy storage units is simply governed by their internal resistance. The hybrid passive storage system works well as an energy source for starting motors but is ineffective as an energy source for electric vehicles (EVs).

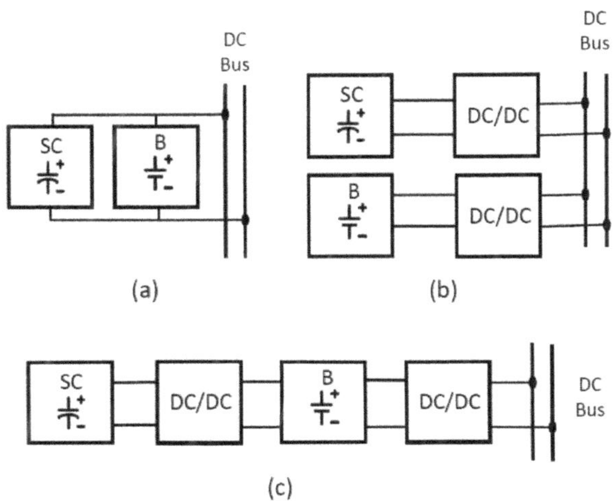

Figure 5. Hybrid energy storage systems (HESS): (**a**) passive, (**b**) active with low-voltage battery and high-voltage supercapacitors, (**c**) active with low-voltage supercapacitor and high-voltage battery.

To optimize the merging of supercapacitors and batteries into a single package, DC/DC converters are needed to manage the power flow between them and the DC bus. Figure 5b represents a hybrid active energy storage system. This topology allows power exchange directly between the source and the supercapacitors without the use of a DC/DC converter. The main power flow from the battery is controlled by a DC/DC converter, which allows the power to be split, but at the same time increases losses due to the converter.

Moreover, such a system imposes restrictions on the acceptable voltage values of the supercapacitors. The output voltage must be higher than the BEMF (back electromotive force) of the electric motor (which necessitates the use of high-voltage supercapacitors [30]). In another topology (Figure 5c), a supercapacitor is connected to the battery through the bidirectional DC/DC converter. The DC/DC converter allows for proper load sharing between two devices, as well as controlling the power flow between them to maintain the proper voltage of the supercapacitors. In general, hybrid energy storage systems (battery–supercapacitor) can be classified based on their connections (Figure 6).

The easiest and most affordable topology is the passive connection of the battery and supercapacitor to the DC bus; the battery and supercapacitor are connected to the DC bus at the same voltage (which is determined by the battery's state of charge value and charge/discharge characteristic). Because of this, EV and PHEV construction typically uses the passive connection.

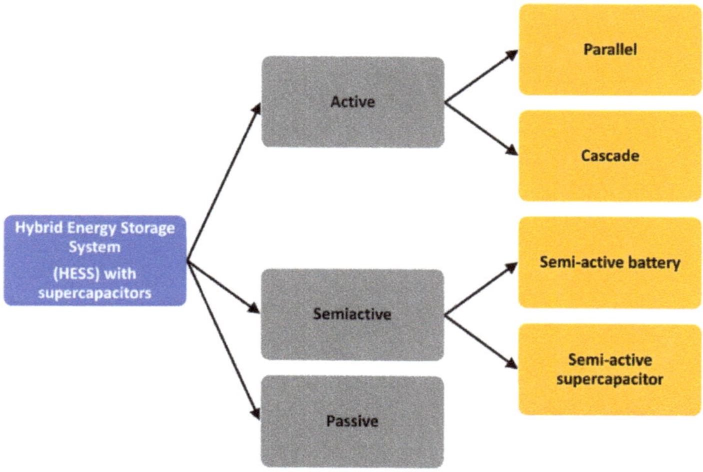

Figure 6. Classification of hybrid energy storage systems with battery and supercapacitors (HESS) (adapted from [22]).

3. Materials and Methods

3.1. Theoretical Considerations on Supercapacitors

Compared to current battery technologies, supercapacitors have a much lower energy density, with the ability to charge and discharge very quickly, and have a high power density. These technical operating characteristics can improve the autonomy of an EV when the supercapacitors and the battery form an HESS. The selection process of a supercapacitor for integration into an HESS is based on the following parameters: voltage, capacitance, energy storage capacity, and energetic losses. The supercapacitor's nominal voltage needs to be as close as possible to the EV's DC bus voltage and satisfy Equation (1) [54]:

$$V_{sc\_max} \leq V_{DCbus\_max} \tag{1}$$

The minimum voltage at which a supercapacitor can operate is closely related to the current I_{c_max} and the power P_0 that the DC/DC converter can handle.

$$V_{sc\_max} \leq \frac{P_0}{I_{c\_max}} \tag{2}$$

The storage capacity of the supercapacitor can be calculated with the relation:

$$E_C = \frac{C_0}{2}\left(V_{sc\_max}^2 - V_{sC_0}^2\right) + \frac{2}{3}K_c\left(V_{sc\_max}^3 - V_{sC_0}^3\right) \tag{3}$$

where V_{sc_max} is the maximum voltage of the supercapacitor, C_0 represents the capacitance of the supercapacitor, V_{sC0} is the initial voltage of the supercapacitor, and K_c is a coefficient that represents the effects of the supercapacitor's diffusion layer. Equations (4) and (5) provide the initial capacitance C_0 for the given coefficient and the braking energy E_B, respectively:

$$C_0 = \frac{2}{\left(V_{sc\_max}^2 - V_{sc\_int}^2\right)}\left[E_B - \frac{2}{3}K_c\left(V_{sc\_max}^3 - V_{sc\_int}^3\right)\right] \tag{4}$$

with V_{sc_int} the intermediate voltage of the supercapacitor.

The braking energy E_B is given by:

$$E_B = \eta_B \int_0^{T_B} P_0(t)dt \tag{5}$$

where η_B is the efficiency of the entire conversion system (including the drive converter, motor, DC/DC converter, and supercapacitor efficiency), the power $P_0(t)$ is the drive shaft power, and T_B is the braking time [55]. The transfer energy (ride-through) E_{RT} available from the supercapacitor is given by Equation (6) and the transfer energy that the supercapacitor package can deliver to the system, when the EV does not accelerate, can be calculated with Equation (7):

$$E_{RT} = \frac{1}{2}C_0\left(V_{sc\_int}^2 - V_{sc\_min}^2\right) + \frac{2}{3}K_c\left(V_{sc\_int}^3 - V_{sc\_min}^3\right) \tag{6}$$

$$E_{RT} = \eta_M \int_0^{T_{RT}} P_0(t)dt \tag{7}$$

where T_{RT} is the transfer time and η_M is the efficiency of the whole system, which depends on the internal resistance of the supercapacitor pack and the power of the drive shaft $P_0(t)$.

To protect the supercapacitors, it is not advised to discharge all of their energy during actual operation. As a result, the maximum permitted discharge energy must be established. Equation (9) can be used to compute the depth of discharge (DoD):

$$DoD = 100\left(1 - \frac{V_{sc\_min}}{V_{sc\_max}}\right) \tag{8}$$

When choosing supercapacitors, it is important to calculate the energy losses of the supercapacitors. Assuming that the supercapacitor is a linear capacitor ($K_C \cong 0$) and neglecting the internal resistance R_{C0}, the charging and discharging currents can be calculated with Equations (9) and (10):

$$i_{c0\_charging} = P_{C0}\left(\frac{C_0}{C_0V_{sc\_min}^2 + 2P_{c0}(t)}\right)^{\frac{1}{2}} \tag{9}$$

$$i_{c0\_discharging} = -|P_{C0}|\left(\frac{C_0}{C_0V_{sc\_0}^2 + 2P_{c0}(t)}\right)^{\frac{1}{2}} \tag{10}$$

where V_{sc_0} is the initial voltage and the charge/discharge power P_{C0} is constant [56]. The power losses of the supercapacitor are calculated using Equations (11) and (12):

$$P_{C\_charging} \cong R_{C0}P_{C0}^2\frac{C_0}{C_0V_{sc0\_min}^2 + 2P_{C0}(t)} \tag{11}$$

$$P_{C\_discharging} \cong R_{C0}P_{C0}^2\frac{C_0}{C_0V_{sc0}^2 + 2P_{C0}(t)} \tag{12}$$

3.2. Modeling and Simulation of the Electric Vehicle

The numerical analysis of the energy efficiency of an HESS for an EV application was carried out by using a numerical model built in the AVL CRUISE software application. The AVL CRUISE application is a complex and dedicated software application for the simulation of different models of propulsion systems for motor vehicles. The initial data, input data, and technical characteristics of the vehicle propulsion system were used to build the numerical model for the simulation process.

For the analysis and study of the theme presented in this article, a virtual EV was built and developed based on the constructive parameters of a Volkswagen (VW) e-Golf electric vehicle. The purpose of the simulations is to highlight the comparative energy performance of the considered energy storage systems (battery system and battery–supercapacitor hybrid system) by analyzing the differences in autonomy and energy consumption. The steps taken to define the computerized simulation process algorithm are the following:

- Creating a new project/version;
- Development of the vehicle model;
- Realization of energy connections;
- Making informational connections;
- Entering the initial technical data into the model;
- Establishing tasks and simulation criteria;
- Establishing the characteristics of the simulation process;
- Running computer simulations;
- Visualization and evaluation of results.

The virtual electric vehicle model of a VW e-Golf was created and developed with the AVL CRUISE software (Figure 7). The constructive data of the vehicle were taken from the technical documentation of the manufacturer.

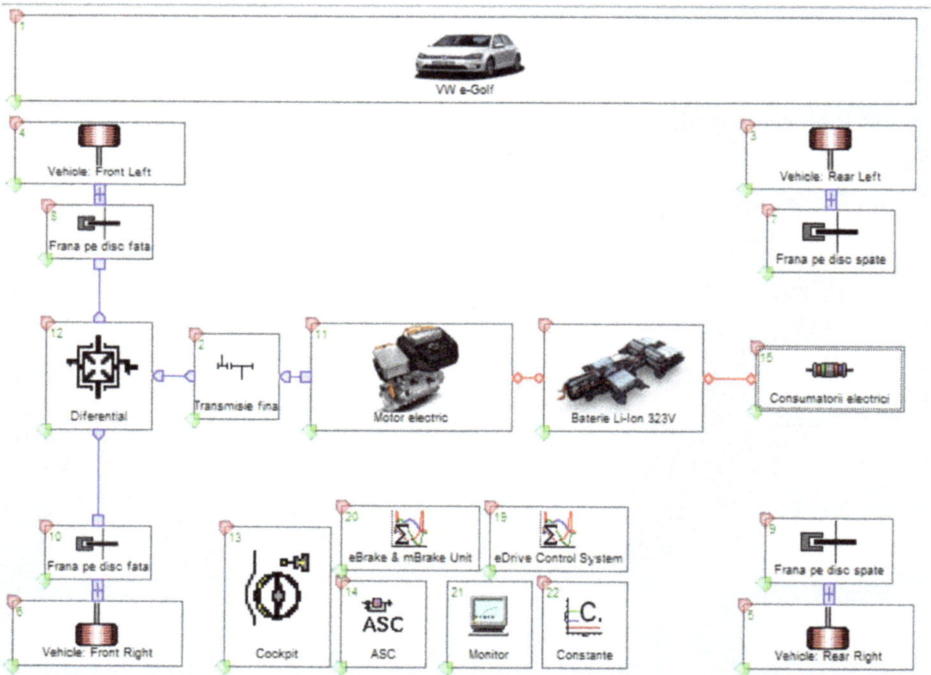

Figure 7. Virtual model of e-Golf electric vehicle without supercapacitor (1: electric vehicle, 2: final transmission, 3: left rear vehicle wheel, 4: left front vehicle wheel, 5: right rear vehicle wheel, 6: right front vehicle wheel, 7: disc brake rear left wheel, 8: disc brake front left wheel, 9: disc brake rear right wheel, 10: disc brake front right wheel, 11: electric motor, 12: differential, 13: passenger, 14: ASC control (anti-slip control), 15: electric consumers, 16: high-voltage electric battery, 17: drive system, 18: braking control system, 19: process monitoring block, 20: model constants).

General information about the electric vehicle's components (1) is provided in Table 1, including constructive parameters, aerodynamic characteristics, overall dimensions, etc.

Table 1. Electric vehicle's component data.

Parameter	Value	Unit
Wheelbase	2629	mm
Curb weight	1615	kg
Total weight	2000	kg
Frontal area	2.5	m^2

The final transmission element (2) is used as a link in the kinematic chain transmission before the differential with the technical data of the considered parameters presented in Table 2.

Table 2. Final transmission component data.

Parameter	Value	Unit
Transmission ratio	6.058	-
Input moment of inertia	0.010	kg·m^2
Output moment of inertia	0.015	kg·m^2
Efficiency	96	%

The wheels (3–6) consider a number of factors and how they affect the running condition (all input data for all vehicle wheels are the same). The data of an e-Golf wheel type 205/55R16 91H can be found in Table 3. Disc brakes (7–10) define the technical characteristics of the braking elements, presented in Table 4.

Table 3. Vehicle's wheel component data.

Parameter	Value	Unit
Moment of inertia	0.1431	kg·m^2
Tire friction coefficient	1	-
Load on wheel	4037.5	N
Static rolling radius/circumference	306.055/1923	mm
Dynamic rolling radius/circumference	308.283/1937	mm

Table 4. Brake component data.

Parameter	Value	Unit
Brake piston surface	1800	mm^2
Friction coefficient	0.25	-
Specific brake factor	1	-
Efficiency	99	%
Moment of inertia	0.02	kg·m^2

In the element e-drive or electric motor (11) the general data of the motor are defined (see Table 5), and in Figure 8 the map of the motor moment is presented.

Table 5. Electric motor component data.

Parameter	Value	Unit
Type	PSM	-
Nominal voltage	323	V
Moment of inertia	1.0×10^{-4}	kg·m^2
Maximum speed (engine)	12,000	1/min

Table 5. *Cont.*

Parameter	Value	Unit
Maximum power (engine)	85	kW
Maximum torque (engine)	270	Nm
Maximum speed (generator)	7000	1/min
Maximum power (generator)	84	kW
Maximum torque (generator)	264	Nm
Efficiency	92.5	%
Initial temperature	20	°C

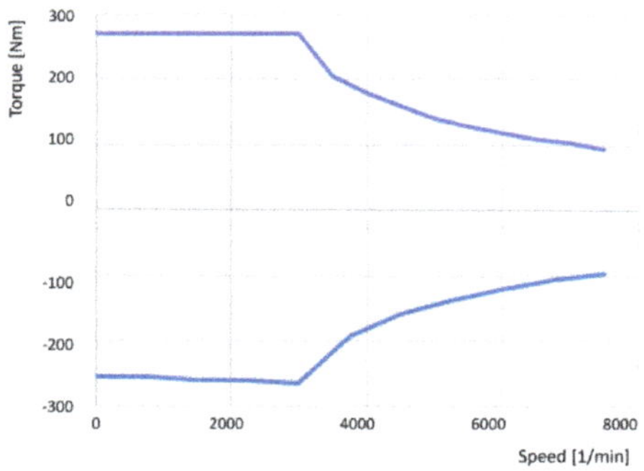

Figure 8. Electric motor torque map.

The differential (12) allows the wheels to rotate at different speeds during the turn and the cockpit component (13) makes the connection between the driver and the vehicle, the connections being made through the data bus connection data bus. The traction force transmission factor to the wheels attached to the running gear is represented by the ASC control (14). All electrical consumers of the vehicle's electrical network are included in the electrical consumers block (15).

The battery component (16) simulates the battery used as a power source for the electric vehicle. The values of the parameters that define the high-voltage Li-ion battery of an e-Golf electric vehicle are presented in Table 6. The battery of the VW e-Golf vehicle is a battery with a 323 V nominal voltage and a weight of 318 kg (represents 21% of the total weight of the unloaded vehicle). The battery is composed of 264 cells of 25 Ah from Panasonic, arranged in 27 modules, having the configuration 88s3p. The technical and functional parameters of a single cell are presented in Table 7.

Table 6. General parameters of e-Golf high-voltage Li-ion battery.

Parameter	Value	Unit
Maximum charge	75	Ah
Nominal voltage	323	V
Energy capacity	24.2	kWh
Maximum voltage	339	V
Minimum voltage	308	V
Initial SoC	95	%
Number of cell/rows	88	-
Number of rows	3	-
Internal resistance (charging/discharging)	0.001	Ω
Operating temperature	33	°C
Specific heat transition	0.4	W/K
Specific heat capacity	795	J/kg·K
Total weight	318	kg

Table 7. Parameters of a single cell.

Parameter	Value	Unit
Maximum charge	25	Ah
Nominal voltage	3.667	V
Maximum voltage	4.2	V
Minimum voltage	3.2	V
Weight	0.724	kg

The drive system (eDrive control system) (17) is a user-defined function used when simulating the ECU (engine control unit) in both engine and generator modes. The brake control system (18) controls the conversion of engine torque to brake pressure. The monitor element (19) monitors selected signals on the data bus connection bus. The constants component (20) allows the definition of constants, which can be used by other components through the data bus.

To carry out the simulations, as well as to exploit the characteristics of the EV, the WLTP test cycle (worldwide harmonized light-duty vehicles test procedure) was considered (Table 8). The WLTP test cycle is a globally harmonized test method for passenger cars and light commercial vehicles and from 1 September 2018, the WLTP testing cycle replaced the new European driving cycle (NEDC). The consumption and CO2 emission values measured according to WLTP are greater than those measured according to NEDC when taking into account the (more) realistic test conditions; for conventional vehicles, a gap of up to 31% was found between the declared and actual fuel consumptions [57].

Table 8. Parameters for the WLTP test cycle [58].

Parameter	Value	Unit
Time	1800	sec
Distance	23.27	km
Average speed	46.5	km/h
Maximum speed	131.3	km/h
Stationary time	12.6	%
Constant functioning	3.7	%
Acceleration	43.8	%
Deceleration	39.9	%
Average positive acceleration	0.41	m/s^2
Maximum positive acceleration	1.67	m/s^2
Average deceleration	−0.45	m/s^2
Minimum deceleration	−1.50	m/s^2

The validation of the model consists in the comparative analysis of the EV e-Golf's autonomy following the simulations and the one provided by the manufacturer (Volkswagen GmbH) of 133.6 km for a 100% charge. For the simulation, the limit values of the battery charge level were chosen to be a maximum of 95% and a minimum of 20%, because a battery will never be charged to 100%, and after reaching the charge level of 20%, the energy performance decreases suddenly. In Figure 9, the charge level (state of charge—SoC) of the battery is represented according to the distance traveled, and under the considered conditions, the autonomy given by the manufacturer would be 113.56 km and the value obtained by the simulation was 114.50 km (error of 0.82%), which was considered to validate the numerical model used to analyze further the energy efficiency of an HESS for an EV using a simulation.

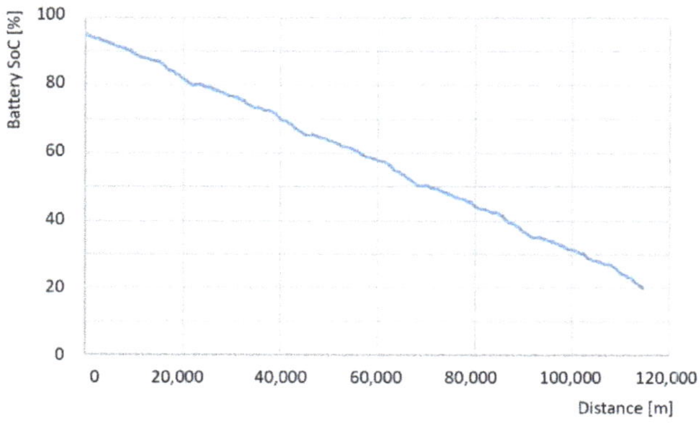

Figure 9. The simulated autonomy of the e-Golf vehicle.

For the supercapacitor and the battery to work harmoniously, the nominal voltage of the supercapacitor and the battery must be as close as possible. On this basis, the technical characteristics of the supercapacitor type BMOD0165 produced by the Maxwell company were chosen for use as the input data of the model, and the numerical model of the VW e-Golf electric vehicle equipped with an HESS (battery—supercapacitor) is represented in Figure 10.

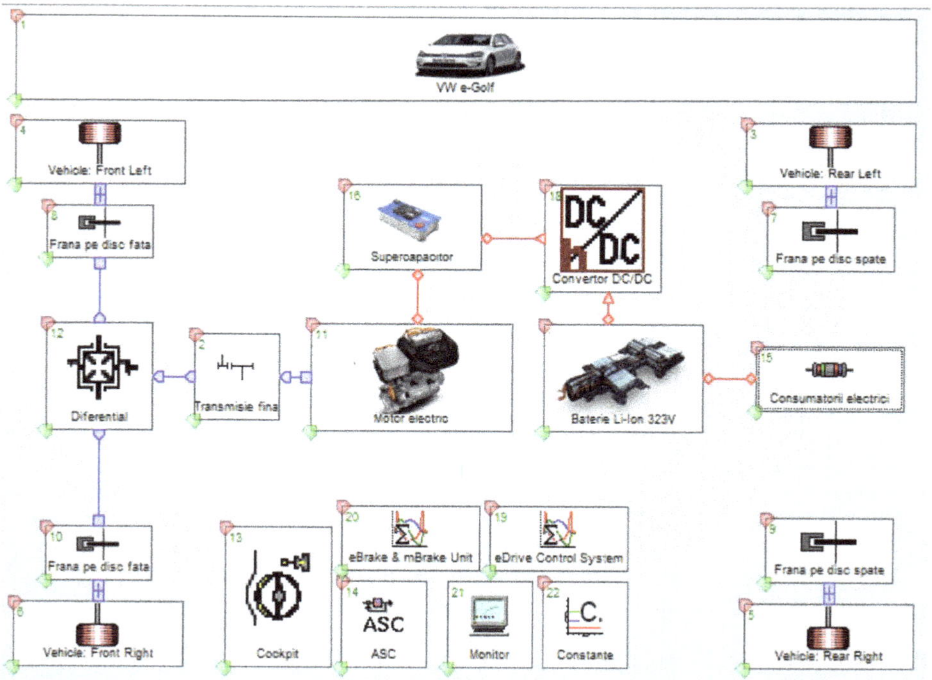

Figure 10. The virtual model of the e-Golf vehicle with a hybrid energy storage system (16—supercapacitor, 18—the DC/DC converter).

In addition to the EV model with a battery, there are two additional constructive elements: the supercapacitor and the DC/DC converter. The supercapacitor (16) is an electrical energy storage device that uses two different mechanisms to store electrical energy and contribute to its total capacity, electrostatic capacitance, based on the electrical double layer, and electrochemical pseudo capacitance. The technical and functional data of the supercapacitor used are presented in Table 9. Since the voltage of the supercapacitor pack must be as close as possible to the battery voltage, the configuration of a module consisting of seven supercapacitors mounted in series was chosen, a configuration that satisfies the requirement of connection with the battery. This results in a total capacitance of 23.57 F and a normal voltage of 340.2 V for the supercapacitor module.

Table 9. Maxwell BMOD0165 supercapacitor data.

Parameter	Value	Unit
Capacitance	165	F
Nominal voltage	48.6	V
Maximum current	1970	A
Maximum voltage	51	V
Minimum voltage	46	V

A DC/DC converter (18) should be implemented for voltage and current control. At the same time, the DC/DC converter can also optimize the motor control to avoid the overvoltage phenomenon, and the specific parameters introduced in the numerical model are presented in Table 10.

Table 10. DC/DC converter component data.

Parameter	Value	Unit
Maximum power	2	kW
Nominal voltage	336.6	V
Maximum voltage	367.2	V
Minimum voltage	255	V

4. Results

Based on the simulations performed on the validated numerical model, the following section will present the results obtained for the virtual EV based on the constructive and functional characteristics of the VW e-Golf vehicle (with and without HESS). The element considered for the comparative analysis was the driving range in the context of an operating regime based on the WLTP test cycle.

In Figure 11, the battery charge level according to the distance traveled when the vehicle is equipped with a hybrid energy storage system is shown. It can be seen that the battery discharges to 20% after traveling 131.3 km, 14.67 km more than the situation where the vehicle is not equipped with a hybrid energy storage system. This difference represents an increase in autonomy of approximately 15%, and Figure 12 more precisely represents the variation in the battery SoC with time.

Figure 12 represents the variation in the battery charge level when the vehicle is equipped with an HESS (red line), and when the vehicle operates only with the battery (blue line). As can be seen in the graph, if the vehicle is equipped with a hybrid system (with battery–supercapacitor), the discharge of the battery is much smoother, which means less charge–discharge variations, which leads to a longer battery life.

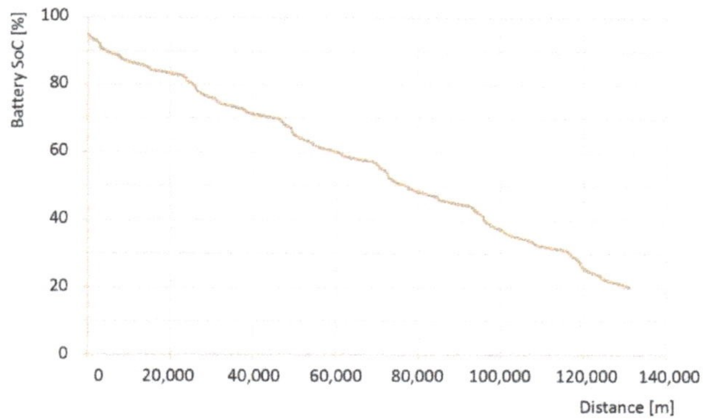

Figure 11. Vehicle autonomy with hybrid energy storage system.

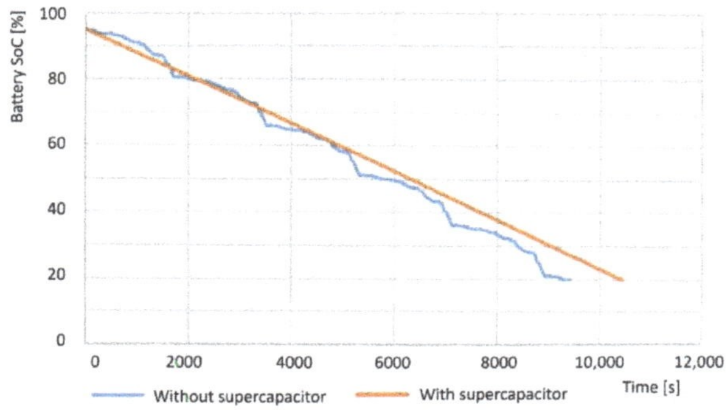

Figure 12. Battery charge level variation with and without supercapacitor.

Figure 13 shows the variation in the electric power of the battery when working in tandem with the supercapacitor pack. To be able to better analyze the variation in the electric power of the battery as a function of time, this graph was made using only one WLTP cycle. The Li-ion battery constantly provides energy for the supercapacitor while it powers the electric motor or absorbs energy from braking and as can be seen from the graph, the electric power of the battery is stable and works at about 2.99 kW. In contrast, the electric power of the battery without a supercapacitor varied between −50 and 30 kW with frequent fluctuations, as seen in Figure 14.

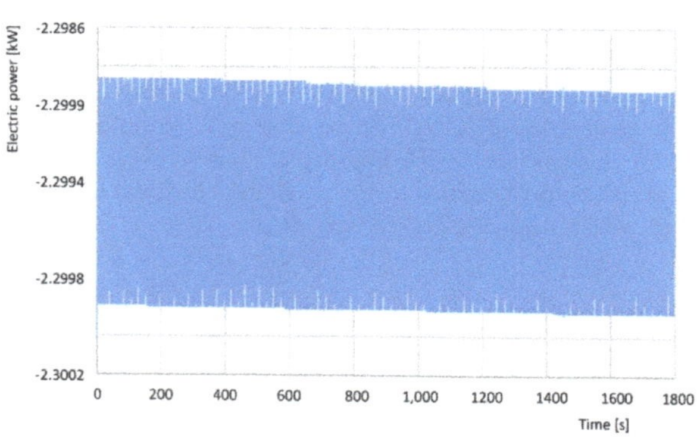

Figure 13. The electric power of the battery with a supercapacitor.

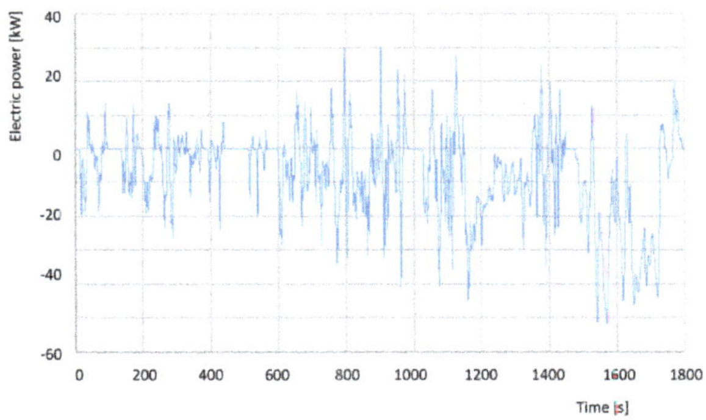

Figure 14. Battery electric power without a supercapacitor.

There is a clear power fluctuation, and the battery is frequently charged and discharged, which inevitably leads to battery degradation. By using a hybrid energy storage system, these power fluctuations are smoothed out (no sudden peaks) and consequently, battery life is extended.

As seen in Figure 15, the variation in the electrical power of the supercapacitor module is very similar to the variation in the electrical power of the battery when there is no HESS. This can be explained by the fact that by using a hybrid energy storage system, the supercapacitor pack takes over most of the battery's load, while the battery is only used to provide the energy needed by the supercapacitor pack. The battery's output energy can be maintained at a constant level with the help of the DC/DC converter, which eliminates significant charge–discharge variations.

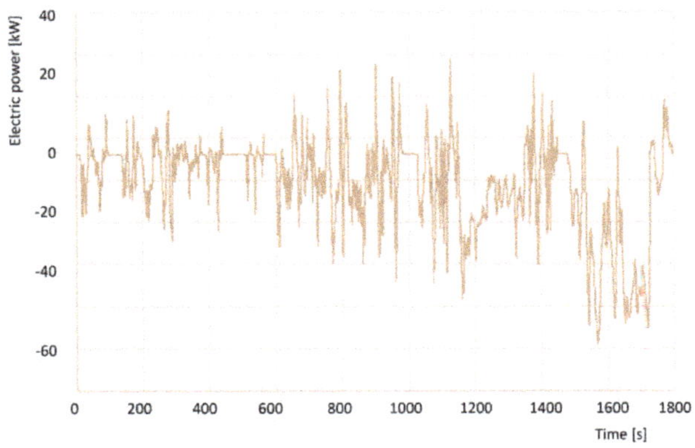

Figure 15. The electrical power of the supercapacitor pack.

Figure 16 represents the variation in the vehicle's electricity consumption when it is equipped with supercapacitors, and when it is not. The energy consumption is lower if the vehicle is equipped with an HESS and the energy consumption values per 100 km are shown in Figures 17 and 18, respectively.

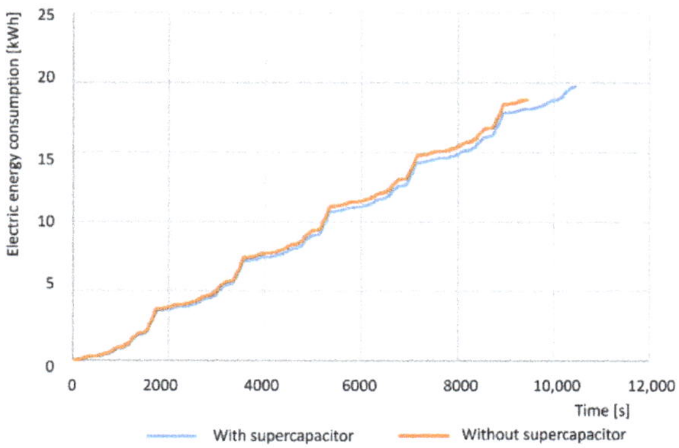

Figure 16. Electricity consumption as function of time.

Electrical_Consumption [kWh]	Distance [m]
15.5407	99999.6
15.5409	100000.0
15.5411	100001.0

Figure 17. Power consumption with SC.

Electrical_Consumption [kWh]	Distance [m]
16.0138	99999.3
16.014	100000.0
16.0142	100001.0

Figure 18. Power consumption without SC.

Based on these values, it was determined that if the vehicle is equipped with supercapacitors, the consumption of electricity decreases by 2.95% per 100 km.

The initial simulations were carried out with only one person on board the vehicle (the driver), weighing 75 kg. Simulations of the model were also carried out to obtain the variation in electricity consumption and vehicle autonomy depending on the number of passengers (with and without an HESS). Figure 19 shows the energy consumption values of the vehicle with and without supercapacitors depending on the number of passengers, and Table 11 calculates the variations in consumption and autonomy depending on the number of passengers. The electricity consumption if the vehicle is equipped with an HESS increases by 0.67% on average for each passenger (of 75 kg) added and by 0.73% on average if the vehicle is not equipped with supercapacitors. From the yield point of autonomy, it decreases by 0.45% on average for each added passenger when the vehicle operates with an HESS and by 0.49% if the vehicle is not equipped with supercapacitors.

Through the statistical analysis of the data in Table 11, there is a strong linear correlation when interpolating the results using statistical analysis. For the energy consumption analysis, R^2 is 0.9990 for the case without SC and 0.9989 for the case with SC, and in the case of the autonomy analysis, R^2 is 1 for the case without SC and 0.9981 for the case with SC. This shows that the use of an HESS with a supercapacitor does not directly (and negatively) influence the performance of the electric vehicle, managing to minimize the effect of its own weight added to the total weight of the vehicle.

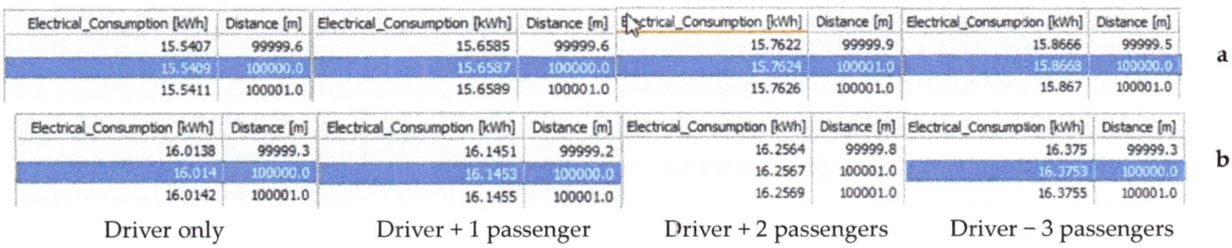

Figure 19. Electricity consumption depends on the number of passengers. (**a**) With supercapacitor; (**b**) without supercapacitor.

Table 11. Variations in energy consumption and autonomy depending on the number of passengers.

Number of Passengers	Energy Consumption (kWh)				Autonomy (km)			
	With SC *	%	Without SC	%	With SC	%	Without SC	%
1	15.5409	-	16.0140	-	131.310	-	114.504	-
2	15.6587	+0.75	16.1453	+0.81	130.719	−0.45 s	113.859	−0.56
3	15.7624	+1.43	16.2567	+1.52	130.128	−0.90	113.352	−1.00
4	15.8668	+2.09	16.3753	+2.26	129.542	−1.34	112.758	−1.52

* SC—supercapacitor.

5. Conclusions

Hybrid energy storage systems (HESSs) have been and are being studied and researched through the lens of the fact that they can increase the energy efficiency of an electric vehicle and, therefore, implicitly the most important parameter, the autonomy traveled for a battery charge.

The large-scale application of such systems faces certain problems related to the high cost of supercapacitors, the thermal behavior being different from that of Li-ion battery cells, high dimensions, and weight, etc. These problems require future studies and research to optimize/solve them.

For the application presented in this article, related to the implementation of an HESS (battery–supercapacitor) for an e-Golf electric vehicle, a virtual model (AVL CRUISE) was created based on its technical, functional, and constructive characteristics. The model was validated based on the data presented by the manufacturer and it was further decided to connect the high-voltage Li-ion battery of the electric vehicle (323 V nominal voltage and 24.2 kWh energy capacity) with a supercapacitor module (340.2 V nominal voltage and 23.57 F capacitance). The results obtained showed that the use of an HESS system leads to a decrease in energy consumption, by 2.95% per 100 km (for WLTP cycle conditions). Simulations of the model were also carried out to obtain the variation in electricity consumption and vehicle autonomy depending on the number of passengers. The electricity consumption if the vehicle is equipped with a hybrid energy storage system increases by 0.67% on average for each passenger (of 75 kg) added and by 0.73% on average if the vehicle is not equipped with supercapacitors.

It should be noted that using a supercapacitor reduces the energy peaks needed by the electric propulsion group (or battery) in the case of abrupt accelerations and decelerations, a common occurrence in urban vehicle operation. This fact has been confirmed by various researchers as directly leading to the extension of the effective life of a battery, which has a direct effect on costs and environmental protection.

A new and interesting direction of research is also given by the results obtained through the prism of the fact that the use of HESS battery–supercapacitor manages to increase the energy efficiency of energy sources by minimizing the effect of the weight added to the hybrid energy source (compared to a conventional battery source). This was also confirmed by recent research carried out on this issue investigating the possibilities and opportunities to use HESS systems in the construction of electric propulsion groups of aircraft [59,60].

Author Contributions: Conceptualization, E.A.K. and F.M.; methodology, E.A.K. and F.M.; software, E.A.K.; validation, E.A.K. and F.M; investigation, E.A.K.; writing—original draft preparation, F.M; writing—review and editing, F.M. All authors have read and agreed to the published version of the manuscript.

Funding: This research received no external funding.

Acknowledgments: This work was possible due to support of AVL List GmbH through university program and with support of TUCN internal grant for publishing program.

Conflicts of Interest: The authors declare no conflict of interest.

References

1. Sepasi, S.; Ghorbani, R.; Liaw, B.Y. A novel on-board state-of-charge estimation method for aged Li-ion batteries based on model adaptive extended Kalman filter. *J. Power Sources* **2014**, *245*, 337–344. [CrossRef]
2. Sepasi, S.; Roose, L.R.; Matsuura, M.M. Extended Kalman Filter with a fuzzy Method for Accurate Battery Pack State of Charge Estimation. *Energies* **2015**, *8*, 5217–5223. [CrossRef]
3. Siang, F.T.; Chee, W.T. A review of energy sources and energy management system in electric vehicles. *Renew. Sustain. Energy Rev.* **2013**, *20*, 82–102.
4. Şahin, M.E.; Blaabjerg, F.; Sangwongwanich, A. A Comprehensive Review on Supercapacitor Applications and Developments. *Energies* **2022**, *15*, 674. [CrossRef]
5. Horn, M.; MacLeod, J.; Liu, M.; Webb, J.; Motta, N. Supercapacitors: A new source of power for electric cars? *Econ. Anal. Policy* **2019**, *61*, 93–103. [CrossRef]
6. Roy, P.K.S.; Bora Karayaka, H.; Yan, Y.; Alqudah, Y. Investigations into best cost battery-supercapacitor hybrid energy storage system for a utility scale PV array. *J. Energy Storage* **2019**, *22*, 50–59. [CrossRef]
7. He, X.; Zhang, X. A comprehensive review of supercapacitors: Properties, electrodes, electrolytes and thermal management systems based on phase change materials. *J. Energy Storage* **2022**, *56*, 106023. [CrossRef]
8. Huang, S.; Zhu, X.; Sarkar, S.; Zhao, Y. Challenges and opportunities for supercapacitors. *APL Mater.* **2019**, *7*, 100901. [CrossRef]
9. Kim, K.; An, J.; Park, K.; Roh, G.; Chun, K. Analysis of a Supercapacitor/Battery Hybrid Power System for a Bulk Carrier. *Appl. Sci.* **2019**, *9*, 1547. [CrossRef]
10. Vukajlović, N.; Milićević, D.; Dumnić, B.; Popadić, B. Comparative analysis of the supercapacitor influence on lithium battery cycle life in electric vehicle energy storage. *J. Energy Storage* **2020**, *31*, 101603. [CrossRef]
11. Xiao, G.; Chen, Q.; Xiao, P.; Zhang, L.; Rong, Q. Multiobjective Optimization for a Li-Ion Battery and Supercapacitor Hybrid Energy Storage Electric Vehicle. *Energies* **2022**, *15*, 2821. [CrossRef]
12. Muralee Gopi, C.V.V.; Vinodh, R.; Sambasivam, S.; Obaidat, I.M.; Kim, H.-J. Recent progress of advanced energy storage materials for flexible and wearable supercapacitor: From design and development to applications. *J. Energy Storage* **2020**, *27*, 101035. [CrossRef]
13. Thakur, A.K.; Choudhary, R.B. High-performance supercapacitors based on polymeric binary composites of polythiophene (PTP)-titanium dioxide (TiO2). *Synth. Met.* **2016**, *220*, 25–33. [CrossRef]
14. Saravanakumar, B.; Purushothaman, K.K.; Muralidharan, G. Interconnected V2O5 nanoporous network for high-performance supercapacitors. *ACS Appl. Mater. Interfaces* **2012**, *4*, 4484–4490. [CrossRef]
15. Thakur, A.K.; Deshmukh, A.B.; Choudhary, R.B.; Karbhal, I.; Majumder, M.; Shelke, M.V. Facile synthesis and electrochemical evaluation of PANI/CNT/MoS2 ternary composite as an electrode material for high performance supercapacitor. *Mater. Sci. Eng. B* **2017**, *223*, 24–34. [CrossRef]
16. Tantawy, N.S.; El-Taib Heakal, F.; Ahmed, S.Y. Synthesis of worm-like binary metallic active material by electroless deposition approach for high-performance supercapacitor. *J. Energy Storage* **2020**, *31*, 101625. [CrossRef]
17. Xu, L.; Wu, J.; Zhou, W.; Jiang, F.; Xu, J. Binder-free hierarchical porous N-doped graphene directly anchored on carbon fiber cloth for high-performance electrochemical energy storage. *J. Energy Storage* **2020**, *31*, 101682. [CrossRef]
18. Song, Z.; Hofmann, H.; Li, J.; Hou, J.; Han, X.; Ouyang, M. Energy management strategies comparison for electric vehicles with hybrid energy storage system. *Appl. Energy* **2014**, *134*, 321–331. [CrossRef]
19. Hu, J.; Liu, D.; Du, C.; Yan, F.; Lv, C. Intelligent energy management strategy of hybrid energy storage system for electric vehicle based on driving pattern recognition. *Energy* **2020**, *198*, 117298. [CrossRef]
20. Xu, H.; Shen, M. The control of lithium-ion batteries and supercapacitors in hybrid energy storage systems for electric vehicles: A review. *Int. J. Energy Res.* **2021**, *45*, 20524–20544. [CrossRef]
21. Yang, B.; Wang, J.; Zhang, X.; Wang, J.; Shu, H.; Li, S.; He, T.; Lan, C.; Yu, T. Applications of battery/supercapacitor hybrid energy storage systems for electric vehicles using perturbation observer based robust control. *J. Power Sources* **2020**, *448*, 227444. [CrossRef]

22. Kouchachvili, L.; Yaïci, W.; Entchev, E. Hybrid battery/supercapacitor energy storage system for the electric vehicles. *J. Power Sources* **2018**, *374*, 237–248. [CrossRef]
23. Hung, K.; Masarapu, C.; Ko, T.; Wei, B. Wide-temperature range operation supercapacitors from nanostructured activated carbon fabric. *J. Power Sources* **2009**, *193*, 944–949. [CrossRef]
24. Ma, S.; Jiang, M.; Tao, P.; Song, C.; Wu, J.; Wang, J.; Deng, T.; Shang, W. Temperature effect and thermal impact in lithium-ion batteries: A review. *Prog. Nat. Sci. Mater. Int.* **2018**, *28*, 653–666. [CrossRef]
25. Durganjali, C.S.; Chawla, V.; Raghavan, H.; Radhika, S. Design, development, and techno-economic analysis of extreme fast charging topologies using Super Capacitor and Li-Ion Battery combinations. *J. Energy Storage* **2022**, *56*, 106140. [CrossRef]
26. Voicu, I.; Louahlia, H.; Gualous, H.; Gallay, R. Thermal management and forced air-cooling of supercapacitors stack. *Appl. Therm. Eng.* **2015**, *85*, 89–99. [CrossRef]
27. Zhou, W.; Liu, Z.; Chen, W.; Sun, X.; Luo, M.; Zhang, X.; Li, C.; An, Y.; Song, S.; Wang, K.; et al. A Review on Thermal Behaviors and Thermal Management Systems for Supercapacitors. *Batteries* **2023**, *9*, 128. [CrossRef]
28. Wieczorek, M.; Lewandowski, M.; Jefimowski, W. Cost comparison of different configurations of a hybrid energy storage system with battery-only and supercapacitor-only storage in an electric city bus. *Bull. Pol. Acad. Sci. Tech. Sci.* **2019**, *67*, 1095–1106.
29. Qi, J.; Su, M. Analysis of Micro-Electric Vehicle with Super Capacitor/Battery Hybrid Energy Storage System. *J. Phys. Conf. Ser.* **2023**, *2459*, 012091. [CrossRef]
30. Lahyani, A.; Sari, A.; Lahbib, I.; Venet, P. Optimal hybridization and amortized cost study of battery/supercapacitors system under pulsed loads. *J. Energy Storage* **2016**, *6*, 222–231. [CrossRef]
31. Hebner, R.; Beno, J.; Walls, A. Flywheel batteries come around again. *IEEE Spectr.* **2002**, *39*, 46–51. [CrossRef]
32. Zhang, J.W.; Wang, Y.H.; Liu, G.C.; Tian, G.Z. A review of control strategies for flywheel energy storage system and a case study with matrix converter. *Energy Rep.* **2022**, *8*, 3948–3963. [CrossRef]
33. Jaafar, A.; Akli, C.R.; Sareni, B.; Roboam, X.; Jeunesse, A. Sizing and energy management of a hybrid locomotive based on flywheel and accumulators. *IEEE Trans. Veh. Technol.* **2009**, *58*, 3947–3958. [CrossRef]
34. Talebi, S.; Nikbakhtian, B.; Toliyat, H. A novel algorithm for designing the PID controllers of high-speed flywheels for traction applications. In Proceedings of the Vehicle Power and Propulsion Conference (VPPC), Arlington, TX, USA, 9–12 September 2007; pp. 574–579.
35. Shen, S.; Veldpaus, F.E. Analysis and control of a flywheel hybrid vehicular powertrain. *IEEE Trans. Control Syst.* **2004**, *12*, 645–660. [CrossRef]
36. Lemofouet, S.; Rufer, A. A hybrid energy storage system based on compressed air and supercapacitors with maximum efficiency point tracking (MEPT). *IEEE Trans. Ind. Electron.* **2006**, *53*, 1105–1115. [CrossRef]
37. Ribeiro, P.F.; Johnson, B.K.; Crow, M.L.; Arsoy, A.; Liu, Y. Energy storage systems for advanced power applications. *Proc. IEEE* **2001**, *89*, 1744–1756. [CrossRef]
38. Ali, M.H.; Wu, B.; Dougal, R.A. An overview of SMES applications in power and energy systems. *IEEE Trans. Sustain. Energy* **2010**, *1*, 38–47. [CrossRef]
39. Ise, T.; Kita, M.; Taguchi, A. A hybrid energy storage with a SMES and secondary battery. *IEEE Trans. Appl. Supercond.* **2005**, *15*, 1915–1918. [CrossRef]
40. Zhang, H.; Ren, J.; Zhong, Y.; Chen, J. Design and test of controller in power conditioning system for superconducting magnetic energy storage. In Proceedings of the International Conference on Power Electronics (ICPE), Daegu, Republic of Korea, 22–26 October 2007; pp. 966–972.
41. Trevisani, L.; Morandi, A.; Negrini, F.; Ribani, P.L.; Fabbri, M. Cryogenic fuel-cooled SMES for hybrid vehicle application. *IEEE Trans. Appl. Supercond.* **2009**, *19*, 2008–2011. [CrossRef]
42. Mukherjee, P.; Rao, V.V. Superconducting magnetic energy storage for stabilizing grid integrated with wind power generation systems. *J. Mod. Power Syst. Clean Energy* **2019**, *7*, 400–411. [CrossRef]
43. Boicea, V.A. Energy storage technologies: The past and the present. *Proc. IEEE* **2014**, *102*, 1777–1794. [CrossRef]
44. Malozemoff, A.P.; Maguire, J.; Gamble, B.; Kalsi, S. Power applications of high-temperature superconductors: Status and perspective. *IEEE Trans. Appl. Supercond.* **2002**, *12*, 778–781. [CrossRef]
45. Maxwell Technologies. 160 Volt Module. Available online: https://maxwell.com/products/ultracapacitors/160v-module/ (accessed on 22 April 2023).
46. Ostadi, A.; Kazerani, M. A comparative analysis of optimal sizing of battery-only, ultracapacitor-only, and battery–ultracapacitor hybrid energy storage systems for a city bus. *IEEE Trans. Veh. Technol.* **2015**, *64*, 4449–4460. [CrossRef]
47. Zhang, C.; Min, H.; Yu, Y.; Wang, D.; Luke, J.; Opila, D.; Saxena, S. Using CPE function to size capacitor storage for electric vehicles and quantifying battery degradation during different driving cycles. *Energies* **2016**, *9*, 903. [CrossRef]
48. Chong, L.W.; Wong, Y.W.; Rajkumar, R.K.; Isa, D. An adaptive learning control strategy for standalone PV system with battery supercapacitor hybrid energy storage system. *J. Power Sources* **2018**, *394*, 35–49. [CrossRef]
49. Rimpas, D.; Kaminaris, S.D.; Aldarraji, I.; Piromalis, D.; Vokas, G.; Papageorgas, P.G.; Tsaramirsis, G. Energy management and storage systems on electric vehicles: A comprehensive review. *Mater. Today Proc.* **2022**, *61*, 813–819. [CrossRef]
50. Yan, X.; Patterson, D. Improvement of drive range, acceleration and deceleration performances in an electric vehicle propulsion system. In Proceedings of the 30th Annual IEEE Power Electronics Specialists Conference, Charleston, SC, USA, 1 July 1999; Volume 2, pp. 638–643.

51. Chau, K.; Wong, Y. Hybridization of energy sources in electric vehicles. *Energy Convers. Manag.* **2001**, *42*, 1059–1069. [CrossRef]
52. Miller, J.R. Engineering electrochemical capacitor applications. *J. Power Sources* **2016**, *326*, 726–735. [CrossRef]
53. Song, Z.; Hofmann, H.; Li, J.; Han, X.; Zhang, X.; Ouyang, M. A comparison study of different semi active hybrid energy storage system topologies for electric vehicles. *J. Power Sources* **2015**, *274*, 400411. [CrossRef]
54. Seaman, A.; Dao, T.S.; McPhee, J. A survey of mathematics-based equivalent-circuit and electrochemical battery models for hybrid and electric vehicle simulation. *J. Power Sources* **2014**, *256*, 410–423. [CrossRef]
55. Li, Y.; Yang, D.; Ruan, X. A systematic method for generating multiple-input dc/dc converters. In Proceedings of the Vehicle Power and Propulsion Conference, Harbin, China, 3–5 September 2008; pp. 1–6.
56. Dixon, J.W.; Ortuzar, M.E. Battery, Ultracapacitor, Fuel Cell, and Hybrid Energy Storage Systems for Electric, Hybrid Electric, Fuel Cell, and Plug-In Hybrid Electric Vehicles: State of the Art. *IEEE Trans. Veh. Technol.* **2010**, *59*, 2806–2814.
57. Sibiceanu, A.R.; Iorga, A.; Nicolae, V.; Ivan, F. *Consideration on the Implications of the WLTC (Worldwide Harmonized Light-Duty Test Cycle) for a Middle-Class Car, CONAT Brașov*; Transilvania University Press: Braşov, Romania, 2016; pp. 203–211. ISSN 2069-0401.
58. EUROCOC. WLTP Cycle Replaces NEDC. Available online: https://www.eurococ.eu/en/blog/wltp-cycle-replaces-nedc/ (accessed on 22 April 2023).
59. Russo, A.; Cavallo, A. Stability and Control for Buck–Boost Converter for Aeronautic Power Management. *Energies* **2023**, *16*, 988. [CrossRef]
60. Russo, A.; Cavallo, A. Supercapacitor stability and control for More Electric Aircraft application. In Proceedings of the 2020 European Control Conference, ECC 2020, Saint Petersburg, Russia, 12–15 May 2020; IEEE: Piscataway, NJ, USA, 2020; pp. 1909–1914.

batteries

MDPI

Article

Hierarchical $CaMn_2O_4$/C Network Framework toward Aqueous Zn Ion Hybrid Capacitors as Competitive Cathodes

Lifen Ding [1], Qingchao Gao [2,3,*] and Changzhou Yuan [3]

1 School of Mechanical and Electronic Control Engineering, Beijing Jiaotong University, Beijing 100044, China; lfding@bjtu.edu.cn
2 Shenzhen Major Industry Investment Group & Tankeblue Semiconductor Co., Ltd., Shenzhen 518108, China
3 School of Materials Science & Engineering, University of Jinan, Jinan 250022, China; mse_yuancz@ujn.edu.cn
* Correspondence: agaoqc_me@163.com

Abstract: Manganese-based materials have received more attention as cathodes for aqueous zinc ion hybrid capacitors (AZIHCs) due to their advantages such as abundant reserves, low cost, and large theoretical capacity. However, manganese-based materials have the disadvantage of poor electrical conductivity. Herein, a solid-phase method was used to synthesize a hierarchical carbon-coated calcium manganate ($CaMn_2O_4$/C) network framework as the cathode for AZIHCs. Thanks to the unique structural/componential merits including conductive carbon coating and hierarchical porous architecture, the achieved $CaMn_2O_4$/C cathode shows an exceptionally long life of close to 5000 cycles at 2.0 A g^{-1}, with a reversible specific capacity of 195.6 mAh g^{-1}. The assembled $CaMn_2O_4$/C-based AZIHCs also display excellent cycling stability with a capacity retention rate of 84.9% after 8000 cycles at 1.0 A g^{-1}, and an energy density of 21.3 Wh kg^{-1} at an output power density of 180.0 W kg^{-1}.

Keywords: $CaMn_2O_4$/C; hierarchical porous; network framework; cathodes; aqueous zinc ion hybrid capacitors

1. Introduction

It is well known that non-renewable chemical energy sources such as coal, natural gas, and oil are decreasing, and the world is facing an energy crisis with rapid economic development. Ionic hybrid capacitors (IHCs) have both the higher power density of supercapacitors together with the higher energy density of secondary batteries, which is a promising energy storage device. The most mature research is on alkali ionic hybrid capacitors (AIHCs), and some of the research has been put into commercial use [1–3]. However, most AIHCs use organic electrolytes, which are not only costly but also highly flammable during use. Aqueous zinc ionic hybrid capacitors (AZIHCs) use water as a solvent, which not only saves cost but also reduces the combustion of the electrolyte [4–10]. In addition, AIHCs are manufactured in a more restrictive environment, whereas AZIHCs are manufactured in a milder environment. AZIHCs have great potential to become large-scale commercially available energy storage devices [11,12].

The main cathode materials for AZIHCs are vanadium-based oxides and manganese-based oxides. In recent years, researchers have studied vanadium-based oxides as electrode materials for AZIHCs. Ma et al. synthesized rod-shaped V_2O_5 as the cathode electrode material for AZIHCs [13], and AC was used as the anode electrode for assembling AZIHCs. By optimizing the mass ratio of the anode together with the cathode, the specific capacity of AZIHCs at 0.1 A g^{-1} is 57.4 mAh g^{-1}, and the maximum energy density was 34.6 Wh kg^{-1}. In addition, manganese-based oxides have received more attention due to their advantages of abundant reserves, low cost, multivalency, and high theoretical capacity [14–16]. However, manganese-based materials have the disadvantage of poor electrical conductivity as electrode materials, which leads to the poor rate performance and cycling performance of such materials [17–20]. Zeng et al. used ultrasonic treatment to prepare MnO_2-CNTs composite electrodes [21]. The MnO_2-CNTs electrode had a high specific capacitance of 254.1 mAh g^{-1} (0.256 A g^{-1}). Gao et al. synthesize ZMO nanoparticles on heterostructures of carbon nanotubes for stable Zn^{2+} storage [22]. Highly conductive CNTs and smaller ZMO can promote fast electron transport and make the ZMO/CNTs composites have high electrical conductivity. The results showed that the prepared ZMO/CNTs materials exhibited excellent cycling stability with an initial specific capacity of 220.3 mAh g^{-1} at 0.1 A g^{-1}.

In this work, hierarchical carbon-coated calcium manganate ($CaMn_2O_4$/C) network framework was synthesized by a simple yet mass-producible solid-phase method using polyvinyl pyrrolidone (PVP) as the carbon source, $Ca(NO_3)_2{\cdot}4H_2O$ together with $Mn(NO_3)_2{\cdot}4H_2O$ as the calcium and manganese sources, respectively. The introduction of C into $CaMn_2O_4$/C increased its electrical conductivity. As a result, the $CaMn_2O_4$/C cathode materials exhibit an exceptionally long lifetime of nearly 5000 cycles at 2.0 A g^{-1} with a reversible specific capacity of 195.6 mAh g^{-1}. The assembled AZIHCs were obtained by using porous carbon (PC) as the anode and $CaMn_2O_4$/C as the cathode as well as a mixed aqueous solution (0.2 M $MnSO_4$ and 2.0 M $ZnSO_4$) as the electrolyte. The constructed PC//$CaMn_2O_4$/C AZIHCs achieve an energy density of 21.3 Wh kg^{-1} at a power density of 180.0 W kg^{-1}, along with modest device leakage current and self-discharge capability, and favorable cycling stability with 84.9% capacity retention after 8000 cycles at 1.0 A g^{-1}.

2. Experimental Sections

2.1. Chemicals

The chemical reagents including $Mn(NO_3)_2{\cdot}4H_2O$, $Ca(NO_3)_2{\cdot}4H_2O$, N-methyl-2-pyrrolidinone (NMP), $ZnSO_4$, $MnSO_4$, and polyvinylpyrrolidone (PVP, K30) are of analytic grade, and from Sinopharm Chemical Reagent Co., Ltd. (Shanghai, China). The chemical reagents are directly used as received without further treatment.

2.2. Materials Synthesis

Synthesize of $CaMn_2O_4$/C: 0.4 g of PVP, 0.258 g of $Ca(NO_3)_2{\cdot}4H_2O$ and 0.752 g of $Mn(NO_3)_2{\cdot}4H_2O$ were well prepared in a mortar, ground thoroughly for 30 min. The milled powder was placed in a muffle furnace, heated up to 750 °C held for 3 h at a rate of 5 °C min^{-1}. The resulting sample was named $CaMn_2O_4$/C.

Synthesis of PC materials: The PC material was fabricated according to the previous report [23]. $Zn(NO_3)_2{\cdot}6H_2O$ and PVP(K30) were dissolved in sufficient distilled water in the mass ratio of 1.5:1 and stirred for 30 min. Then, the resulting solution was placed in an oven (at 90 °C) and dried completely. The powder was annealed at 800 °C for 1 h (atmosphere: N_2, rate: 5 °C min^{-1}). The annealed sample and KOH were dissolved in deionized water in the ratio of 1:3 by mass and stirred for 30 min. The resulting solution was dried (at 90 °C) completely. The dried mixture is then activated at 800 °C for 3 h (atmosphere: N_2, rate: 5 °C min^{-1}). The excess KOH in the product is neutralized by aqueous hydrochloric acid. The product was washed with plenty of deionized water and dried at 60 °C for 12 h (in an oven). The sample was named PC.

2.3. Material Characterization

Phase characterization of products was carried out using X-ray diffraction (XRD, Rigaku Ultima IV powder X-ray diffractometer with Cu Ka radiation). The chemical compositions were tested by X-ray photoelectron spectroscopy (XPS, The model number is VG Multilab 2000 and the origin is England). Raman spectra (HR Evolution, Japanese) were collected using 532 nm excitation. The microstructures and morphologies of samples were characterized using transmission electron microscopy (TEM), field-emission scanning electron microscopy (FESEM, JEOL-6300F), and high-resolution TEM (HRTEM, the model is JEM-2100 and the manufacturer is Nippon Electronics Co., Ltd., Mumbai, India).

2.4. Electrochemical Measurements

Preparation of $CaMn_2O_4$/C cathode: Briefly, a mixture of poly(vinylidene fluoride) (PVDF): $CaMn_2O_4$/C: acetylene black was added to the mortar in the mass ratio of 1:7:2, and a reasonable amount of NMP was added dropwise, and the paste was ground for 40 min to form a homogeneous paste. The slurry was evenly coated on a 12 mm diameter carbon paper, placed in an oven, and dried overnight at 110 °C. The amount of the $CaMn_2O_4$/C on the carbon paper was about 1.5 mg.

Preparation of PC anode: Typically, a mixture of PVDF:acetylene black:PC was added to the mortar in a mass ratio of 1:2:7. A reasonable amount of NMP was added dropwise and the paste was ground for half an hour to form a homogeneous slurry. The slurry was uniformly coated on a stainless-steel mesh of 12 mm in diameter and then placed in an oven for drying overnight at 60 °C. The mass loading of PC on the stainless-steel mesh was around 1.0 mg.

Glass fiber (Whatman GF/D) was used for the diaphragm, zinc foil was used as the reference electrode, and a mixture of 0.2 M $MnSO_4$ and 2.0 M $ZnSO_4$ in aqueous solution was used for the electrolyte. The CR2016 button half-cell was assembled to study the electrochemical performance of $CaMn_2O_4$/C. The voltage range was 0.4–1.9 V (vs. Zn/Zn^{2+}).

$CaMn_2O_4$/C was used as the cathode along with PC as the anode to assemble the AZIHCs. The separator and electrolyte were kept the same as those for the electrochemical evaluation of $CaMn_2O_4$/C, as mentioned above. The devices were tested over a voltage range of 0–1.8 V. Galvanostatic intermittent titration technique (GITT) and typical charge/discharge tests were conducted on an 8-channel Land Test System (CT2001A). Cyclic voltammetry (CV) was performed on the electrochemical workstation (IviumStat. h, Eindhoven, The Netherlands) at 25 °C.

3. Results and Discussion

3.1. Synthesis and Structural Analysis

The microstructures of the as-obtained $CaMn_2O_4$/C are characterized by FESEM. As illustrated in Figure 1a, the hierarchical porous network framework is evident for the $CaMn_2O_4$/C, which is constructed with nano-building blocks of about 0.2 μm in size (Figure 1b,c). The co-existence of Ca, Mn, and O species can be visualized based on the elemental mapping images (Figure 1d–f), moreover, the C mapping image authenticates the uniform distribution in the $CaMn_2O_4$/C composite framework (Figure 1g). In addition, by EDS spectroscopy data, the $CaMn_2O_4$/C composites had a C relative atomic ratio is about 8.29 at.% (Figure S1). As shown in Figure S2, we performed a Raman spectroscopic evaluation of $CaMn_2O_4$/C. The G and D bands of carbon are located at 1607.2 cm^{-1} and 1333.9 cm^{-1}, respectively, indicating the presence of graphitic structures and defects/disturbances. The intensity ratio (i.e., I_D/I_G) of the D and G bands can be used to estimate the degree of graphitization, which is 1.35 for $CaMn_2O_4$/C.

The more detailed structure of $CaMn_2O_4$/C was characterized as well by TEM and HRTEM techniques, as shown in Figure 2a,b. Clearly, a uniform carbon layer of about 3.53 nm in thickness is located upon the surface of the well-crystalline $CaMn_2O_4$, and the well-defined lattice fringes with a spacing of 0.300 nm, which corresponds to the (320) lattice plane of $CaMn_2O_4$, is apparent.

The crystal structures of the resulting $CaMn_2O_4$/C were checked by XRD. As illustrated in Figure 3, typical XRD peaks here match well with the standard data of the $CaMn_2O_4$ (JCPDS PDF# 76-0516). In specific, the distinct diffraction peaks at 2θ = 18.3, 31.0, 32.9, 35.0, 39.2, 40.3, 40.5, and 43.5° are related to (200), (111), (320), (211), (031), (131), (240) and (231) planes, respectively [24,25]. Moreover, no other discernable reflections are found for the potential impurities, suggesting that the $CaMn_2O_4$/C is successfully prepared by such a facile synthetic method.

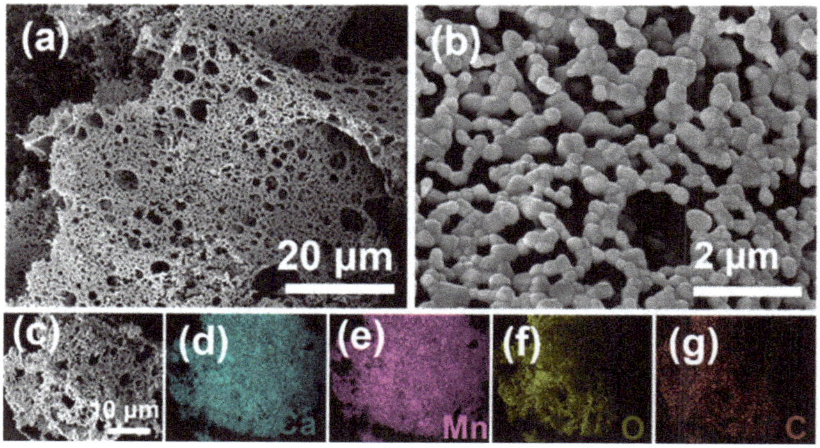

Figure 1. (**a–c**) FESEM images, and corresponding elemental ((**d**) Ca; (**e**) Mn; (**f**) O and (**g**) C) mapping images of $CaMn_2O_4$/C.

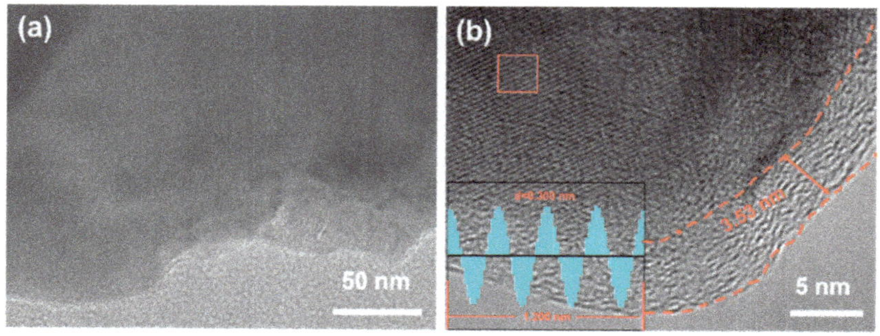

Figure 2. (**a**) TEM and (**b**) HRTEM images of $CaMn_2O_4$/C. The inset for the intensity profile for the measured interlayer spacing.

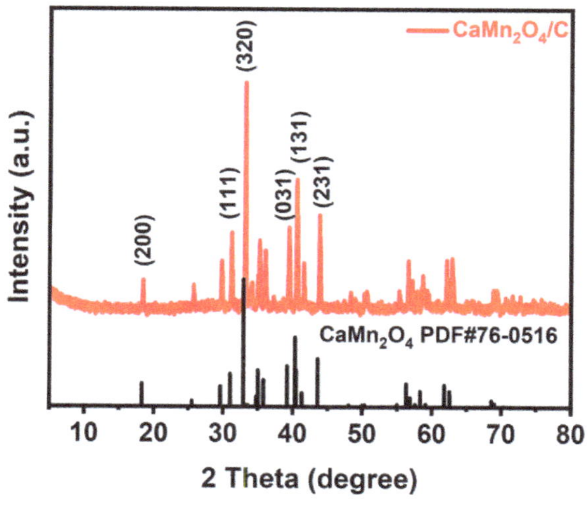

Figure 3. XRD pattern of $CaMn_2O_4$/C.

XPS determinations of $CaMn_2O_4$/C were performed to enable further analysis of the chemical state of the surface elements. As illustrated in Figure 4a, the characteristic peaks of C 1s, Mn 2p, O 1s, and Ca 2p can be observed. Figure 4b shows the XPS spectrum of Mn 2p, and the characteristic peaks at 641.8 and 653.0 eV correspond to Mn $2p_{3/2}$ and Mn $2p_{1/2}$ [25,26]. As illustrated in Figure 4c, the O 1s spectrum of $CaMn_2O_4$/C can be divided into two peaks at 529.2 (metal-oxygen bond) and 531.2 (O-H) eV [25,26]. The Ca spectrum of $CaMn_2O_4$/C, as presented in Figure 4d, is fitted as two peaks corresponding to Ca $2p_{3/2}$ (350.3 eV) and Ca $2p_{3/2}$ (346.8 eV), respectively [27]. The C 1s spectrum (Figure 4e), can be decomposed into three peaks at 284.5, 285.8, and 285.0 eV, corresponding to C-C, O-C=C, and C-O, respectively [26].

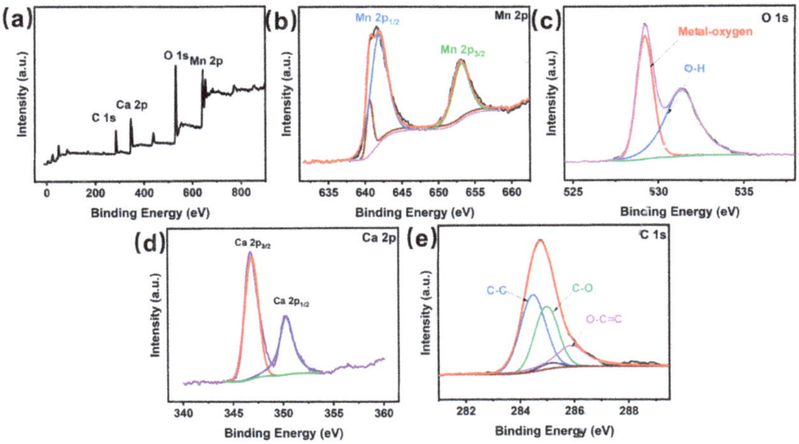

Figure 4. High-resolution core-level spectra of (**a**) XPS survey spectrum, (**b**) Mn 2p, (**c**) O 1s, (**d**) Ca 2p, and (**e**) C 1s, respectively, for the $CaMn_2O_4$/C sample.

3.2. Electrochemical Evaluation of the $CaMn_2O_4$/C Cathode

As illustrated in Figure 5, the electrochemical performance of $CaMn_2O_4$/C is collected. The initial three CV curves of $CaMn_2O_4$/C are shown in Figure 5a. The CV curves for the initial cycle under a scanning rate of 0.1 mV s^{-1} are slightly different from the subsequent ones, which may be relevant to the progressive activation of the electrode in subsequent cycles. For the anodic scan of the second cycle, two peaks centered at 1.60 as well as 1.54 V appear and overlap in the CV curve, which corresponds to the progressive electrochemical extraction of Zn^{2+} [28]. In the cathodic scan, two distinct peaks were observed (around 1.38 as well as 1.23 V), which could be ascribed to the gradual embedding of Zn^{2+} [29]. The constant current charge/discharge plots of $CaMn_2O_4$/C at 0.1 A g^{-1} are profiled in Figure 5b. $CaMn_2O_4$/C displays a discharge-specific capacity of 248.1 mAh g^{-1} (at 0.1 A g^{-1}). The rate behaviors of the $CaMn_2O_4$/C are summarized, as shown in Figure 5c. The $CaMn_2O_4$/C cathode provides reversible specific capacities of 260.9, 242.3, 219.9, 198.3, 135.1, and 112.2 mAh g^{-1} at 0.4, 0.6, 0.8, 1.0, 2.0, and 3.0 A g^{-1}, respectively, equivalenting to a capacity retention of 43%. When the current density is again returned to 0.4 A g^{-1}, a reversible specific capacity as high as 275.0 mAh g^{-1} is remained by $CaMn_2O_4$/C cathode. This excellent rate performance, as summarized in Figure 5d, is also comparable to many other cathode materials recently reported in the literature, such as $Ca_{0.28}MnO_2 \cdot 0.5\ H_2O$ [25], ZMO@$Ti_3C_2T_x$ [27], ZMO QD@C [30], ZMO [31], and K, Fe-ZMO [32]. More surprisingly, our fabricated $CaMn_2O_4$/C cathode exhibits robust electrochemical stability with an exceptionally long lifetime of 5000 cycles at 2.0 A g^{-1} with a retained reversible capacity of 195.6 mAh g^{-1}, as illustrated in Figure 5e.

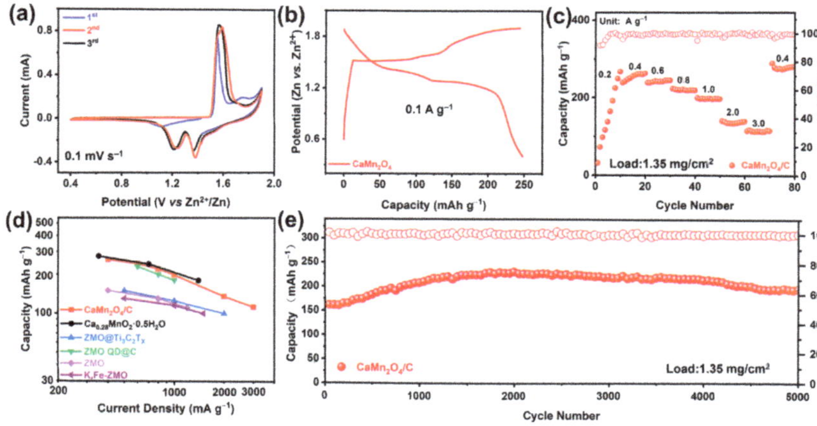

Figure 5. (**a**) CV curves and (**b**) the first charge/discharge plots (0.1 A g^{-1}), (**c**) rate behaviors, (**d**) comparison in rate capacities with other cathodes reported in the literature in recent years, and (**e**) long cycling performance (1.0 A g^{-1}) of the $CaMn_2O_4$/C cathode.

The GITT tests were utilized to investigate the diffusion coefficient (D_{Zn}) of Zn^{2+} during the charging/discharging processes of $CaMn_2O_4$/C (Figure S3). The GITT was performed after three charging and discharging cycles of the $CaMn_2O_4$/C half-cell. The GITT was carried out with the relaxation time of 30 min at 0.2 A g^{-1} and the charging and discharging time of 10 min. The D_{Zn} values of $CaMn_2O_4$/C are calculated according to the equation [20]:

$$D_{Zn} = \frac{4L^2}{\pi\tau}\left(\frac{m_b V_M}{M_B S}\right)^2\left(\frac{\Delta E_S}{\Delta E_t}\right)^2 \quad (1)$$

where m_b, V_M, and M_B represent the mass, molar volume of the active material, and molar mass, respectively. S corresponds to the surface area, τ corresponds to the relaxation time, ΔEs for the steady-state potential change of the current pulse, and ΔE_t for the iR drop after relaxation time. The D_{Zn} values of $CaMn_2O_4$/C were then calculated to be in the range of 10^{-12} and 10^{-10} cm^2 s^{-1} over the discharge-charge processes (Figure 6a,b). This is also consistent with its better cycling and multiplication properties [33].

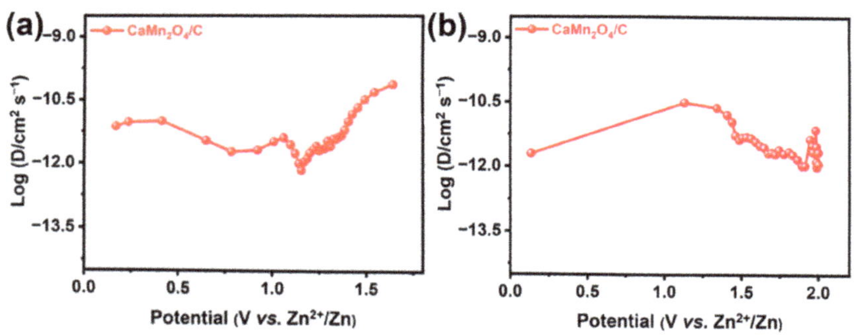

Figure 6. Specific D_{Zn} values over the (**a**) discharge and (**b**) charging processes of the $CaMn_2O_4$/C cathode.

In order to deeply analyze the storage kinetics of Zn^{2+}, CV measurements of $CaMn_2O_4$/C were carried out under different scanning rates of 0.2–1.0 mV s^{-1}. Two reduction peaks and one oxidation peak can be clearly observed (Figure 7a). The profile of the CV curves of $CaMn_2O_4$/C does not appear significantly with increasing scan rate, indicating that it has good reaction kinetics. The capacitance effect of the cathode is calculated from the relationship between the current (i) as well as the scan rate (v) by the equation: $i = av^b$

(b as well as a are constants). Generally, the value of b is equal to 0.5 for ion diffusion-controlled processes, and the case of $b = 1$ corresponds to the surface capacitance-controlled processes, and the value of b is greater than 0.5 but less than 1, which indicates a mixed mechanism. The b values of peak 1, peak 2, and peak 3 are 0.71, 0.73 and 0.72 (Figure 7b) [24], which indicates the pseudocapacitive contribution dominated Zn^{2+}-storage process for the $CaMn_2O_4$/C composite cathode. Specific pseudocapacitive contribution of the cathode can be distinguished by the equation: $i = k_1v + k_2v^{1/2}$, where k_1v and $k_2v^{1/2}$ represent the one from the capacitive part and the current of the diffusion-controlled part. Figure 7c shows the capacity share contributed by the diffusion control process and the capacitive effect (red region) at 0.8 mV s^{-1}, highlighting a capacitive contribution as large as about 91.8% for the $CaMn_2O_4$/C electrode. More strikingly, the capacitive contribution can be up to 92.7% when the sweep rate is up to 1.0 mV s^{-1} (Figure 7d). It is the remarkable capacitive contrition of the $CaMn_2O_4$/C here that ensures its faster reaction kinetics toward efficient zinc storage [34,35].

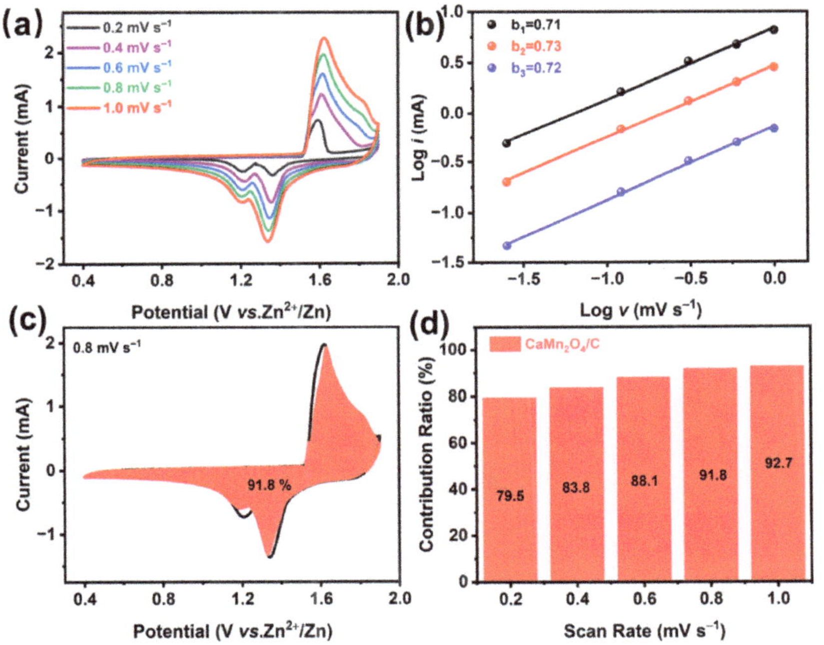

Figure 7. (**a**) CV curves, (**b**) corresponding logi vs. logv plots, (**c**) CV curves with the pseudocapacitive contribution (red region), and (**d**) diagram for the pseudocapacitive contributions at different scan rates for the $CaMn_2O_4$/C cathode.

3.3. Electrochemical Properties of PC//$CaMn_2O_4$/C AZIHCs

The unique PC//$CaMn_2O_4$/C AZIHCs were assembled with the $CaMn_2O_4$/C as the cathode and the PC as the anode, as schematically illustrated in Figure 8a, and their electrochemical performance was comprehensively measured. Figure 8b shows the CV curves of AZIHCs at 5.0–40.0 mV s^{-1} (the voltage range: 0–1.8 V). The device CV curves at 5.0–40.0 mV s^{-1} are close to an ideal rectangular shape, showing the characteristic capacitive nature of the hybrid device. The profile of the CV curve does not appear to change with increasing scanning rate, indicating modest polarization of the AZIHCs occurring. Figure 8c exhibits the constant-current charge/discharge plots of the PC//$CaMn_2O_4$/C AZIHCs. These curves are almost straight lines, showing the good capacitive characteristics of the AZIHCs. It is worth noting that the reversible capacitance of the cell can be as high as 80.0–17.8 F g^{-1} at 0.2–2.0 A g^{-1}. The device provides an energy density of 21.3 Wh kg^{-1}

at an output power density of 180.0 W kg^{-1}. When charged to 1.8 V at 0.1 A g^{-1}, the device leakage current was only 7 μA (Figure 8d) and the open-circuit voltage (after 20 h of static operation) dropped by 0.75 V, equivalenting to a voltage loss of only 41.6% (Figure 8e). As shown in Figure 8f, the AZIHCs exhibit excellent cycling stability with 84.9% capacity retention (at 1.0 A g^{-1}) after 8000 cycles, equivalenting to a capacitance degradation of 0.046% per cycle, indicative of superb electrochemical stability of our fabricated PC//$CaMn_2O_4$/C AZIHCs.

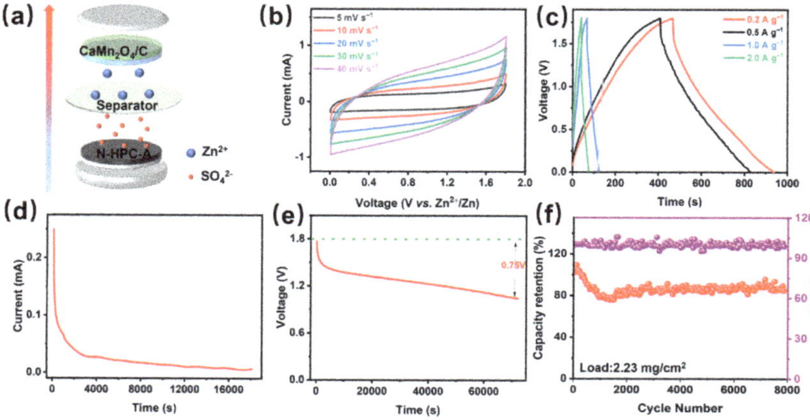

Figure 8. Electrochemical performance of N-HPC-A//$CaMn_2O_4$/C AZIHCs: (**a**) schematic diagram of device assembly, (**b**) CV curves, (**c**) constant current charge-discharge plots from 0.2 to 2.0 A g^{-1}, (**d**) leakage current diagram, (**e**) self-discharge diagram, and (**f**) cycling performance.

4. Conclusions

In summary, in our work, we explored a solid-phase synthesis avenue to smartly fabricate a hierarchical $CaMn_2O_4$/C network framework and utilize it as the competitive cathode for AZHICs. Benefiting from its unique structural/componential advantages, the $CaMn_2O_4$/C cathode was endowed with superb electrochemical stability and high-rate capacities. Typically, the cathode displayed an exceptionally long cycle life of nearly 5000 cycles at 2.0 A g^{-1}, with a reversible specific capacity of 195.6 mAh g^{-1}. Utilizing the PC as the anode, and $CaMn_2O_4$/C as the cathode, the assembled PC//$CaMn_2O_4$/C AZIHCs exhibited capacity retention of 84.9% (at 1.0 A g^{-1}) after 8000 cycles and a power density of 180.0 W kg^{-1} at an energy density of 21.3 Wh kg^{-1}. All the data here featured the good prospects of our prepared porous $CaMn_2O_4$/C network framework for practical applications.

Supplementary Materials: The following supporting information can be downloaded at: https://www.mdpi.com/article/10.3390/batteries9120586/s1, Figure S1: EDS spectrum of $CaMn_2O_4$/C and individual atoms percentage; Figure S2: Raman spectra of the $CaMn_2O_4$/C cathode; Figure S3: GITT curves overcharge and discharge processes of the $CaMn_2O_4$/C cathode.

Author Contributions: Conceptualization, L.D. and Q.G.; methodology, L.D.; software, L.D.; validation, L.D., Q.G. and C.Y.; formal analysis, L.D.; investigation, L.D.; resources, L.D.; data curation, Q.G. and C.Y.; writing—original draft preparation, L.D.; writing—review and editing, Q.G. and C.Y.; visualization, L.D.; supervision, Q.G.; project administration, Q.G. All authors have read and agreed to the published version of the manuscript.

Funding: This research received no external funding.

Data Availability Statement: Data are contained within the article or Supplementary Material.

Conflicts of Interest: The authors declare no conflict of interest.

References

1. Gopalakrishnan, A.; Sharma, C.S. High-performance dual carbon Li-ion hybrid capacitor constructed from N, S—Co-doped candle soot derived carbon nanoparticles anode and porous carbon cathode. *J. Energy Storage* **2022**, *55*, 105788. [CrossRef]
2. Wang, G.Y.; Wang, X.H.; Sun, J.F.; Zhang, Y.M.; Hou, L.R.; Yuan, C.Z. Porous carbon nanofibers derived from low-softening-point coal pitch towards all-carbon potassium ion hybrid capacitors. *Rare Met.* **2022**, *22*, 3706. [CrossRef]
3. Zhou, J.; Hu, H.Y.; Li, H.Q.; Chen, Z.P.; Yuan, C.Z.; He, X.J. Advanced carbon-based materials for Na, K, and Zn ion hybrid capacitors. *Rare Met.* **2023**, *42*, 719. [CrossRef]
4. Bommireddy, P.R.; Karnam, J.B.; Park, S. Ni-Co PBA-decorated CNTs as battery-type cathode materials for potassium-ion hybrid capacitors. *J. Energy Storage* **2023**, *62*, 106870. [CrossRef]
5. Rajkumar, P.; Thirumal, V.; Radhika, G.; Gnanamuthu, R.; Subadevi, R.; Sivakumar, M.; Yoo, K.; Kim, J. Eco-friendly production of carbon electrode from biomass for high performance lithium and zinc ion capacitors with hybrid energy storage characteristics. *Mater. Lett.* **2023**, *354*, 135320. [CrossRef]
6. Boruah, B.D.; Mathieson, A.; Wen, B.; Jo, C.; Deschler, F.; Volder, M.D. Photo-rechargeable Zinc-Ion capacitor using 2D graphitic carbon nitride. *Nano Lett.* **2020**, *20*, 5967–5974. [CrossRef]
7. Maughan, P.A.; Tapia-Ruiz, N.; Bimbo, N. In-situ pillared MXene as a viable zinc-ion hybrid capacitor. *Electrochim. Acta* **2020**, *341*, 136061. [CrossRef]
8. Chen, X.; Zhang, H.; Gao, Y.; Liu, J.-H.; Cao, X.; Zhan, C.; Wang, S.; Wang, J.; Dou, S.-X.; Cao, D. Zinc-ion hybrid supercapacitors: Design strategies, challenges, and perspectives. *Carbon Neutralization* **2022**, *1*, 159–188. Available online: https://onlinelibrary.wiley.com/doi/pdfdirect/10.1002/cnl2.22 (accessed on 30 October 2023). [CrossRef]
9. Eskusson, J.; Thomberg, T.; Lust, E.; Jänes, A. Electrochemical characteristics of Zn-ion hybrid supercapacitors based on aqueous solution of different electrolytes. *J. Electrochem. Soc.* **2022**, *169*, 020512. [CrossRef]
10. Wang, Y.; Sun, S.; Wu, X.; Liang, H.; Zhang, W. Status and opportunities of zinc ion hybrid capacitors: Focus on carbon materials, current collectors, and separators. *Nano-Micro Lett.* **2023**, *15*, 78. Available online: https://link.springer.com/content/pdf/10.1007/s40820-023-01065-x.pdf (accessed on 30 October 2023). [CrossRef]
11. Li, Y.; Zhang, X.; Lu, T.; Zhang, Y.; Li, X.; Yu, D.; Zhao, G. Boosting the capacitance of aqueous zinc-ion hybrid capacitors by engineering hierarchical porous carbon architecture. *Batteries* **2023**, *9*, 429. Available online: https://www.mdpi.com/2313-0105/9/8/429/pdf?version=1692264370 (accessed on 30 October 2023). [CrossRef]
12. Tang, H.; Yao, J.; Zhu, Y. Recent developments and future prospects for zinc-ion hybrid capacitors: A review. *Adv. Energy Mater.* **2021**, *11*, 2003994. [CrossRef]
13. Jagadale, A.D.; Rohit, R.C.; Shinde, S.K.; Kim, D. Materials development in hybrid zinc-ion capacitors. *ChemNanoMat* **2021**, *7*, 1082–1098. [CrossRef]
14. Ma, X.; Wang, J.; Wang, X.; Zhao, L.; Xu, C. Aqueous V_2O_5/activated carbon zinc-ion hybrid capacitors with high energy density and excellent cycling stability. *J. Mater. Sci.-Mater. Electron.* **2019**, *30*, 5478–5486. [CrossRef]
15. Pramanik, A.; Chattopadhyay, S.; Maiti, S.; De, G.; Mahanty, S. Hollow-porous nanospheres of $ZnMn_2O_4$ spinel: A high energy density cathode for rechargeable aqueous battery. *Mater. Chem. Phys.* **2021**, *263*, 124373. [CrossRef]
16. Aristote, N.T.; Deng, X.; Zou, K.; Gao, X.; Momen, R.; Li, F.; Deng, W.; Hou, H.; Zou, G.; Ji, X. General overview of sodium, potassium, and zinc-ion capacitors. *J. Alloys Compd.* **2022**, *913*, 165216. [CrossRef]
17. Wang, J.; Wang, J.-G.; Liu, H.; You, Z.; Li, Z.; Kang, F.; Wei, B. A highly flexible and lightweight MnO_2/graphene membrane for superior zinc-ion batteries. *Adv. Funct. Mater.* **2021**, *31*, 2007397. Available online: https://onlinelibrary.wiley.com/doi/pdf/10.1002/adfm.202007397 (accessed on 30 October 2023). [CrossRef]
18. Huang, C.; Wang, Q.; Zhang, D.; Shen, G. Coupling N-doping and rich oxygen vacancies in mesoporous $ZnMn_2O_4$ nanocages toward advanced aqueous zinc ion batteries. *Nano Res.* **2022**, *15*, 8118–8127. [CrossRef]
19. Yang, S.; Zhang, M.; Wu, X.; Wu, X.; Zeng, F.; Li, Y.; Duan, S.; Fan, D.; Yang, Y.; Wu, X. The excellent electrochemical performances of $ZnMn_2O_4$/Mn_2O_3: The composite cathode material for potential aqueous zinc ion batteries. *J. Electroanal. Chem.* **2019**, *832*, 69–74. [CrossRef]
20. Chen, F.; Wang, Q.; Yang, X.; Wang, C.; Zang, H.; Tang, Y.; Li, T.; Geng, B. Construction of hollow mesoporous $ZnMn_2O_4$/C microspheres with carbon nanotubes embedded in shells for high-performance aqueous zinc ions batteries. *Nano Res.* **2023**, *16*, 1726–1732. [CrossRef]
21. Cui, K.; Sun, M.; Zhang, J.; Xu, J.; Zhai, Z.; Gong, T.; Hou, L.; Yuan, C. Facile solid-state synthesis of tetragonal $CuFe_2O_4$ spinels with improved infrared radiation performance. *Ceram. Int.* **2022**, *48*, 10555–10561. [CrossRef]
22. Jiang, H.; Zhang, Y.; Pan, Z.; Xu, L.; Zheng, J.; Gao, Z.; Hu, T.; Meng, C.: Wang, J. Layered $Ca_{0.28}MnO_2{\cdot}0.5H_2O$ as a high performance cathode for aqueous zinc-ion battery. *Small* **2020**, *16*, 2000597. Available online: https://onlinelibrary.wiley.com/doi/pdf/10.1002/smll.202002852 (accessed on 30 October 2023).
23. Gao, F.; Mei, B.; Xu, X.; Ren, J.; Zhao, D.; Zhang, Z.; Wang, Z.; Wu, Y.; Liu, X.; Zhang, Y. Rational design of $ZnMn_2O_4$ nanoparticles on carbon nanotubes for high rate and durable aqueous zinc-ion batteries. *Chem. Eng. J.* **2022**, *448*, 137742. Available online: https://www.sciencedirect.com/science/article/pii/S1385894722032296 (accessed on 30 October 2023). [CrossRef]
24. Gao, Q.; Li, T.; Liu, C.; Sun, J.; Liu, Y.; Hou, L.; Yuan, C. Hierarchically porous N-doped carbon framework with enlarged interlayer spacing as dual-carbon electrodes for potassium ion hybrid capacitors. *Carbon Neutr.* **2023**, *2*, 18. [CrossRef]

25. Tao, Y.; Li, Z.; Tang, L.; Pu, X.; Cao, T.; Cheng, D.; Xu, Q.; Liu, H.; Wang, Y.; Xia, Y. Nickel and cobalt Co-substituted spinel $ZnMn_2O_4$@ N-rGO for increased capacity and stability as a cathode material for rechargeable aqueous zinc-ion battery. *Electrochim. Acta* **2020**, *331*, 135296. [CrossRef]
26. Yang, C.; Han, M.; Yan, H.; Li, F.; Shi, M.; Zhao, L. In-situ probing phase evolution and electrochemical mechanism of $ZnMn_2O_4$ nanoparticles anchored on porous carbon polyhedrons in high-performance aqueous Zn-ion batteries. *J. Power Sources* **2020**, *452*, 227826. [CrossRef]
27. Cai, K.; Luo, S.; Cong, J.; Li, K.; Yan, S.; Hou, P.; Wang, Q.; Zhang, Y.; Liu, X.; Lei, X. Synthesis and optimization of $ZnMn_2O_4$ cathode material for zinc-ion battery by citric acid sol-gel method. *J. Electrochem. Soc.* **2022**, *169*, 030531. [CrossRef]
28. Shi, M.; Wang, B.; Shen, Y.; Jiang, J.; Zhu, W.; Su, Y.; Narayanasamy, M.; Angaiah, S.; Yan, C.; Peng, Q. 3D assembly of MXene-stabilized spinel $ZnMn_2O_4$ for highly durable aqueous zinc-ion batteries. *Chem. Eng. J.* **2020**, *399*, 125627. [CrossRef]
29. Gao, Z.; Lu, G.-G.; Cao, L.-C.; Zhu, Z.-X.; Li, Y.-X.; Wei, F.-X.; Ji, Z.; Sui, Y.-W.; Qi, J.-Q.; Meng, Q.-K.; et al. Rationally designed Mn_2O_3@ $ZnMn_2O_4$/C core–shell hollow microspheres for aqueous zinc-ion batteries. *Dalton Trans.* **2023**, *52*, 1768–1776. Available online: http://pubs.rsc.org/en/content/articlepdf/2023/DT/D2DT03652E (accessed on 30 October 2023). [CrossRef]
30. Mallick, S.; Choutipalli, V.S.K.; Bag, S.; Subramanian, V.; Raj, C.R. Defect engineered ternary spinel: An efficient cathode for an aqueous rechargeable zinc-ion battery of long-term cyclability. *ACS Appl. Mater. Interfaces* **2022**, *14*, 37577–37586. [CrossRef]
31. Deng, S.; Tie, Z.; Yue, F.; Cao, H.; Yao, M.; Niu, Z. Rational design of $ZnMn_2O_4$ quantum dots in a carbon framework for durable aqueous zinc-ion batteries. *Angew. Chem. Int. Ed.* **2022**, *61*, e202115877. [CrossRef]
32. Soundharrajan, V.; Sambandam, B.; Kim, S.; Islam, S.; Jo, J.; Kim, S.; Mathew, V.; Sun, Y.-k.; Kim, J. The dominant role of Mn^{2+} additive on the electrochemical reaction in $ZnMn_2O_4$ cathode for aqueous zinc-ion batteries. *Energy Stor. Mater.* **2020**, *28*, 407–417. [CrossRef]
33. Shao, T.; Zhang, Y.; Cao, T.; Yang, Y.; Li, Z.; Liu, H.; Wang, Y.; Xia, Y. Structural regulation of $ZnMn_2O_4$ cathode material by K, Fe-double doping to improve its rate and cycling stability for rechargeable aqueous zinc-based batteries. *Chem. Eng. J.* **2022**, *431*, 133735. Available online: https://www.sciencedirect.com/science/article/pii/S1385894721053092 (accessed on 30 October 2023). [CrossRef]
34. Cheng, C.; Wu, D.X.; Gong, T.Y.; Yan, Y.S.; Liu, Y.; Ji, X.W.; Hou, L.R.; Yuan, C.Z. Internal and external cultivation design of zero-strain columbite-structured MNb_2O_6 toward lithium-ion capacitors as competitive anodes. *Adv. Energy Mater.* **2023**, *13*, 202302107. [CrossRef]
35. Chen, L.; Yang, Z.; Qin, H.; Zeng, X.; Meng, J. Advanced electrochemical performance of $ZnMn_2O_4$/N-doped graphene hybrid as cathode material for zinc ion battery. *J. Power Sources* **2019**, *425*, 162–169. [CrossRef]

batteries

Review

Tuning of Ionic Liquid–Solvent Electrolytes for High-Voltage Electrochemical Double Layer Capacitors: A Review

Yan Wang [1,2,3], Kaiyuan Xue [1], Changzeng Yan [4], Yuehui Li [5], Xingyun Zhang [1], Kailimai Su [1], Pengjun Ma [1], Shanhong Wan [1,*] and Junwei Lang [1,2,3,*]

1 Laboratory of Clean Energy Chemistry and Materials, Lanzhou Institute of Chemical Physics, Chinese Academy of Sciences, Lanzhou 730000, China; wangyan@licp.cas.cn (Y.W.); 2021023631@qdu.edu.cn (K.X.); 2021023612@qdu.edu.cn (X.Z.); suklm@licp.cas.cn (K.S.); mapengjun@licp.cas.cn (P.M.)

2 Shandong Laboratory of Advanced Materials and Green Manufacturing at Yantai, Yantai 264000, China

3 Qingdao Center of Resource Chemistry & New Material, Qingdao 266000, China

4 College of Smart Energy, Shanghai Jiao Tong University, Shanghai 200240, China; czyan@sjtu.edu.com

5 State Key Laboratory of Oxo Synthesis and Selective Oxidation, Lanzhou Institute of Chemical Physics, Chinese Academy of Sciences, Lanzhou 730000, China; yhli@licp.cas.cn

* Correspondence: shwan@licp.cas.cn (S.W.); jwlang@licp.cas.cn (J.L.)

Abstract: Electrochemical double-layer capacitors (EDLCs) possess extremely high-power density and a long cycle lifespan, but they have been long constrained by a low energy density. Since the electrochemical stability of electrolytes is essential to the operating voltage of EDLCs, and thus to their energy density, the tuning of electrolyte components towards a high-voltage window has been a research focus for a long time. Organic electrolytes based on ionic liquids (ILs) are recognized as the most commercially promising owing to their moderate operating voltage and excellent conductivity. Despite impressive progress, the working voltage of IL–solvent electrolytes needs to be improved to meet the growing demand. In this review, the recent progress in the tuning of IL- based organic electrolyte components for higher-voltage EDLCs is comprehensively summarized and the advantages and limitations of these innovative components are outlined. Furthermore, future trends of IL–solvent electrolytes in this field are highlighted.

Keywords: EDLCs; electrolytes; IL–solvent; high-voltage; component tuning

1. Introduction

Renewable forms of energy, including solar, wind, hydroelectric, geothermal, and biomass, are essential to move forward the sustainable economy [1]. However, these renewable energies have the disadvantages of intermittency and require the participation of energy storage devices [2]. Reversible energy storage and discharge technology has become indispensable, requiring the development of systems that deliver both high energy and high-power capability. Electrical energy can be stored electrochemically in batteries and capacitors. Various types of batteries include lithium-ion batteries (Li-ion), sodium–sulfur batteries (Na–S), lead–acid (Pb–acid) batteries, and flow batteries. Among the various types of battery systems, lithium-ion batteries hold the dominant position in the market with their energy density of up to 300 Wh/kg. Nevertheless, their power density and lifetime are constrained by the mechanism of store charge, that is, the change is converted into electrical energy through an electrochemical redox reaction. Moreover, the batteries always face the problem of short life and slow charging rate due to the limitations of volume charge, phase transition, and solid diffusion rate during charging and discharging.

Electrochemical double-layer capacitors (EDLCs) [3,4], also referred to as supercapacitors, store and release energy based on charge adsorption and desorption on electrode surfaces (Figure 1b). Therefore, they exhibit advantages such as exceedingly high power density (>10 kW kg^{-1}), rapid charging/discharging, and an extremely long cycle life. The devices do not undergo Faraday reactions during operation, making their charge and discharge processes very effective, and the cycle life significantly surpasses other energy storage devices. Thus, EDLCs are considered as a promising renewable energy storage device and have become the preferred technology for applications that need to provide a rapid and sustained energy supply [5]. Moreover, EDLCs are suitable for incorporation into grid energy storage to conserve power units and for frequency regulation, as well as to enhance power quality and smoothness. As a result of the application of these features, the global EDLC market continues to grow, which can be observed in Figure 1c. Among the regional markets, the Asia–Pacific region has great potential for market growth in the upcoming years, whereas Europe and North America also hold a considerable market share.

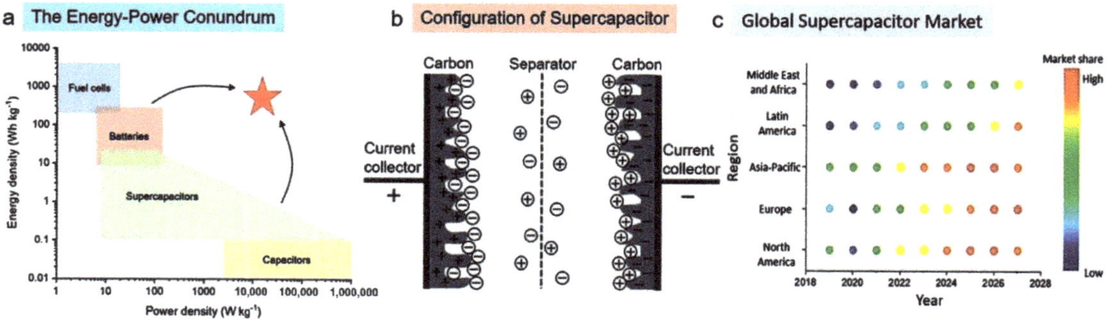

Figure 1. (**a**) A Ragone plot depicting various energy storage systems; (**b**) fundamental design and operational principles of an EDLC; (**c**) global distribution of supercapacitor market shares by region and year [3].

Typically, the energy density of an EDLC is determined by its electrodes and electrolytes [6]. Based on the equation of energy density ($E = 1/2\ CV^2$, where C is the capacitance of the electrodes and V represents the operating voltage window) [7], one strategy is to develop novel electrode materials to deliver higher capacitance. Carbon-based materials have become the most widely studied active materials [8,9] due to their excellent chemical and thermal stability. According to the basic structure and shape of carbon materials, three main types of carbon-based materials for symmetrical EDLCs are as follows: (I) graphene [10,11] and carbon nanotubes [12,13]; (II) carbide-derived carbon and aerogels [14,15]; and (III) activated carbon (AC) [16,17]. Nevertheless, the reported specific capacitance for alternative materials typically ranges between 100 and 150 F g^{-1}, which is not significantly superior to activated-carbon-based electrodes. Strategies for modifying carbon-based materials have been employed in high-voltage EDLCs, encompassing methods like heteroatom doping, selection of diverse raw precursor materials, activation techniques for AC, and controlling the growth process of CNT to prevent the formation of amorphous carbon. However, the synthesis process of these carbon materials for high-voltage uses is considered to be complex and costly, which is not suitable for industrial application. Apart from increasing the specific capacitance of electrodes [18], expanding the operating voltage window (V) is another efficient technique for enhancing the energy density. Meanwhile, the operating voltage of EDLCs is closely associated with the properties of the electrolyte, which are mainly limited by the electrochemical stability of the electrolyte. Due to this consideration, the development of alternative electrolyte salts, or the tuning of electrolyte composition, represents one of the keys to achieving high-energy EDLCs nowadays [19–21].

Ionic liquids (ILs) are salt-like materials formed by ionic bonds between cations and anions. Due to their wide electrochemical potential windows, high thermal and chemical stability, non-volatility, and non-combustibility, ILs can significantly help EDLCs obtain higher energy densities than aqueous electrolytes. Therefore, ILs as electrolytes in EDLCs have garnered considerable interest [22]. To date, extensive research has been conducted on IL-based electrolytes, including pure ILs, binary mixtures of ILs, and ILs combined with organic solvent electrolytes. Among them, solvent-free IL–electrolytes can provide the highest operating voltage window in the range of 3–4 V for the carbon–electrode combination [23]. However, due to large ion sizes, their high viscosity and low ionic conductivity limit the dynamic behaviors of the EDLCs by ionic diffusion. Moreover, the high cost of neat ILs also poses a challenge for the commercialization of EDLCs. IL–organic solvent electrolytes, featuring a moderate operating voltage window, are the most extensively commercialized type of electrolyte for EDLCs at present. However, the operating voltage of EDLC devices using commercial tetra–ethylamine tetrafluoroborate ($TEABF_4$)/acetonitrile electrolyte is limited to 2.8 V. Thus, the development of a novel IL-based electrolyte with a voltage up to 3.0 V or higher has been a research hotspot in this field.

Recently, many studies have focused on developing suitable IL–solvent electrolyte components for EDLCs to attain increased voltage and enhanced energy density. IL–solvent electrolytes were mentioned in some reviews [21,24,25] as the key materials of high-energy-density EDLCs, but so far, no review has summarized in great detail the tuning of the components of IL–solvent for the high voltage requirement. Therefore, this review will focus on the tuning strategies for high-voltage IL–solvent electrolytes for EDLCs that have been executed so far, focusing on the tuning of concentration, solvent composition, and the solute salt component of electrolytes based on common ILs, as well as the design principle. At the same time, we also hope to provide reference strategies for the further development of advanced electrolytes in the future.

2. Electrolytes for EDLCs

Electrolytes are the important components of EDLCs, affecting the electrochemical performance, operation life, and safety. Electrolytes with high voltage and ionic conductivity have demonstrated a substantial enhancement in the overall performance of energy storage devices [26]. The cell voltage is mainly limited by the electrochemical stability window (ESW) of the electrolyte, which is defined as the range between the reduction potential and oxidation potential. The ESW is influenced by the characteristics of ions and solvents, along with the interactions between ion–ion, ion–solvent, and electrode–electrolyte surfaces. Several studies have suggested that the decomposition of the electrolyte can be attributed to the presence of catalytic active centers, particularly functional groups on the carbon electrode [27].

Electrolytes primarily consist of solute salts and solvents, in which the solute salts provide ions and the solvents facilitate pathways for the transport of these electrolyte ions. The electrolytes can be classified into liquid electrolytes and solid electrolytes [28,29], and the liquid electrolytes can be further classified into aqueous electrolytes and non-aqueous electrolytes according to the types of solvents. The liquid electrolytes used in EDLCs include aqueous and organic electrolytes and pure ILs (Table 1). Aqueous electrolytes possess inherent advantages, including safety, low cost, high conductivity (~1 S cm^{-1}), minimum ion-pairing possibility, and environmental friendliness [30]. The ionic conductivity of aqueous electrolytes significantly surpasses that of non-aqueous electrolytes, resulting in superior rate performance. Nevertheless, the stable window of water is limited to only 1.23 V, which imposes constraints on the practical application of aqueous electrolyte [31–33]. "Water–in–salt" (WIS) electrolyte refers to an electrolyte with an exceptionally high concentration that approaches saturation levels. The number of "free" water molecules is lower than that of the solute molecules, so WIS electrolytes can increase the operating voltage of aqueous EDLCs [34–37]. Until now, the working potential voltages of symmetric EDLCs assembled by AC electrode and WIS electrolyte are still $\leq$3.0 V [38]. In contrast, organic electrolytes [39,40] have wider stable electrochemical potential windows of 2.5–3.0 V and therefore are more preferred than aqueous electrolytes.

Table 1. Electrolytes used in EDLCs.

Types	Electrolytes	Operating Voltage (V)	References
Aqueous electrolyte	6M KOH,6M NaOH, 1M HCl, 0.5M H_2SO_4	0.9	[33]
	1M NaCl, 1M KCl, 0.5M Na_2SO_4, 0.5M K_2SO_4	1.7	[34]
	0.5M H_2SO_4	1.2	[34]
	LiCl (1M, 10M, 20M)	1.4, 1.5, 2.35	[34]
	KCH_3COO	2.0	[38]
	WIS (21M LiTFSI)	2.2	[35]
	WIS (21M LiTFSI)	2.6	[36]
	WIS (31.3M LiTFSI)	2.4	[37]
Solvent-free ILs	$EMImBF_4$	3.5	[41]
	EMImFSI	2.0	[42]
	EMImTFSI	3.0	[43]
	$Pyr_{14}TFSI$	3.7	[44]
	MPImFSI	3.5	[45]
	MPPyrFSI	3.5	[45]
	MeoMPyrFSI	3.5	[45]
	$Pip_{14}B(CN)4$	3.7	[46]
	$Pyr_{14}B(CN)4$	3.7	[46]
Mixed ILs	EMImTFSI + $EMImBF_4$	3.5	[47]
	$Pyr_{14}FSI$ + $PIP_{13}FSI$	3.7	[48]
	$EMImBF_4$ + $TMABF_4$	3.5	[41]
	EMImTFSI + MPPyrTFSI	3.5	[49]
IL–solvent	$Pyr_{14}TFSI$/AN	3.1	[50]
	$Pyr_{14}BF_4$/AN	3.1	[50]
	$Pyr_{14}BF_4$/PC	3.2	[51]
	$TEABF_4$/AN or PC	2.5–2.8	[52]
	$SBPBF_4$/AN or PC	2.8–3.2	[53]
	($TEMABF_4$; $DEDMABF_4$; $TMPABF_4$; $TMEABF_4$)/AN	3.0	[54]

ILs are unique ionic compounds composed of organic cations and organic/inorganic anions. They are also described as molten salts with melting points below 100 °C [55]. ILs are also known for their remarkable properties, including the ability to maintain liquid state over a wide temperature range, negligible vapor pressure, and high thermal stability. The higher electrochemical stability and thermal stability of ILs make them well-suited for high voltage and high-temperature supercapacitors. The cations of ILs can be classified as imidazolium [42,43], pyrrolidinium [44,45], pyridinium, piperidium, quaternary ammonium [54], and quaternary phosphonium. So far, ILs utilizing pyrrolidinium, ammonium, and imidazolium cations are the most promising and widely utilized types. In general, the pyrrolidinium-based ILs have higher electrochemical stability than imidazole-based ILs, while the imidazole-based ILs show higher conductivity and lower viscosity than pyrrolidinium-based ILs [56]. These features make pyrrolidinium-based ILs suitable for greater energy output, while imidazole-based ILs have fast charge/discharge performance. The conventional ILs in EDLCs are based on cations of 1–ethyl–3–methylimidazolium (EMIm) [41], N-propyl-N-methylpyrrolidinium (Pyr_{13}), 1–butyl–1–methylpyrrolidinium (Pyr_{14}) [50], 1–butyl–3–methylimidazolium (BMIm) [57]. Additionally, the most commonly used anions include bis(trifluoromethanesulfonyl)imide ($TFSI^-$), hexafluorophosphate (PF_6^-), bis(fluorosulfonyl)imide (FSI^-), and tetrafluoroborate (BF_4^-). These reported ILs exhibited operating potential windows up to 3.5 V or ever higher when used alone as electrolytes [58]. Furthermore, the structure of the most commonly used anions and cations for ILs is depicted in Figure 2a [59]. In addition, some new ILs (Figure 2b), such as azepanium [60], acetate- and lactate-based piperidinium [46], piperazinium [61], and bis(oxalate)borate-based ILs [62] have been reported for EDLCs (Figure 2b). Relevant reviews have summarized the regulation of the cation and anion structures of ILs used in EDLCs [63,64]. The adjusting methods of the structures include the optimization of side chain lengths for cations and incorporation of functional groups into the cations, such as substituting F with the electron-withdrawing substituent (–CN). Mixed ILs with different anions or cations could alter the initial ion accumulation due to the difference in ion sizes, thereby adjusting the thickness of the electric double layer on the electrode interface and the distribution of ions within the nanopores, which lead to the excellent electrochemical performances of EDLCs, such as wide operating voltage, high capacitance, and long cycle life [41,47–49]. However, solvent-free single ILs and mixed ILs are not utilized in commercial devices owing to their drawbacks such as high expense, limited transport properties, and excessive viscosity. Therefore, the reasonable selection and manipulation of IL-based electrolytes to exploit their advantages and solve their shortcomings will be the future direction of building EDLCs with high energy density.

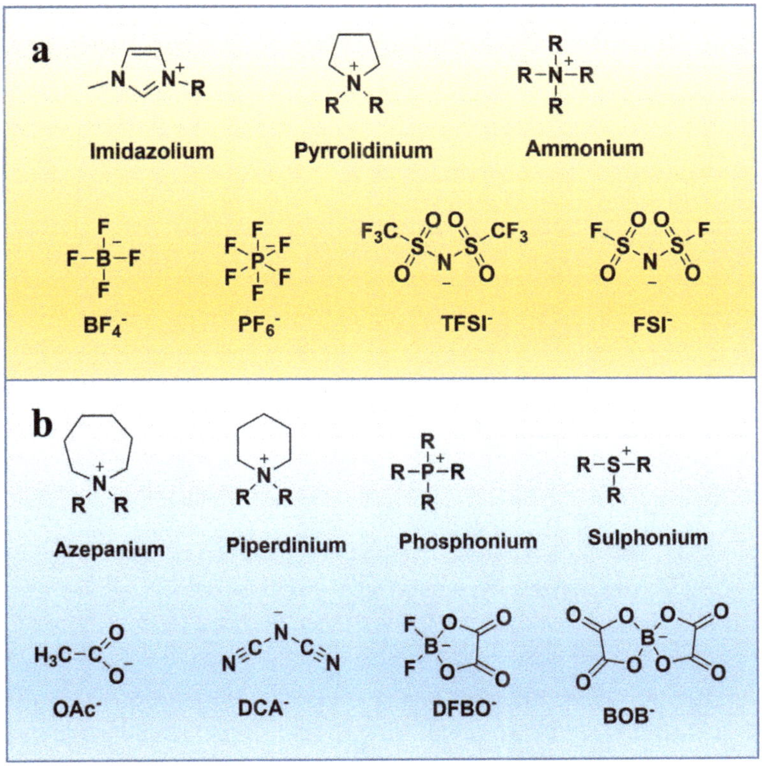

Figure 2. Commonly used cations and anions (**a**) and other cations and anions with new structure (**b**) in ILs.

The cations and anions in ILs are bound by potent Coulombic forces and van der Waals forces, resulting in the high viscosity, low conductivity, and low ion migration rate of ILs. Thus, adding organic solvents to ILs would weaken the Coulombic force and van der Waals force to alleviate the drawbacks of low conductivity and high viscosity of pure ILs. In IL–solvent electrolyte, ILs provide a similar role to other salts, supplying cations and anions in organic solutions. The operating voltage windows of EDLCs with these IL–solvent electrolytes are 2.7–3.0 V [65], which are generally lower than that of pure-IL electrolytes (Table 1). Therefore, EDLCs using IL–solvent electrolytes have moderate operating voltage windows between aqueous electrolytes and pure-IL electrolytes and have excellent electrical conductivity; thus, they are the most widely commercialized currently. The majority of commercially available organic electrolytes are solutions of 1 M ammonium salt, often $TEABF_4$, dissolved in organic solvents like propylene carbonate (PC) or acetonitrile (AN) [66]; their operating voltages are 2.7–2.8 V. Compared with PC-based electrolytes, AN-based IL electrolytes are more commonly used due to their higher conductivity and superior performance at low temperatures. However, the voltage window of the IL–solvent electrolyte system still has a lot of room for improvement. Therefore, increasing the operating voltage of EDLCs using IL–solvent electrolytes is still the research direction in this field, and it is also the focus of this review. There have been some reviews on the structure regulations of ILs and their application in EDLCs [67–71], so many reports on the use of innovative pure ILs as electrolytes for EDLCs are out of the scope of this article.

3. Tuning of IL–Solvent Electrolytes for EDLCs

At high operating voltage, the electrolyte components may decompose on the surface of the carbon-based active material, resulting in an irreversible electrochemical process for the device [72,73]. Therefore, the stability of each component of the electrolyte determines the operating voltage limit. Over recent years, many efforts have been made to develop advanced IL–solvent electrolyte systems to meet the necessary criteria for achieving high-energy EDLCs. At present, for the existing IL solute salts, the methods of tuning IL–solvent electrolyte components to achieve higher voltage can be divided into four categories (Figure 3): (1) increasing the concentration of IL solute salts; (2) selecting single-substitute solvents with high electrochemical stability to replace AN and PC; (3) adding co-solvent to AN or PC to improve the electrochemical stability or conductivity; and (4) introducing solute salt additives.

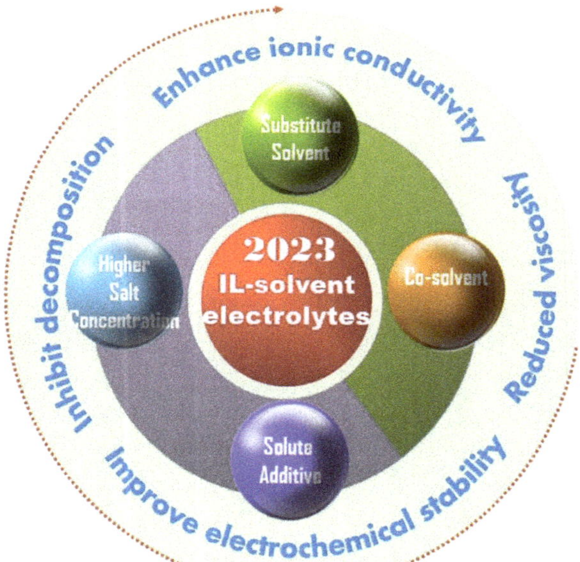

Figure 3. The methods of increasing the voltage of IL–solvent electrolytes as well as the corresponding improved physicochemical properties.

3.1. *Higher Salt Concentration for High Voltage*

The concentration of solute salt in commercial electrolyte is usually 1.0 M. The energy density of EDLCs is determined by the quantity of electrolyte ions per unit area of the electrode surface. The IL concentration plays a pivotal role in regulating viscosity, ionic liquid conductivity, and the potential window. Higher salt concentration can not only increase the number of adsorbed/desorbed ions, which may increase the specific capacity (C), but also has been considered as a method to break through the voltage limits of EDLCs [74], because less free solvent molecules will reduce the redox reaction on the electrode. For example, the use of 2.0 M N,N–dimethyl pyrrolidinium tetrafluoroborate ($Pyr_{11}BF_4$)/AN [75], enabled the EDLC to achieve a high operating voltage of 3.4 V with a capacitance retention of 88% after 500 h. In addition, EDLCs using higher-salt-concentration electrolytes also surpassed the current leading devices in terms of high-temperature performance (retaining 91% capacitance after floating 500 times at 3.0 V and 60 °C) (Figure 4a,b). Meanwhile, post-mortem gas chromatography/mass spectrometry (GC–MS) indicated that the higher performances of 2M $Pyr_{11}BF_4$/AN were related to its high thermal and electrochemical stability. In addition, a Pyr_{14}TFSI/PC (1:1 wt.%) mixture (equivalent to a concentration of 2.3 M) [51] provided the operating voltage of the EDLCs at 3.5 V, which was similar to that of EDLCs using neat Pyr_{14}TFSI. The electrolyte system also ensured excellent cycle stability, with a 5% decrease in capacitance after 100,000 cycles at 3.5 V. Another report of 2.3 M $Pyr_{14}BF_4$/PC electrolyte [76] indicated that the electrolyte allowed the cells to work at a high operating voltage of 3.2 V in a float test, and the conductivity and viscosity were comparable to those of the traditional 1.0 M Et_4NBF_4/PC electrolyte. Moreover, due to the high electrochemical stability of electrolyte, the high cycle stability of EDLCs at 3.2 V was confirmed by charge-discharge measurement. $EMImBF_4$ with small contents (from 0.005 M to 1.0 M) of different organic solvents (AN, PC or γ–butyrolactone) [77] were studied as electrolytes in EDLCs at high potentials ($\geq$3.2 V). Jia et al. [52] systematically studied the effect of $EMImBF_4$ concentrations (including 1.0 M, 2.0 M, 4.0 M, and 6.54 M) on the potential window, ionic conductivity, and viscosity of the $EMImBF_4$/AN electrolyte. The results show that the ESWs of FRGO-based EDLCs increased with increasing IL concentrations: 1M $EMImBF_4$/AN (3.4 V), 2M $EMImBF_4$/AN (3.4 V), and 4M $EMImBF_4$/AN (3.5 V). The increase in specific capacitances had a similar trend: 1M $EMImBF_4$/AN (94.4 F g^{-1}), 2M $EMImBF_4$/AN (147.8 F g^{-1}), and 4M $EMImBF_4$/AN (158.3 F g^{-1}). The maximum specific capacitance offered by the 4M $EMImBF_4$/AN was due to the fact that the capacitance of EDLCs increased with the increase in voltage across electrodes. The reason for the low ESW of dilute solution was that the "freer" AN molecule was prone to irreversible redox reactions on the electrode surface. With rising IL concentrations, the viscosity of the electrolytes increased, whereas the ionic conductivity decreased. The peak was reached when the concentration of $EMImBF_4$ was 2.0 M (Figure 4c–f).

Scalia et al. [78] used N–butyl–N–methyl–pyrrolidinium 4,5-dicyano-2–(trifluoromethyl) imidazole (Pyr_{14}TDI) as the solute salt for EDLCs for the first time and evaluated the performances of the electrolytes with different Pyr_{14}TDI/PC ratios from 1:3 to 3:1 (*w*/*w*). By adjusting the concentrations of the solution, a wide liquid range and high conductivities were achieved. The highest cell voltage of 3.3 V was attained with a Pyr_{14}TDI/PC ratio of 3:1. Meanwhile, although the voltage window of the most diluted electrolyte Pyr_{14}TDI:PC (1:3) was limited to 3.0 V, it delivered the highest specific energy and power. Balducci et al. [79] explored the impact of salt concentrations on the theoretical and practical energy of EDLCs with PC-based $Pyr_{14}BF_4$ and trimethylsulfonium bis[(trifluoromethyl)sulfonyl]imide (Me_3TFSI). The study indicates that 1.5 M $Pyr_{14}BF_4$/PC and 3.8 M Me_3TFSI/PC were promising for higher operative voltages of 3.49 V and 2.88 V, respectively.

Based on the examples given above, a high concentration of IL–solvent electrolyte does have a significant effect on raising the voltage, but the reasons for the formation of this electrochemical stability in high-concentration electrolytes have not been clearly studied. Moreover, there are still some shortcomings to limit their commercial application, such as low conductivity, high viscosity, and high cost.

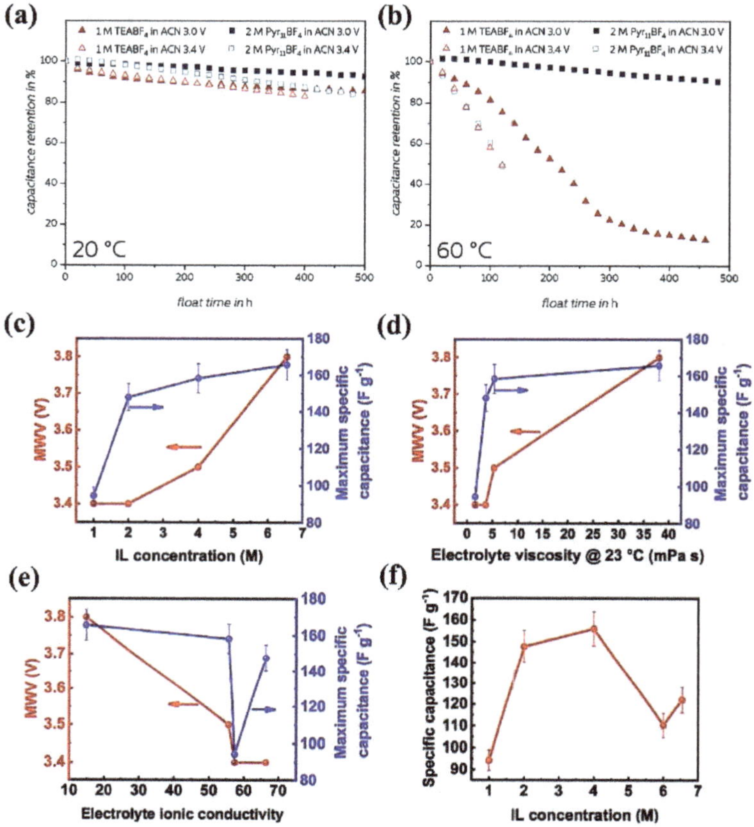

Figure 4. Comparisons of the floating stability of 1 M $TEABF_4$/AN and 2 M $Pyr_{11}BF_4$/AN at 20 °C (**a**) and 60 °C (**b**) [75]. The maximum operating voltages (MWVs) and maximum specific capacitance related to (**c**) the concentration of IL, (**d**) the viscosity of the electrolyte, and (**e**) the ionic conductivity of electrolytes. The variation curve of specific capacitance with IL concentration at 3.4 V (**f**) [52].

3.2. The Use of Single-Substitute Solvent with High Electrochemical Stability

AN and PC are the commonly used organic solvents in the electrolyte for EDLCs. With regard to improving the stable voltage window of IL-based electrolytes, several organic solvents with higher electrochemical stability were proposed and tested as the alternative solvents of AN and PC [80], such as ethyl isopropyl sulfone, propionitrile, adiponitrile, and butylene carbonate. Ethyl isopropyl sulfone (EiPS) has a high boiling point (265 °C) and a relatively low melting point (−8 °C) and low reactivity with water, which makes it a potential organic solvent for high-voltage EDLCs [81,82]. Balducci et al. [83] further studied the electrochemical performance of EDLCs using $TEABF_4$/EiPS at high voltage and different temperatures from 20 to 80 °C. The results show that 0.5 M $TEABF_4$/EiPS electrolyte featured high electrochemical and thermal stability, allowing EDLCs to operate at high voltages from 3.0 V to 3.4 V at 60 °C and 80 °C. At 20 °C, 90% and 74% of the initial capacitance of EDLCs with $TEABF_4$/EiPS were retained after 500 h of floating at 3.4 V and 3.6 V, respectively. This superior stability is one of the highest reports of laboratory-scale $TEABF_4$-based EDLCs and is much higher than AN-based $TEABF_4$ electrolyte. In addition, the devices retained 68% of their original capacitance after floating at 3.4 V at 60 °C for 500 h, while devices using $TEABF_4$/electrolytes were unstable under these conditions. Based on a postmortem GC–MS analysis of the $TEABF_4$/EiPS electrolyte, no soluble decomposition products were found from the solvent, and the post-mortem

X-ray photoelectron spectra (XPS) analysis of the electrodes revealed the accumulation of non-soluble decomposition compounds on the electrode surface at high temperatures. The electrochemical reactions on the electrode surface were confirmed, including cation degradation via Hofmann elimination, the fluorination of the carbon surface, the in situ formation of HF, and the decomposition of EiPS. Based on the characterization results, the high electrochemical stability was attributed to the deposition of EiPS decomposition products on the electrode surface, which might act as a passivation layer to inhibit further degradation (Figure 5).

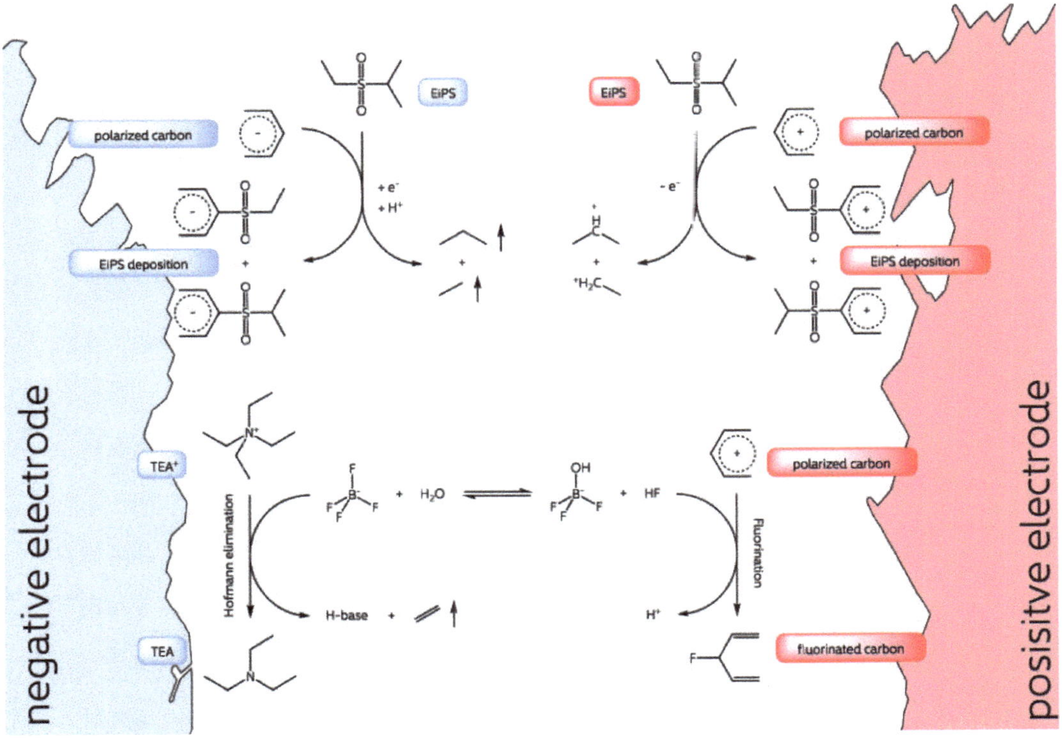

Figure 5. Proposed aging processes occurred on the interphase between $TEABF_4$/EiPS electrolyte and activated carbon electrodes at elevated temperatures [83].

Butylene carbonate (BC) is similar to PC in reduction stability and has higher oxidative stability than PC. Therefore, BC has higher electrochemical stability than PC and is a potential high-voltage alternative solvent of PC. BC has some isomers, including BC, 2,3–BC, and 1,2–BC. The boiling point of 2,3–butylene carbonate (BC) [84] is higher (243 °C) than that of PC. A spiro-(1,10)-bipyrrolidinium tetrafluoroborate ($SBPBF_4$)/2,3–BC electrolyte showed a voltage window of 3.5 V, far exceeding the PC-based electrolyte (2.7 V). The interface stability at high voltage was explained by the fact that the 4th and 5th positions provided a high-level protection from the external environment. This protective effect made these sites less vulnerable to attack by compounds that decompose the electrolyte (such as H_2O). Furthermore, 1,2–Butylene carbonate, as the isomer of 2,3–BC, also worked as alternative for PC [85]. The use of 1.5 M $Pyr_{14}BF_4$/1,2–BC allowed EDLCs to operate at 3.15 V, and the capacitance retention reached 90% after 500 h of floating.

Propionitrile (PN) has a higher boiling point (97 °C) and a lower melting point (−93 °C) than AN [86]. Nguyen et al. [87] reported the utilization of 1.0 M $SBPBF_4$/PN and N, N-dimethyl pyrrolidinium tetrafluoroborate ($DMPBF_4$)/PN as EDLC electrolytes and compared their performances with 1.0 M $DMPBF_4$/AN and $SBPBF_4$/AN electrolyte. The PN-based electrolytes exhibited high electrochemical stability, low viscosity, good ionic conductivity, and excellent anodic potential limit. The symmetrically stable electrochemical windows of alternative electrolytes helped to protect these electrolytes from decomposition when charged to high oxidation potentials. Consequently, EDLCs using 1.0 M $DMPBF_4$/PN and $SBPBF_4$/PN electrolytes were cycled over 10,000 times at 3.5 V and 1000 h at 3.3 V. The $DMPBF_4$/PN electrolyte enabled EDLCs to achieve the highest power density (22.9 kW kg^{-1}) and energy density (49.3 Wh kg^{-1}). Otherwise, adiponitrile (ADN) also worked as an alternative solvent of AN due to its high boiling point (295 °C) and high flashpoint (163 °C). The voltage of EDLCs using 1.75 M $Pyr_{14}BF_4$/ADN electrolyte [88] ranged from 3.2 V to 3.7 V, and the capacitance retention of devices reached 81% at 3.7 V after 100,000 cycles (at 5 A g^{-1}). The operating voltage of EDLCs reached 3.4 V when galvanostatic charge-discharge processes were performed at room temperature as well as 60 °C, while that of EDLCs reached 3.0 V in the float tests. The temperature-programmed desorption and nitrogen adsorption analysis of the aged electrodes proved that the primary factor contributing to the degradation of EDLC performance was the formation of surface functional groups, which led to the gradual blockage of pores. Nevertheless, electrolytes using these nitrile solvents exhibit elevated viscosity and reduced conductivity compared to AN, thus constraining the power performance of the EDLCs.

It is noted that the use of some new solvents with high electrochemical stability indeed improves the voltage window of the IL–solvent electrolytes. However, due to their high viscosity and low ionic conductivity, this is usually accompanied by a reduction in power output.

3.3. Adding Co-Solvents to AN or PC for High Voltage

Co-solvents are often employed to adjust macroscopic properties of electrolyte components, including viscosity, dielectric constant, and melting point [89]. Electrolytes with multi-solvent systems demonstrate broad tolerances owing to the co-solvent's relatively lower viscosity and higher dielectric constant [90]. AN works as the widely used solvent for EDLC electrolyte, as the co-solvent can improve the electrochemical stability of AN-based electrolyte. As mentioned earlier, EDLCs utilizing $TEABF_4$–EiPS electrolyte achieved an operating voltage window of 3.4 V [83]. Balducci et al. [53] used AN and EiPS as mixed solvents to prepare 1.0 M $TEMABF_4$/EiPS/AN electrolyte. It was found that the operating voltage of the electrolytes using EiPS and AN (75:25 and 50:50) as co-solvent was 3.0 V, and the electrolyte system with EiPS/AN (75:25) showed excellent cycle and floating stability. In fact, EDLCs using the electrolyte with a 75:25 ratio exhibited only a 3% loss of initial capacitance after a 500 h cycles test at 3.0 V. Lang et al. [91] found that incorporating methyl acetate (MA) into $EMImBF_4$/AN can improve the performance of EDLCs at low temperature. The optimal pairing of an APDC electrode with the $EMImBF_4$/AN/MA electrolyte allowed the EDLCs to operate normally at −50 °C and exhibited an operating voltage of 3.5 V, a maximum power density of 26 kW kg^{-1}, a maximum energy density of 80 Wh kg^{-1}, and a long cycle lifespan of 10,000 cycles. Subsequent studies found that adding a fluorinated ether, 1,1,2,2–tetrafluoroethyl–2,2,3,3–tetrafluoropropyl ether (TTE), into $TEABF_4$/AN electrolyte could improve the stability of the electrolyte. As a result, a high working voltage of 3.6 V was assembled, which was much higher than the typical working voltage of 2.7 V for AN-based electrolytes without TTE [92]. In addition, the leakage current of $TEABF_4$–TTE/AN was smaller than that of electrolyte without TTE at the same charging voltage, and no decomposition of electrolyte was observed at 3.6 V. When charged to 3.6 V, the open-circuit voltage of ELDC with $TEABF_4$–TTE/AN decayed by 2.03 V after 24 h, which was lower than that of the EDLC without TTE (2.60 V). A mechanistic analysis indicated that cell self-discharge with the TTE additive was slowed down, which could be attributed to the inhibition of the faradaic reaction process caused by electrolyte decomposition.

As a common solvent for EDLCs with organic electrolyte, PC has higher electrochemical stability and better safety than AN, but its electrical conductivity is lower than AN. Thus, the electrochemical stability and conductivity of IL/PC-based electrolytes also need to be improved by adding other solvents. A common linear ether, 1,2–dimethoxyethane (DME), which has favorable chemical/electrochemical stability and quite low viscosity, has also been as a co-solvent for PC to enhance the transport behavior of ions. The EDLCs assembled with $EMImBF_4$–PC–DME (PC: DME = 1:1) electrolyte [93] had a voltage of 3.0 V and the property of fast charge-discharge. Ethylene carbonate (EC) has lower viscosity and a higher dielectric constant than PC. Dimethyl carbonate (DMC) has a linear structure, and its viscosity and melting point are significantly lower than those of ethylene carbonate (EC). For instance, the dielectric constants of propylene carbonate (PC), DMC, and EC are 65, 3.1, and 90 V/m, respectively, while their viscosities are 2.5, 0.6, and 1.9 mPa s, respectively. Therefore, the adding of DMC or EC into PC-based electrolyte is beneficial to the preparation of advanced electrolytes with high electrical conductivity and low viscosity [94]. The electrolytes containing $SBPBF_4$ with a solvent mixture of PC/DMC/EC (1:1:1 in volume) and PC/DMC (1:1 in volume) were studied for EDLCs [95]. The PC/DMC system exhibited the lowest viscosity (3.14 mPa s), while the PC/DMC/EC system showed the highest conductivity (18.08 mS cm^{-1}). Additionally, $SBPBF_4$ using a solvent mixture of the PC/DMC/EC system exhibited a stable voltage of 3.2 V and superior rate performance. In addition, the monofluorination of carbonates can increase relative permittivity and kinematic and dynamic viscosity. The dynamic viscosity of fluoroethylene carbonate (FEC) at 40 °C (4.1 mPa s) is higher than that of PC at 25 °C (2.53 mPa), and the relative dielectric constant of FEC at 40 °C (78.4) is higher than that of PC at 25 °C (64.92). Jänes et al. [96] introduced FEC as the co-solvent to construct FEC/PC solvent mixtures in various volume ratios. The EDLCs assembled with 1M $(C_2H_5)_3CH_3NBF_4$/FEC:PC (1:19) (5 vol.% FEC) and microporous TiC–CDC electrodes reached a voltage of 3.0 V. Therefore, the co-solvent in the above IL–PC system could reduce the viscosity or freezing point of electrolyte and slightly increase the voltage window.

From the above examples, it can be seen that co-solvents are mainly eaters, and there are few types of co-solvents reported at present. The co-solvent can slightly enhance the voltage window of the IL–AN (PC) electrolyte, and the main promotion effect is the improvement of other electrochemical properties, such as the increase in conductivity to improve the rate performance or low–temperature performance, which also are the important factors of EDLCs.

3.4. The Use of Solute Salt Additives for High Voltage

The ions of the solute salt play a direct role in forming the electric double layer, so the selection of solute salts is crucial for developing advanced EDLCs [97,98]. Encouragingly, it is of practical significance to study the IL–solvent electrolyte additives that can improve the operating voltage for improving energy density. F-containing compounds can form an effective passivation layer on the electrode surface, thereby minimizing the decomposition of solvent or AN polymerization. Back in 2005, Ishikawa et al. [99] reported that the cathodic stability of $EMImBF_4$ could be expanded by adding $LiBF_4$. Compared with $EMImBF_4$ and $TEMABF_4$/PC, $EMImBF_4$+$LiBF_4$/PC had a higher electrochemical stability window. They later confirmed that the enhancement of cathodic stability in the $LiBF_4$/$EMImBF_4$ system was associated with an "inhibiting effect" caused by the presence of Li^+ ions, rather than the "coating effect" from the formation of SEI [100]. The protective impact of Li^+ ions helped in curbing the side reactions on the AC electrode. Li et al. [101] used $LiFP_6$ as an additive to prepare the $EMImBF_4$/$LiPF_6$/PC/DMC electrolyte, which had a low viscosity of 2.89 mPa s, a high ionic conductivity of 20.72 mS cm^{-1}, and high electrochemical stability. The inhibiting effect brought by Li^+ effectively hindered the reduction of cations in IL. The advanced 3V-class EDLCs were successfully constructed using this electrolyte, and their comprehensive performance was obviously better than that of EDLCs using pure $EMImBF_4$ or $EMImBF_4$/solvent, with a capacitance retention of 95.8% after 20,000 cycles at 3.0 V. The improvement of the overall performance was primarily attributed to the improved ion transport capacity, the balance charge storage of the electrodes, and the passivation of the electrode surface, which were caused by the combined action of Li salt and carbonate solvent.

Krause et al. [102] reported fluorinated additives such as ethyl(2,2,2–trifluoroethyl) carbonate (ETFEC), (2,2,2–trifluorothyl) methyl carbonate (TFEMC), lithium difluoro(oxalate) borate (LiDFOB), bis(2,2,2–trifluorothyl) carbonate (bTFEC), and tris(2,2,2–trifluoroethul) phophite (TTFEPi) for the $SBPBF_4$/AN system. The results demonstrate that LiDFOB and ETFEC could improve the stability at high voltage (>2.7 V). In particular, when 2 wt% LiDFOB additive was added to $SBPBF_4$/AN, the resulting EDLC cells revealed good capacitance retention and reduced internal resistance in the voltage ranges of 3.5 to 4.0 V (Figure 6). In addition, EDLCs using TTFEPi showed lower self-discharge performance. The electrochemical decomposition degree of the internal components of the cell was reflected by the steady-state current value. The cell with TTFEPi additive showed the lowest current value in the test, indicating that TTFEPi might reduce aluminum corrosion at high voltage to improve the coulombic efficiency.

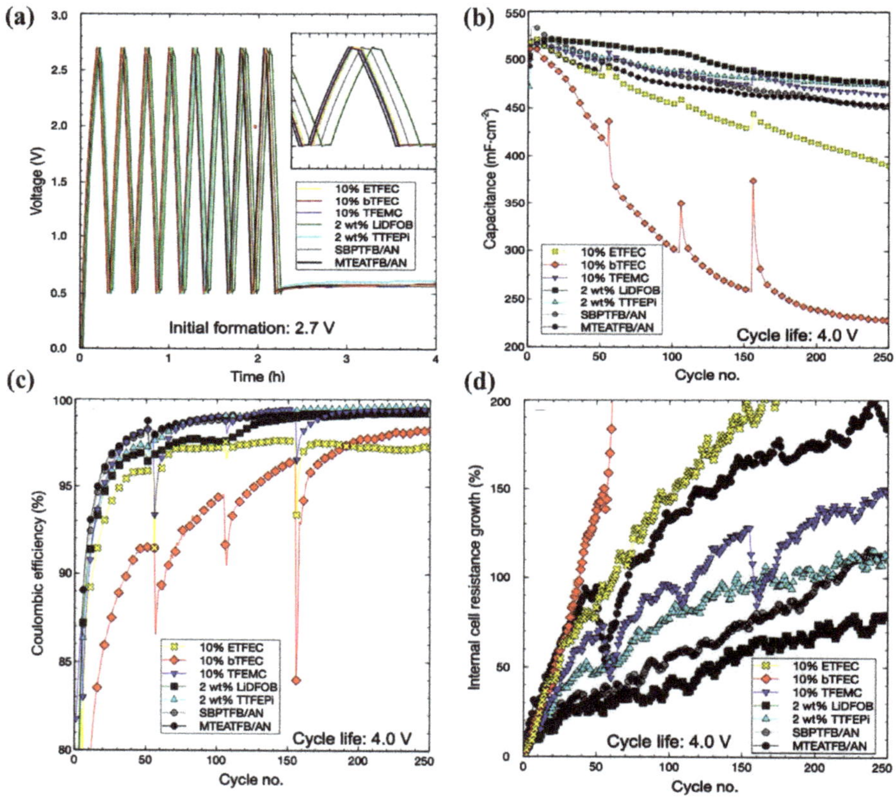

Figure 6. (**a**) Eight constant current charge/discharge cycles at a current of 5 mA across a voltage range of 0.5 to 2.7 V. (**b**) Capacitance of EDLCs during cycle life testing across a range of 0.5 to 4.0 V. (**c**) Coulombic efficiency of EDLCs under a voltage range of 0.5 to 4.0 V cycle life. (**d**) Increase in internal resistance as determined by current interrupting a voltage range of 0.5 to 4.0 V in cycle life test [102].

Besides the F-containing compounds, different kinds of ILs can also be used as the additives for IL-based electrolytes. In addition to the organic solvent with high dielectric constant and minimal viscosity, incorporating ILs with other cations can also improve the electric conductivity and reduce the viscosity of IL-based electrolytes. Adding different contents of $EMImBF_4$ to 0.2 M $SBPBF_4$/PC/DME [103] electrolyte reduced the interfacial resistance and bulk resistance, thereby increasing the ionic conductivity. In particular, the electrolyte with 15 vol% $EMImBF_4$ showed superior capacitance, electrochemical stability, ionic conductivity, and low interfacial resistance. Moreover, when 15 vol% $EMImBF_4$ was added, the XPS spectra confirmed the thinnest SEI film. Otherwise, adding $EMImBF_4$ or EMImTFSI as additive into $TEABF_4$/PC [104] also improved the ionic conductivity and stability by inhibiting electrolyte decomposition. The ESWs of the electrolytes with IL additive were wider than that of the conventional electrolyte, which enabled EDLCs to operate stably at 3.7 V. When the amounts of $EMImBF_4$ and EMImTFSI were 7 wt.%, the specific capacitances of the initial value of EDLC after 10,000 cycles were 97.5% and 91.3%, respectively. By mixing N–butyl–N–methylpyrrolidinium bis(trifluoromethanesulfonyl)imide (Pyr_{14}TFSI) and trimethyl propylammonium bis(trifluoromethanesulfonyl)imide (TMPATFSI) with an equivalent mole ratio into AN at 1.0 M concentration [105], the binary IL electrolyte exhibited a low viscosity of 0.692 mPa s^{-1}, an exceptional electric conductivity of 44.3 mS cm^{-1}, and a wide electrochemical stability window of 4.82 V. EDLCs constructed with AC electrodes and this binary ILs electrolyte achieved an elevated operating voltage of 3.1 V, a maximum power density of 32.16 kW kg^{-1}, and a maximum energy density of 28.30 Wh kg^{-1}. These metrics were notably superior compared to those of commercially available EDLCs using organic electrolytes.

Unlike the common ILs, which can improve the conductivity and reduce the viscosity of the electrolyte by cationic regulation, functional ILs with large-sized groups can enlarge the operating voltage window by adjusting the polarization limit of the electrode. The operating voltage of EDLCs is determined by the upper limit potential (P_U) and lower limit potential (P_L), which correspond to the potential of oxidation and reduction of electrolyte, respectively. By adding a silica-grafted ILs additive into the EMImTFSI/PC electrolyte, Yan et al. [106] developed an effective strategy to adjust the polarization potential of electrodes and improved the potential limits of the positive and negative electrodes of EDLCs (Figure 7a). Compared with the electrolyte without additives, the electrolyte with 10% SiO_2–IL–TFSI exhibited a more symmetric potential range for two electrodes, and the P_{0V} was adjusted to ~0 V vs. Ag. In this case, P_U and P_L were expanded to 1.6 and −1.6 V vs. Ag in a three-electrode system, and the voltage of the AC-based EDLCs increased from 2.8 to 3.2 V. The energy density was increased by 39% with the improvement in cycle stability. Subsequently, a polymeric imidazole IL with ether branch (PVEImTFSI) was synthesized and worked as an additive to EMImTFSI/PC. It was found that PVEImTFSI could also enlarge the limit potential ranges of carbon-based EDLC [107]. The addition of PVEImTFSI improved the electrochemical stability and effectively adjusted the upper limit potential (P_U) and lower limit potential (P_L) of the electrolytes. The P_U and P_L of electrolytes with 10% PVEImTFSI reached 1.7 and −1.5 V vs. Ag. The working voltage of EDLCs with 10% PVEImTFSI was increased to 3.2 V compared with EDLCs at 2.8 V without additives. As a result, the energy density was increased by approximately 41.5%. The cycling stability of EDLCs with electrolyte containing 10% PVEImTFSI at 3.0 V and 3.2 V was also significantly enhanced. In addition, the equivalent circuit model was simulated and the Bode plots were proposed according to the results of electrochemical impedance spectroscopy (EIS). Both simulated and measured date show that the resistance of each element was reduced by adding PVEImTFSI, which improved the ion dynamic and reduced the energy barrier of the ion diffusion, thereby improving the electrochemical performance of EDLCs (Figure 7b).

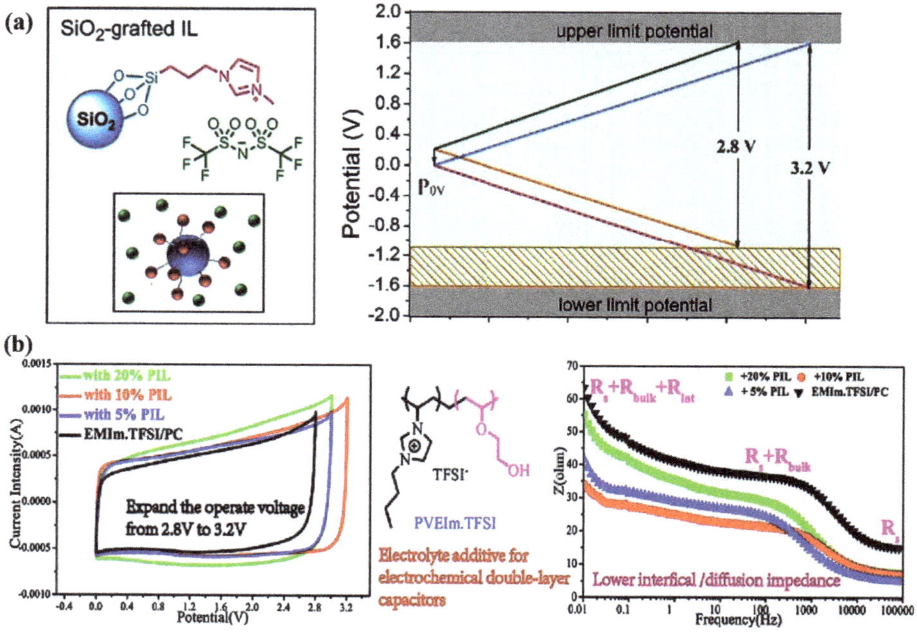

Figure 7. (**a**) The structure of SiO_2–IL–TFSI and the schematic of increasing the operational potential [106]; (**b**) chemical structure of PVEImTFSI and its positive effects on voltage and impedance [107].

Moreover, novel electrolytes for EDLCs were prepared by adding cetyltrimethylammonium-bromide-grafted Ti_3C_2 MXene (Ti_3C_2–CTAB) into $EMImBF_4$/PC and Et_4NBF_4/PC [108]. Remarkably, the ionic conductivity of electrolyte with the Ti_3C_2–CTAB additive increased by 38%, and the solid–liquid surface energy decreased from 18.03 mN m^{-1} to 12.37 mN m^{-1}. The hydrogen bonds and electrostatic force generated between Ti_3C_2-CTAB and $EMImBF_4$ promoted the dissociation and ion transfer of electrolyte ion pairs. Comparing with free-additive electrolyte, the EDLCs assembled with 0.5% additive showed higher rate capability (76.4% vs. 45.8%), higher specific capacitance (29.6 F g^{-1} vs. 21.6 F g^{-1}), and longer cycling life after 9000 cycles at 3.0 V (83.6% vs. 59.8%). The addition of Ti_3C_2–CTAB reduced the interfacial resistance of electrolyte/electrode, which improved the electrochemical performance. Consequently, EDLCs based on Ti_3C_2–CTAB/$EMImBF_4$/PC obtained a high energy density of 28.3 Wh kg^{-1} and a power density of 18.3 kW kg^{-1}.

It can be seen from the results reported above that there are many kinds of solute additives, including F-containing compounds, ILs, and ionic compounds with large-sized groups. The effect of adding these additives is immediate, so the addition of solute additives is a very effective and relatively simple way to achieve high voltage. However, designing or selecting suitable additives is still a challenge.

4. Conclusions and Future Perspectives

EDLCs have become important energy storage devices to improve the power quality of renewable energy because of their fast energy storage and high power output. As mentioned above, the electrochemical stability window is an important factor restricting the energy density of EDLCs. The development of electrolytes with wider electrochemical windows is an important direction for constructing high-energy-density EDLCs. IL–solvent electrolytes are the most commercially promising for EDLCs due to their moderate conductivity and electrochemical window. In the past few years, many efforts have been dedicated to the study of new solvents and electrolyte salts to achieve high-voltage EDLCs.

This review provides insights to expand the working window from the IL–solvent component level. In summary, tuning the components of the IL–solvent electrolyte is an effective method to broaden the voltage window of EDLCs with IL–solvent electrolytes. Various excellent organic IL electrolytes have been developed, including those that have increased IL concentration, those that have novel organic solvents with high electrochemical stability, and those that have co-solvents and solute salt additives in their electrolytes. The as–discussed high-voltage IL–solvent electrolyte components and their electrochemical performances are summarized in Table 2. Studies show that the voltage of EDLCs with the use of these electrolytes increased compared to that of the initial. The results indicate that the electrolyte salts, solvents, and additives have strong effects on the electrochemical voltage window and the properties of the interface. Despite the fact that some progress has been made in the research of IL–solvent electrolytes, there are still some works to be carried out in order to meet their practical applications. The future research of IL–solvent electrolytes will involve the following considerations:

Table 2. The methods to achieve high voltage of EDLCs with IL–solvent electrolytes.

Tuning Methods	Electrolytes	Voltage (V)	Ref.	Challenges
High salt concentration	2M $Pyr_{11}BF_4$/AN	3.4	[75]	High viscosity and high cost limit the commercial application
	2.3M Pyr_{14}TFSI/PC	3.5	[51]	
	$EMIMBF_4$/AN or PC or γ-butyrolactone (0.005–1.0 M solvent)	3.2	[77]	
High stable solvent	$TEABF_4$/EiPS	3.4	[83]	Accompanied by a reduction in power output
	$SBPBF_4$/2,3-BC	3.5	[84]	
	$Pyr_{14}BF_4$/1,2-BC	3.15	[85]	
	$SBPBF_4$/PN	3.5	[87]	
	$DMPBF_4$/PN	3.3	[87]	
	$Pyr_{14}BF_4$/ADN	3.4	[88]	
Co-solvent	$TEMABF_4$/EiPS/AN	3.0	[53]	The increase of voltage window is limited
	$EMImBF_4$/MA/AN	3.5	[91]	
	$TEABF_4$/TTE/AN	3.6	[92]	
	$SBPBF_4$/DME/PC	3.2	[95]	
	$(C_2H_5)_3CH_3NBF_4$/FEC/PC	3.0	[96]	
Solute salt additive	$EMImBF_4$/$LiPF_6$/PC/DMC	3.0	[101]	The designing and selecting suitable additives lack systematic guidance
	$SBPBF_4$/LiDFOB/AN	4.0	[102]	
	$TEABF_4$/$EMImBF_4$/PC	3.7	[104]	
	Pyr_{14}TFSI/TMPATFSI/AN	3.1	[105]	
	EMImTFSI/SiO_2-IL/PC	3.2	[106]	
	EMImTFSI/PVEImTFSI/PC	3.2	[107]	
	$EMImBF_4$/Ti_3C_2CTAB/PC	3.0	[108]	

(1) It is necessary to develop high-purity, newly low-cost IL salts with high electrochemical stability windows. Firstly, the structure innovations of IL electrolyte salts with high surface ion density are carried out by the cation structure regulation strategy. Secondly, the purity of IL should be taken into account. It is well known that a small number of halides or water may influence the physicochemical properties of electrolytes and the behavior of EDLCs significantly. Thirdly, the cost is the major consideration while using IL in commercial applications. It is necessary to explore the development of simple and atom-economic synthesis methods with low-cost raw materials and post-treatment purification steps, which make the IL a cheap solute. Finally, the strong effect of innovation on IL–solvent electrolytes on the structure and the formation of the double layer is necessary to study. Among the several aspects, the influences of the solvation process and ion mobility need to be analyzed, which will indicate the advantages and limitations of innovative electrolyte components.

(2) We must continue to screen novel organic co-solvents with high dielectric constants, wide temperature ranges, low costs, and low toxicities to improve the electrochemical stability of EDLCs. Nevertheless, in some AN systems, the increase in voltage is obtained at the expense of power reduction. Therefore, it is necessary to carry out research on the tuning technology of multi-solvent components and explore the synergistic mechanism between the components. Considering that the high electrolyte concentration and highly stable alternative solvents in the above strategies can increase the voltage window and that additives and co-solvents can also increase the conductivity, the combination of these strategies can develop a high-concentration ionic liquid–mixed solution electrolyte to achieve a synergistic effect in high-performance EDLCs.

(3) The selection and design of solute additives still remain a major challenge. Large-cation ILs are the main research content of large-size salts as IL additives, and the rarely involved anions may be a future direction. Meanwhile, it is necessary to analyze the influence of solute additives to the functional groups on the electrode surface, which will allow us to deeply understand the advantages of electrolyte components and the enhanced mechanisms.

(4) As the microscopic reaction of the electrode–electrolyte interface occurs in the Stern layer of the electric double layer, which is determined by the ion dynamics and physicochemical interactions, theoretical study of the formation, thickness, and composition of the Stern layer will be the focus of research in the field of high voltage. The ionic interaction and kinetics of the IL–solvent system at the electrolyte/electrode interface should be systematically studied using molecular dynamics simulations. The influence mechanisms of ion size and solvation on the formation and thickness of the Stern layer in the electric double layer also need to be systematically studied via simulation. In addition, it is necessary to introduce an effective computational screening method to identify ILs, solvents, and additives suitable for EDLCs.

(5) A clear understanding of the electrochemical degradation mechanism of the IL–solvent electrolyte of EDLCs at high voltage is still lacking at present, which gives important guidance for the design of IL–solvent components with a wide operating window. The inhibition of electrochemical reactions on the electrode surface caused by the electrolyte components requires a deep understanding so that a highly electrochemical stable electrolyte and an electrode material with a stable surface can be designed.

Author Contributions: Y.W.: investigation, writing—original draft, methodology, funding acquisition; K.X.: investigation, formal analysis; X.Z.: investigation, formal analysis; C.Y.: funding acquisition, formal analysis; Y.L.: funding acquisition; K.S.: review—editing; P.M.: supervision—review, funding; S.W.: supervision—review; J.L.: supervision, funding acquisition, writing—review and editing. All authors have read and agreed to the published version of the manuscript.

Funding: This research was funded by the Taishan Scholars Program, No. tsqnz20230629; the Western Young Scholars Foundations of Chinese Academy of Sciences; the Science Fund of Shandong Laboratory of Yantai Advanced Materials and Green Manufacturing, No. AMGM2022A02, AMGM2023A04; the Collaborative Innovation Alliance Fund for Young Science and Technology Worker, No. HZJJ23-5; and the FAW Volkswagen China Environmental Protection Foundation Automotive Environmental Protection Innovation Leadership Program.

Data Availability Statement: Not applicable.

Conflicts of Interest: The authors declare no conflicts of interest.

References

1. Hu, X.; Zuo, D.; Cheng, S.; Chen, S.; Liu, Y.; Bao, W.; Deng, S.; Harris, S.J.; Wan, J. Ultrafast materials synthesis and manufacturing techniques for emerging energy and environmental applications. *Chem. Soc. Chem.* **2023**, *52*, 1103. [CrossRef]
2. Zhang, Z.; Ding, T.; Zhou, Q.; Sun, Y.; Qu, M.; Zeng, Z.; Ju, Y.; Li, L.; Wang, K.; Chi, F. A review of technologies and applications on versatile energy storage systems. *Renew. Sustain. Energy Rev.* **2021**, *148*, 111263. [CrossRef]
3. Huang, J.; Xie, Y.; You, Y.; Yuan, J.; Xu, Q.; Xie, H.; Chen, Y. Rational design of electrode materials for advanced supercapacitors: From lab research to commercialization. *Adv. Funct. Mater.* **2023**, *33*, 2213095. [CrossRef]

4. Zhao, J.; Burke, A.F. Review on Supercapacitors: Technologies and performance evaluation. *J. Energ. Chem.* **2021**, *59*, 276–291. [CrossRef]
5. Pershaanaa, M.; Bashir, S.; Ramesh, S.; Ramesh, K. Every bite of supercap: A brief review on construction and enhancement of supercapacitor. *J. Energy Storage* **2022**, *50*, 104699. [CrossRef]
6. Pathak, M.; Bhatt, D.; Bhatt, R.C.; Bohra, B.S.; Tattari, G.; Rana, S.; Arya, M.C.; Sahoo, N.G. High Energy Density Supercapacitors: An Overview of Efficient Electrode Materials, Electrolytes, Design, and Fabrication. *Chem. Rec.* **2024**, *24*, e202300236. [CrossRef]
7. Simon, P.; Gogotsi, Y. Materials for electrochemical capacitors. *Nat. Mater.* **2008**, *7*, 845–854. [CrossRef]
8. Song, G.; Li, C.; Wang, T.; Hu, K.L.F.; Cheng, S.; Hondo, E.; Liu, S.; Kawi, S. Hierarchical hollow carbon particles with encapsulation of carbon nanotubes for high performance supercapacitors. *Small* **2023**, *20*, 2305517. [CrossRef]
9. Luo, L.; Lan, Y.; Zhang, Q.; Deng, J.; Luo, L.; Zeng, Q.; Gao, H.; Zhao, W. A review on biomass-derived activated carbon as electrode materials foe energy storage supercapacitors. *J. Energy Storage* **2022**, *55*, 105839. [CrossRef]
10. Wu, Y.; Ye, J.; Jiang, G.; Ni, K.; Shu, N.; Taberna, P.; Zhu, Y.; Simon, P. Electrochemical characterization of single layer graphene/electrolyte interface: Effect of solvent on the interfacial capacitance. *Angew. Chem. Int. Ed.* **2021**, *60*, 13317–13322. [CrossRef]
11. Tareen, A.K.; Khan, K.; Iqbal, M.; Zhang, Y.; Long, J.; Nazeer, F.; Mahmood, A.; Mahmood, N.; Shi, Z.; Ma, C.; et al. Recent advanced in novel graphene: New horizons in renewable energy storage technologies. *J. Mater. Chem. C* **2022**, *10*, 11472–11531. [CrossRef]
12. Das, S.; Manuraj, M.; Rakhi, R.B.; Ajayaghosh, A. High-frequency electrochemical double layer capacitor based in carbon nanotubes ink coated eggshell membrane electrodes. *J. Energy Storage* **2022**, *45*, 103799. [CrossRef]
13. Liu, J.; Yuan, Y.; Fang, H.; Xu, Y.; Sun, W.; Chen, S.; Wang, Y.; Lv, L. Redox-active tetramino-benzoquinone π-π stacking and H-bonding onto multiwalled carbon nanotubes toward a high-performance asymmetric supercapacitor. *ACS Appl. Energy Mater.* **2022**, *5*, 8112–8122. [CrossRef]
14. Martín-Illán, J.; Sierram, L.; Ocón, P.; Zamora, F. Electrochemical double-layer capacitor based on carbon@ covalent organic framework aerogels. *Angew. Chem. Int. Ed.* **2022**, *61*, e202213106. [CrossRef]
15. Shalkh, J.S.; Shalkh, N.S.; Mishra, Y.K.; Pawar, S.S.; Parveen, N.; Shewale, P.M.; Sabale, S.; Kanjanaboos, P.; Praserthdam, S.; Lokhande, C.D. The implementation of graphene-based aerogel in the field of supercapacitor. *Nanotechnology* **2021**, *32*, 362001.
16. Najib, S.; Erdem, E. Current progress achieved in novel materials for supercapacitor electrodes: Mini review. *Nanoscale Adv.* **2019**, *1*, 2817–2827. [CrossRef] [PubMed]
17. Schütter, C.; Pohlmann, S.; Balducci, A. Industrial requirements of materials for electrical double layer capacitors: Impact on current and future applications. *Adv. Energy Mater.* **2019**, *9*, 1900334. [CrossRef]
18. Tao, R.; Yang, T.; Zhang, J.; Wu, Z.; Qiu, L. Design strategies of covalent organic framework-based electrodes for supercapacitor application. *Chem. Commun.* **2023**, *59*, 3175–3192. [CrossRef] [PubMed]
19. Balducci, A. Electrolytes for high voltage electrochemical double layer capacitors: A perspective article. *J. Power Sources* **2016**, *326*, 534–540. [CrossRef]
20. Simon, P.; Gogotsi, Y. Perspectives for electrochemical capacitors and related devices. *Nat. Mater.* **2020**, *19*, 1151–1163.
21. Lim, J.M.; Jang, Y.S.; Nguyen, H.V.T.; Kim, J.S.; Yoon, Y.; Park, B.J.; Seo, D.H.; Lee, K.; Han, Z.; Ostrikov, K.; et al. Advances in high-voltage supercapacitors for energy storage systems: Materials and electrolyte tailoring to implementation. *Nanoscale Adv.* **2023**, *5*, 615–626. [CrossRef]
22. Kim, E.; Han, J.; Ryu, S.; Choi, Y.; Yoo, J. Ionic liquid electrolyte for electrochemical energy storage devices. *Materials* **2021**, *14*, 4000. [CrossRef] [PubMed]
23. Ray, A.; Saruhan, B. Application of ionic liquids for batteries and supercapacitors. *Materials* **2021**, *14*, 2942. [CrossRef] [PubMed]
24. Miao, L.; Song, Z.; Zhu, D.; Li, L.; Gan, L.; Liu, M. Ionic liquids for supercapacitive energy storage: A mini-review. *Energy Fuels* **2021**, *35*, 8443–8455. [CrossRef]
25. Gerdroodbar, A.E.; Alihemmati, H.; Safavi-Mirmahaleh, S.; Golshan, M.; Damircheli, R.; Eliseeva, S.N.; Salami-Kalajahi, M. A review on transport pathways and coordination chemistry between ions and electrolytes in energy storage devices. *J. Energy Storage* **2023**, *74*, 109311. [CrossRef]
26. Xia, L.; Yu, L.; Hu, D.; Chen, G.Z. Electrolytes for electrochemical energy storage. *Mater. Chem. Front.* **2017**, *1*, 584–618. [CrossRef]
27. Venâncio, R.; Vicentini, R.; Pinzón C., M.J.; Corrêa, D.A.; Miranda, A.N.; Queiroz, A.C.; Degasperi, F.T.; Siqueira, L.J.A.; Da Silva, L.M.; Zanin, H. Combining electrochemical, molecular simulation and operando techniques to investigate the stability of electrodes and organic electrolytes used in EDLCs. *Energy Storage Mater.* **2023**, *62*, 102943. [CrossRef]
28. Zhang, J.; Wang, J.; Zuo, D.; Xu, J.; Wang, Q.; Zhang, H. A supramolecular gel polymer electrolyte based on poly(vinyl alcohol) and tannic acid for flexible electrical double layer capacitors. *J. Energy Storage* **2023**, *72*, 108618. [CrossRef]
29. Gajewski, P.; Lewandowska, A.; Szczesniak, K.; Voelkel, A.; Marcinkowska, A. Optimization of ionogel polymer electrolytes composition for their best performance in electric double layer capacitor. *ChemElectroChem* **2022**, *27*, e202200745. [CrossRef]
30. Bhat, M.Y.; Hashmi, S.A.; Khan, M.; Choi, D.; Qurashi, A. Frontiers and recent developments on supercapacitor's materials, design, and applications: Transport and power system applications. *J. Energy Storage* **2023**, *58*, 106104. [CrossRef]
31. Lu, S.; Song, Y.; Guo, K.; Chen, X.; Xu, J.; Zhao, L. Effect of aqueous electrolytes on the electrochemical behaviors of ordered mesoporous carbon composites after KOH activation as supercapacitors electrodes. *J. Electroanal. Chem.* **2018**, *818*, 58–67. [CrossRef]

32. Zhang, X.; Shen, C.; Kao, E.; Warren, R.; Zhang, R.; Teh, K.S.; Zhong, J.; Wei, M.; Li, B.; Chu, Y.; et al. Titanium Disulfide Coated Carbon Nanotube Hybrid Electrodes Enable High Energy Density Symmetric Pseudocapacitors. *Adv. Mater.* **2018**, *30*, 1704754. [CrossRef]
33. Tian, Z.; Deng, W.; Wang, X.; Liu, C.; Li, C.; Chen, J.; Xue, M.; Li, R.; Pan, F. Superconcentrated aqueous electrolyte to enhance energy density for advanced supercapacitors. *Funct. Mater. Lett.* **2017**, *10*, 1750081. [CrossRef]
34. Xiao, D.; Zhang, L.; Li, Z.; Dou, H.; Zhang, X. Design strategies and research progress for Water-in-Salt electrolytes. *Energy Storage Mater.* **2022**, *44*, 10–28. [CrossRef]
35. Xiao, D.; Wu, Q.; Liu, X.; Dou, Q.; Liu, L.; Yan, B.; Yu, H. High-voltage and high-capacitance aqueous supercapacitor using carbon nanorods electrodes and water-in-salt electrolyte. *ChemElectroChem* **2019**, *6*, 439–443. [CrossRef]
36. Martins, V.L.; Mantovi, P.S.; Torresi, R.M. Suppressing early capacitance fade of electrochemical capacitors with water-in-salt electrolytes. *Electrochim. Acta* **2021**, *372*, 137854. [CrossRef]
37. Lannelongue, P.; Bouchal, R.; Mourad, E.; Bodin, C.; Olarte, M.; Vot, S.; Favier, F.; Fontaine, O. "Water-in-Salt" for Supercapacitors: A Compromise between Voltage, Power Density, Energy Density and Stability. *J. Electrochem. Soc.* **2018**, *165*, A657–A663. [CrossRef]
38. Samanta, P.; Ghosh, S.; Kundu, A.; Samanta, P.; Murmu, N.C.; Kulia, T. Review a strategic way of high-performance energy storage device development with envirmentally viable "Water-in-salt" electrolytes. *J. Energy Chem.* **2023**, *78*, 350–373. [CrossRef]
39. Laheäär, A.; Jänes, A.; Lust, E. $NaClO_4$ and $NaPF_6$ as potential non-aqueous electrolytes salts for electrical double layer capacitor application. *Electrochim. Acta* **2012**, *82*, 309–313. [CrossRef]
40. Chidiac, J.; Timperman, L.; Anouti, M. Role of FTFSI anion asymmetry on physical properties of AFTFSI (A=Li, Na and K) based electrolytes and consequences on supercapacitor application. *ChemPhysChem* **2021**, *22*, 1863–1879. [CrossRef]
41. Wang, X.; Mahandzhiyski, A.Y.; Arstad, B.; Van Aken, K.L.; Mathis, T.S.; Gallegos, A.; Tian, Z.; Ren, D.; Sheridan, E.; Grimes, B.A.; et al. Selective charging behavior in an ionic mixture electrolyte supercapacitor system for higher energy and power. *J. Am. Chem. Soc.* **2017**, *139*, 18681–18687. [CrossRef] [PubMed]
42. Handa, N.; Sugimoto, T.; Yamagata, M.; Kikuta, M.; Kono, M.; Ishikawa, M. A neat ionic liquid electrolyte based on FSI anion for electric double layer capacitor. *J. Power Sources* **2008**, *185*, 1585–1588. [CrossRef]
43. Largeot, C.; Portet, C.; Chmiola, J.; Teberna, P.L.; Gogotsi, Y.; Simon, P. Relation between the ion size and pore size for an electric double-layer capacitor. *J. Am. Chem. Soc.* **2008**, *130*, 2730–2731. [CrossRef]
44. Rennie, A.J.R.; Martins, V.L.; Torresi, R.M.; Hall, P.J. Ionic liquids containing sulfonium cations as electrolytes for electrochemical double layer capacitors. *J. Phys. Chem. C* **2015**, *119*, 23865–23874. [CrossRef]
45. Lee, J.H.; Ryu, J.; Lee, A.S.; Na, W.; Yoon, H.; Kim, W.; Koo, C.M. High-voltage ionic liquid electrolytes based on ether functionalized pyrrolidinium for electric double-layer capacitors. *Electrochim. Acta* **2016**, *222*, 1847–1852. [CrossRef]
46. Martins, V.L.; Rennie, A.J.R.; Sanchez-Ramirez, N.; Torresi, R.M.; Hall, P.J. Improved performance of ionic liquid supercapacitors by using tetracyanoborate anions. *ChemElectroChem* **2018**, *5*, 598–604. [CrossRef] [PubMed]
47. Aken, K.L.V.; Beidaghi, M.; Gogotsi, Y. Formulation of ionic-liquid electrolyte to expand the voltage window of supercapacitors. *Angew. Chem. Int. Ed.* **2015**, *54*, 4806–4809. [CrossRef]
48. Lin, R.; Taberna, P.L.; Fantini, S.; Presser, V.; Pérez, C.R.; Malbosc, F.; Rupesinghe, N.L.; Teo, K.B.K.; Gogosti, Y.; Simon, P. Capacitive energy storage from −50 to 100 °C using an ionic liquid electrolyte. *J. Phys. Chem. Lett.* **2011**, *2*, 2396–2401. [CrossRef]
49. Newell, R.; Faure-Vincent, J.; Iliev, B.; Schubert, T.; Aradilla, D. A new high performance ionic liquid mixture electrolyte for large temperature range supercapacitor application (−70 °C to 80 °C) operating at 3.5 V cell voltage. *Electrochim. Acta* **2018**, *267*, 15–19. [CrossRef]
50. Krummacher, J.; Schütter, C.; Passerini, S.; Balducci, A. Characterization of different conductive salts in CAN-based electrolytes for electrochemical double-layer capacitors. *ChemElectroChem* **2013**, *4*, 353–361. [CrossRef]
51. Balducci, A. High voltage electrochemical double layer capacitor containing mixtures of ionic liquids and organic carbonate as electrolytes. *Electrochem. Commun.* **2011**, *13*, 814–817. [CrossRef]
52. Wong, S.I.; Lin, H.; Sunarso, J.; Wong, B.T.; Jia, B. Optimization of ionic-liquid based electrolyte concentration for high-energy density graphene supercapacitors. *Appl. Mater. Today* **2020**, *18*, 100522. [CrossRef]
53. Schütter, C.; Bothe, A.; Balducci, A. Mixtures of acetonitrile and ethyl isopropyl sulfone as electrolytes foe electrochemical double layer capacitors. *Electrochim. Acta* **2020**, *331*, 135421. [CrossRef]
54. Koh, A.R.; Hwang, B.; Roh, K.C.; Kim, K. The effect of the ionic size of small quaternary ammonium BF_4 salts on electrochemical double layer capacitors. *Phys. Chem. Chem. Phys.* **2014**, *16*, 15146–15151. [CrossRef] [PubMed]
55. Lei, Z.; Chen, B.; Koo, Y.M.; Macfarlane, D.R. Introduction: Ionic liquids. *Chem. Rev.* **2017**, *117*, 6633. [CrossRef] [PubMed]
56. Choudhary, G.; Dhariwal, J.; Saha, M.; Trivedi, S.; Banjare, M.K.; Kanaoujiya, R.; Behera, K. Ionic liquids: Environmentally sustainable materials for energy conversion and storage applications. *Environ. Sci. Pollut. Res.* 2023. [CrossRef] [PubMed]
57. Wang, X.; Zhou, H.; Sheridan, E.; Walmsley, J.C.; Ren, D.; Chen, D. Geometrically confined favorable ion packing for high gravimetric capacitance in carbon-ionic liquid supercapacitors. *Energy Environ. Sci.* **2016**, *9*, 232–239. [CrossRef]
58. Ali, E. Supercapacitors utilising ionic liquids. *Energy Storage Mater.* **2017**, *9*, 47–69.
59. Lethesh, K.C.; Bamgbopa, M.O.; Susantyoko, R.A. Prospects and design insights of neat ionic liquids as supercapacitor electrolytes. *Front. Energy Res.* **2021**, *9*, 741772. [CrossRef]

60. Pohlmann, S.; Olschläger, T.; Giidrich, P.; Jacquemin, J.A.V.J.; Balducci, A. Mixtures of azepanium based ionic liquids and propylene cabonates as high voltage electrolytes for supercapacitors. *Electrochim. Acta* **2015**, *153*, 426–432. [CrossRef]
61. Kasprzak, D.; Stepniak, I.; Galinski, M. Acetate- and lactate-based ionic liquids: Synthesis, characterisation and electrochemical properties. *J. Mol. Liq.* **2018**, *264*, 233–241. [CrossRef]
62. Nguyen, H.V.; Lee, S.; Kwak, K.; Lee, K.K. Bis(oxalate)borate-containing electrolytes for high voltage electric double-layer capacitors: A comparative study. *Electrochim. Acta* **2019**, *321*, 134649. [CrossRef]
63. Mishra, A.; Shetti, N.P.; Basu, S.; Reddy, K.R.; Aminabhavi, T.M. Chapter 7- Recent developments in ionic liquid-based electrolytes for energy storage supercapacitors and rechargeable batteries. In *Green Sustainable Process for Chemical and Environmental Engineering and Science*; Elsevier Inc.: Amsterdam, The Netherlands, 2020; pp. 199–221.
64. Feng, J.; Wang, Y.; Xu, Y.; Sun, Y.; Tang, Y.; Yan, X. Ion regulation of ionic liquid electrolytes for supercapacitors. *Energy Environ. Sci.* **2021**, *14*, 2859. [CrossRef]
65. Krummacher, J.; Schütter, C.; Hess, L.H.; Balducci, A. Non-aqueous electrolytes for electrochemical capacitors. *Curr. Opin. Electrochem.* **2018**, *9*, 64–69. [CrossRef]
66. Zhong, C.; Deng, Y.D.; Hu, W.B.; Qiao, J.L.; Zhang, L.; Zhang, J.J. A review of electrolyte materials and compositions for electrochemical supercapacitors. *Chem. Soc. Rev.* **2015**, *44*, 7484–7539. [CrossRef]
67. Shahzad, S.; Shah, A.; Kowsari, E.; Iftikhar, F.J.; Nawab, A.; Piro, B.; Akhter, M.S.; Rana, U.A.; Zou, Y. Ionic liquids as environmentally benign electrolytes for high-performance supercapacitors. *Glob. Chall.* **2019**, *3*, 1800023. [CrossRef]
68. Pan, S.; Yao, M.; Zhang, J.; Li, B.; Xing, C.; Song, X.; Su, P.; Zhang, H. Recognition of ionic liquids as high-voltage electrolytes for supercapacitors. *Front. Chem.* **2020**, *8*, 261. [CrossRef] [PubMed]
69. Stettner, T.; Balducci, A. Protic ionic liquids in energy storage devices: Past, present and future perspective. *Energy Storage Mater.* **2021**, *40*, 402–444. [CrossRef]
70. Sun, L.; Zhuo, K.; Chen, Y.; Du, Q.; Zhang, S.; Wang, J. Ionic liquid-based redox active electrolytes for supercapacitors. *Adv. Funct. Mater.* **2022**, *32*, 2203611. [CrossRef]
71. Zhou, T.; Gui, C.; Sun, L.; Hu, Y.; Lyu, H.; Wang, Z.; Song, Z.; Yu, G. Energy application of ionic liquids: Recent developments and future prospects. *Chem. Rev.* **2023**, *123*, 12170–12253. [CrossRef]
72. Bittner, A.M.; Zhu, M.; Yang, Y.; Waibel, H.F.; Konuma, M.; Starke, U.; Weber, C.J. Ageing of electrochemical double layer capacitors. *J. Power Sources* **2012**, *203*, 262–273. [CrossRef]
73. Kurzweil, P.; Schottenbauer, J.; Schell, C. Past, present and future of electrochemical capacitors: Pseudocapacitance, aging mechanisms and service life estimation. *J. Energy Storage* **2021**, *35*, 102311. [CrossRef]
74. Zheng, J.P. The limitations of energy density of battery/double-layer capacitor asymmetric cells. *J. Electrochem. Soc.* **2003**, *150*, A484–A492. [CrossRef]
75. Köps, L.; Kreth, F.A.; Bothe, A.; Balducci, A. High voltage electrochemical capacitors operating at elevated temperature besed on 1,1-dimethylpyrrolidinium tetrafluoroborate. *Energy Storage Mater.* **2022**, *44*, 66–72. [CrossRef]
76. Pohlmann, S.; Balducci, A. A new conducting salt for high voltage propylene carbonated-based electrochemical double layer capacitors. *Electrochim. Acta* **2013**, *110*, 221–227. [CrossRef]
77. Palm, R.; Kurig, H.; Tõnurist, K.; Jänes, A.; Lust, E. Influence of different organic solvent additives on 1-ethyl-3-methylimidazolium tetrafluoroborate electrolyte based electrical double layer capacitors. *J. Electrochem. Soc.* **2013**, *160*, A1741–A1745. [CrossRef]
78. Scalia, A.; Varzi, A.; Moretti, A.; Ruschaupt, P.; Lamberti, A.; Tress, E.; Passerini, S. Electrolytes based on N-butyl-N-methyl-pyrrolidium 4,5-dicyano-2-(trifluoromethyl) imidazole for high voltage electrochemical double layer capacitors. *ChemElectroChem.* **2019**, *6*, 552–557. [CrossRef]
79. Brandt, A.; Balducci, A. Theoretical and practical energy limitations of organic and ionic liquid-based electrolytes for high voltage electrochemical double layer capacitors. *J. Power Sources* **2014**, *250*, 343–351. [CrossRef]
80. Schütter, C.; Husch, T.; Korth, M.; Balducci, A. Toward new solvents for EDLCs: From computational screening to electrochemical validation. *J. Phys. Chem. C* **2015**, *119*, 13413–13424. [CrossRef]
81. Chiba, K.; Ueda, T.; Yamaguchi, Y.; Oki, Y.; Shimodate, F.; Naoi, K. Electrolyte systems for high withstand voltage and durability I. linear sulfones for electric double-layer capacitors. *J. Electrochem. Soc.* **2011**, *158*, A872. [CrossRef]
82. Senoh, H.; Sakaebe, H.; Sano, H.; Yao, M.; Kuratani, K.; Takeichi, N.; Kiyobayashi, T. Sulfone-based electrolyte solutions for rechargeable magnesium batteries using 2,5-dimethoxy-1,4-benzoquinone positive electrode. *J. Electrochem. Soc.* **2014**, *161*, A1315. [CrossRef]
83. Köps, L.; Kreth, F.A.; Leitenschneider, D.; Schutjajew, K.; Glä, R.; Oschatz, M.; Balducci, A. Improving the stability of supercapacitors at high voltages and high temperature by implementation of ethyl isopropyl sulfone as electrolyte solvent. *Adv. Energy Mater.* **2023**, *13*, 2203821. [CrossRef]
84. Chiba, K.; Ueda, T.; Yamaguchi, Y.; Oki, Y.; Saiki, F.; Naoi, K. Electrolyte systems for high withstand voltage and durability II: Alkylated cyclic carbonates for electric double-layer capacitors. *J. Electrochem. Soc.* **2011**, *158*, A1320–A1327. [CrossRef]
85. Hess, L.H.; Balducci, A. 1,2-butylene carbonate as solvent for EDLCs. *Electrochimica Acta* **2018**, *281*, 437–444. [CrossRef]
86. Rumble, J.R.; David, R.L.; Thomas, J.B. *CRC Handbook of Chemistry and Physics: A Ready Reference Book of Chemical and Physical Data*, 99th ed.; CRC Press: Boca Raton, FL, USA, 2018.
87. Nguyen, H.V.T.; Faheem, A.B.; Kwak, K.; Lee, K. Propionitrile as a single organic solvent for high voltage electric double-layer capacitors. *J. Power Sources* **2020**, *463*, 228134. [CrossRef]

88. Bothe, A.; Pourhosseini, S.E.M.; Ratajczak, P.; Beguin, F.; Balducci, A. Towards understanding the impact of operating voltage on the stability of adiponitrile-based electrical double-layer capacitors. *J. Power Sources* **2021**, *496*, 229841. [CrossRef]
89. Tabarov, F.S.; Galimzyanov, R.R.; Krechetov, I.S.; Kalashnik, A.T.; Galimzyanov, T.R.; Boboev, I.R.; Lisitsin, A.V.; Stakhanova, S.V. Vinylene carbonate, toluene and diethyl ether as electrolyte additives for a wide-temperature range operating of EDLCs. *J. Power Sources* **2023**, *560*, 232658. [CrossRef]
90. Yang, H.; Wang, Z.; Qi, Y.; Zhang, C.; Huang, Y.; Yan, J.; Cen, K.; Xiong, G.; Bo, Z.; Ostrikov, K. A strong-weak binary solvation structure for unimpeded low-temperature ion transport in nanoporous energy storage materials. *J. Mater. Chem. A* **2023**, *11*, 16995–18006. [CrossRef]
91. Lang, J.; Zhang, X.; Liu, L.; Yang, B.; Yang, J.; Yan, X. Highly enhanced energy density of supercapacitors at extremely low temperatures. *J. Power Sources* **2019**, *423*, 271–279. [CrossRef]
92. Yao, J.; Shi, M.; Li, W.; Han, Q.; Wu, M.; Yang, W.; Wang, E.; Lu, X. Fluorinated ether-based electrolyte for supercapacitors with increased working voltage and suppressed self-discharge. *ChemElectroChem* **2022**, *9*, e202200223. [CrossRef]
93. Lu, H.; He, L.; Li, X.; Zhang, W.; Che, J.; Liu, X.; Hou, Z.; Du, H.; Qu, Y. Ionic liquid-solvent mixture of propylene carbonate and 1,2-dimethoxyethane as electrolyte for electric double-layer capacitor. *J. Mater. Sci. Mater. Electron.* **2019**, *30*, 13933–13938. [CrossRef]
94. Nambu, N.; Kobayashi, D.; Sasaki, Y. Physical and electrolytic properties of different cyclic carbonates as solvents for electric double-layer capacitors. *Electrochemistry* **2013**, *81*, 814–816. [CrossRef]
95. Yu, X.; Wang, J.; Wang, C.; Shi, Z. A novel electrolyte used in high working voltage application for electrical double-layer capacitor using spiro-(1,1′)-bipyrridinium tetrafluoroborate in mixtures solvents. *Electrochim. Acta* **2015**, *182*, 1166–1174. [CrossRef]
96. Jänes, A.; Thomberg, T.; Eskusson, J.; Lust, E. Fluoroethylene carbonate as co-solvent for propylene carbonate based electrical double layer capacitors. *J. Electrochem. Soc.* **2013**, *160*, A1025. [CrossRef]
97. Vatamanu, J.; Borodin, O.; Olguin, M.; Yushin, G.; Bedrov, D. Charge storage at the nanoscale: Understanding the trends from the molecular scale perspective. *J. Mater. Chem. A* **2017**, *5*, 21049–21076. [CrossRef]
98. Azmi, S.; Klimek, A.; Frackowiak, E. Why electrochemical capacitor electrolytes should not be ignored? *Electrochim. Acta* **2023**, *452*, 142347. [CrossRef]
99. Murayama, I.; Yoshimoto, N.; Egashira, M.; Morita, M.; Kobayashi, Y.; Ishiikawa, M. Characteristics of electric double layer capacitors with an ionic liquid electrolyte containing Li ion. *Electrochemistry* **2005**, *73*, 600–602. [CrossRef]
100. Fukuda, Y.; Tanaka, R.; Ishikawa, M. Beneficial effects of a Li salt on electrode behavior in an ionic liquid for electric double layer capacitors. *Electrochemistry* **2007**, *75*, 589–591. [CrossRef]
101. Li, X.; Li, H.; Feng, R.; He, L.; Lu, H. Electrochemical double-layer capacitor containing mixtures of ionic liquid, lithium salt, and organic solvent as an electrolyte. *Front. Mater.* **2021**, *8*, 633460. [CrossRef]
102. Krause, F.C.; Jones, J.; Smart, M.C.; Chin, K.B.; Brandon, E.J. Screening electrolytes designed for high voltage electrochemical capacitors. *Electrochim. Acta* **2021**, *374*, 137898. [CrossRef]
103. Kim, M.K.K.; Kim, S. Conducting and interface characterization of carbonate-type organic electrolyte containing $EMImBF_4$ as an additive against activated carbon electrode. *Carbon Lett.* **2015**, *16*, 51–56. [CrossRef]
104. Kwon, H.; Jang, S.; Kang, Y.; Roh, K. The effect of ILs as co-salts in electrolytes for high voltage supercapacitors. *Sci. Rep.* **2019**, *9*, 1180. [CrossRef] [PubMed]
105. Bo, Z.; Zhang, X.; Huang, Z.; Huang, Y.; Yan, J.; Cen, K.; Yang, H. Binary ionic liquids hybrid electrolyte based supercapacitors with high energy & power density. *RSC Adv.* **2023**, *13*, 15762. [PubMed]
106. Dou, Q.; Lian, C.; Lei, S.; Chen, J.; Liu, H.; Yan, X. Silica-grafted ionic liquid for maximizing the operational voltage of electrical double-layer capacitors. *Energy Storage Mater.* **2019**, *18*, 253–259. [CrossRef]
107. Wang, Y.; Xue, K.; Zhang, X.; Zhang, X.; Ma, P.; Yang, B.; Xu, S.; Lang, J. High-voltage electrochemical double layer capacitors enabled by polymeric ionic liquid. *Electrochim. Acta* **2023**, *441*, 141829. [CrossRef]
108. Jiang, X.; Wu, X.; Xie, Y.; Wang, Z.; Huang, J.; Qu, Y.; Mu, D.; Zhang, X.; Yang, W.; Zhang, H. Additive engineering enables ionic-liquid electrolyte-based supercapacitors to deliver simultaneously high energy and power density. *ACS Sustain. Chem. Eng.* **2023**, *11*, 5685–5695. [CrossRef]

batteries

MDPI

Article

[SBP]BF_4 Additive Stabilizing Zinc Anode by Simultaneously Regulating the Solvation Shell and Electrode Interface

Xingyun Zhang [1,2,†], Kailimai Su [2,†], Yue Hu [1,2], Kaiyuan Xue [1,2], Yan Wang [2,3], Minmin Han [1,*] and Junwei Lang [2,3,*]

1 College of Mechanical and Electrical Engineering, College of Materials Science and Engineering, Qingdao University, Qingdao 266071, China; 2021023612@qdu.edu.cn (X.Z.); huyue@qdu.edu.cn (Y.H.); 2021023631@qdu.edu.cn (K.X.)
2 Laboratory of Clean Energy Chemistry and Materials, State Key Laboratory of Solid Lubrication, Lanzhou Institute of Chemical Physics, Chinese Academy of Sciences, Lanzhou 730000, China; suklm@licp.cas.cn (K.S.); wangyan@licp.cas.cn (Y.W.)
3 Shandong Laboratory of Advanced Materials and Green Manufacturing at Yantai, Yantai 264000, China
* Correspondence: hanminmin1@qdu.edu.cn (M.H.); jwlang@licp.cas.cn (J.L.)
† These authors contributed equally to this work.

Abstract: The zinc anode mainly faces technical problems such as short circuits caused by the growth of dendrite, low coulomb efficiency, and a short cycle life caused by side reactions, which impedes the rapid development of aqueous zinc-ion batteries (AZIBs). Herein, a common ionic liquid, 1,1-Spirobipyrrolidinium tetrafluoroborate ([SBP]BF_4), is selected as a new additive for pure $ZnSO_4$ electrolyte. It is found that this additive could regulate the solvation sheath of hydrated Zn^{2+} ions, promote the ionic mobility of Zn^{2+}, homogenize the flux of Zn^{2+}, avoid side reactions between the electrolyte and electrode, and inhibit the production of zinc dendrites by facilitating the establishment of an inorganic solid electrolyte interphase layer. With the 1% [SBP]BF_4-modified electrolyte, the Zn | | Zn symmetric cell delivers an extended plating/stripping cycling life of 2000 h at 1 mA cm^{-2}, which is much higher than that of the cell without additives (330 h). As a proof of concept, the Zn||V_2O_5 battery using the [SBP]BF_4 additive shows excellent cycling stability, maintaining its specific capacity at 97 mAh g^{-1} after 2000 cycles at 5 A g^{-1}, which is much greater than the 46 mAh g^{-1} capacity of the non-additive battery. This study offers zinc anode stabilization through high-efficiency electrolyte engineering.

Keywords: aqueous zinc-ion batteries; ionic liquids; electrolyte; additive; zinc anode

1. Introduction

Due to the escalating severity of energy crises and environmental pollution caused by human technological progress and industrialization, the advancement of renewable power conversion/storage technologies has been a focus point and hot topic in the area of power storage [1,2]. Considering their elevated theoretical specific capacity (820 mAh g^{-1}), low redox voltage (−0.76 V vs. SHE), and abundant resources (three times that of lithium metal) of the zinc metal, as well as the cost-effectiveness, high safety, and high ionic conductivity of water-based electrolytes, aqueous zinc-ion batteries (AZIBs) have significant application value and prospects in areas ranging from mobile electronics and electric automobiles to the industrial sector [3,4]. Typically, AZIBs utilize manganese-based materials with large tunnel structures or vanadium-based layered materials as the cathode, with zinc alloys serving as the anode and normal or mildly acidic water-based compounds like $ZnSO_4$, $ZnCl_2$, and $Zn(NO_3)_2$ serving as the electrolyte [5]. At present, there are numerous challenges in the cathode, anode, and electrolyte aspects of AZIBs, and relevant research is in the transitional stage from scientific research to commercialization [6–8].

The zinc anode faces the main technical problems such as short circuits caused by the growth of dendrite, low coulomb efficiency, and a short cycle life caused by side reactions [9]. Among them, zinc dendrite growth is associated with the uneven deposition and solvation of zinc ions, while the side reactions are corrosion, hydrogen evolvement, and obturation of the zinc anode [10]. Therefore, ensuring the reliability of the zinc anode is essential to promoting the practical application of AZIBs. It is not difficult to find that all the reactions occurring at the zinc anode in AZIBs involve strong interactions between Zn^{2+} and H_2O molecules. Firstly, the $Zn(H_2O)_6{}^{2+}$ solvated structure has a high energy barrier, resulting in difficult desolvation, high charge transfer resistance, and slow kinetics during charge and discharge. Secondly, $[Zn(H_2O)_6]^{2+}$ can provide a large number of active H_2O molecules on the interface of the zinc anode, thus causing various side reactions [11,12]. Thirdly, the zinc anode cannot easily form the solid electrolyte interface (SEI) film in an aquatic environment, and a bare zinc anode is prone to side reactions including erosion and hydrogen evolution. Thus, stabilizing the zinc anode requires mitigation of the solvation effect of Zn^{2+} with water and suppression of the activity of water molecules on the anode surface [13].

Researchers have adopted several strategies to increase the stabilization of the zinc anode, including (1) optimizing the electrode structure, (2) constructing functional interfacial layers on the zinc electrode skin, (3) optimizing the separator, and (4) adjusting the composition and concentration of the electrolyte or introducing additives [14–16]. However, the first three methods require a significant amount of time and involve high process costs [17]. High-concentration electrolytes also face challenges such as high cost, limited conductivity, and low-temperature salt precipitation. Introducing additives into electrolytes can directly optimize the solvation structure of Zn^{2+}, regulate the local current distribution, and improve the electrode/electrolyte interface properties [18]. In comparison, the introduction of additives is the simplest and most direct method for stabilizing the zinc anode and is compatible with existing battery manufacturing processes.

Electrolyte additives can be categorized as ionic additives (Na^+, Mg^{2+}, Mn^{3+}, etc.), inorganic additives (oxides and inorganic acids), organic additives (small molecules, polymers, ionic liquids, etc.), and metal additives (lead and tin). Ionic additives containing heavy metals can cause environmental and electrolyte side reaction issues [19,20]. Inorganic and metal additives have received less research attention due to their limited solubility. Organic additives have been a popular topic as they can optimize the solvation structure of zinc ions, regulate uniform nucleation, induce epitaxial deposition, and even produce SEI films in situ [21]. They can also lessen the movement of water and zinc ions by creating hydrogen bonds with water molecules, thereby mitigating water-related side reactions. However, common organic additives have safety hazards such as flammability and toxicity [22]. Consequently, there is an urgent to explore new excellent performance additives that are highly stable, have a wide voltage window, are safe, and have low flammability.

Ionic liquids (ILs) consist of asymmetric organic cations and organic/inorganic anions, which have special characteristics such as a low melting point, good thermal and chemical stability, strong ionic conductivity, a broad electrochemical window, non-volatility, and non-combustibility and have been widely applied in energy storage batteries, metal deposition, organic synthesis, and other fields [23]. In terms of non-aqueous zinc-based batteries, ILs can be used alone as the electrolyte (EMI-DCA [24]) or as the solvent for zinc salt (EMImOTf [25]), and the resulting Zn//graphite battery showed no signs of dendrites or short circuits after hundreds of cycles. In the area of aqueous alkaline zinc batteries, adding a small amount of IL (EMI-DCA [24]) to the KOH electrolyte can improve the secondary nucleation process of zinc, prevent uneven zinc deposition, and inhibit dendrite formation. As for AZIBs, only EMIES [26], [Ch]OAC [27], [BMIM]OTF [28], Me3EtNOTF [29], DES [30], and PIL [31] have been reportedly applied as electrolyte additives, with related work only providing preliminary verification of their feasibility and excellent effects. In summary, other high-performance IL additives need to be screened to provide experimental data for the subsequent design of ideal IL additives with multiple functions such as stabilizing the zinc anode, increasing the voltage window, and improving low-temperature performance [32].

In this paper, a common IL, [SBP]BF_4, was chosen as a novel addition to pure $ZnSO_4$ electrolyte, and its impact on the stability of the zinc anode and the performance of the whole battery was studied systematically [33,34]. In order to avoid side reactions between the electrolyte and electrode, it is discovered that the [SBP]BF_4 is preferentially adsorbed on the zinc anode and facilitates the creation of an inorganic SEI layer. On the other hand, the solvation sheath of hydrated Zn^{2+} ions may be controlled by the [SBP]BF_4, promote the ionic mobility of Zn^{2+}, homogenize the flux of Zn^{2+}, and thereby inhibit the formation of zinc dendrites. The findings of the experiment demonstrated that a minimal concentration of [SBP]BF_4 (1%) may effectively prevent the production of zinc dendrites and produce uniform zinc deposition. Consequently, the zinc anode delivered an extended plating/stripping cycling life of 2000 h in the Zn||Zn symmetric cell with additive at 1 mA cm^{-2}, which is much higher than that of the cell without additives (where the life was only 330 h). Due to the critical dual functions originating from [SBP]BF_4, with 2000 cycles at 5 A g^{-1}, the Zn V_2O_5 battery maintained its specific capacity of 97 mAh g^{-1}, which is significantly higher than the non-additive battery's 46 mAh g^{-1} capacity; this indicates outstanding cycling stability.

2. Experimental Section

2.1. Materials

The following chemicals were procured from various suppliers: Pyrrolidine, 1,4-dibromobutane, and ammonium fluoroborate (NH_4BF_4) were acquired from Shanghai Macklin Chemical Co., Ltd., Shanghai, China. Potassium carbonate (K_2CO_3) and acetonitrile (ACN) were sourced from Sinopharm Chemical Reagents Co., Ltd., Shanghai, China. Vanadium pentoxide nanosheets were purchased from the same company. Zinc sulfate monohydrate ($ZnSO_4{\cdot}H_2O$, AR) was obtained from Aladdin Reagent, Shanghai, China. Acetylene black was procured from Sinopharm Chemical Reagents Co., Ltd., China. Polyvinylidene fluoride (PVDF) was acquired from Sinopharm Chemical Reagents Co., Ltd., China. Finally, 1-methyl-2-pyrrolidone (NMP) was purchased from Sinopharm Chemical Reagents Co., Ltd., China.

2.2. Preparation of [SBP]BF_4

Pyrrolidine (1.1 mol, 78.2 g) was combined with 1,4-dichlorobutane (1.1 mol, 139.7 g) and potassium carbonate (2.1 mol, 290 g) in acetonitrile (ACN, 1.5 L). The resultant mixture was stirred under reflux at 60 °C for 20 h. Upon cooling, the mixture was filtered and the filtrate was subjected to vacuum drying to yield 170 g of a crude chloridate salt. This crude product was then treated with 50% hydrofluoroboric acid (HBF_4, 0.9 mol, 155 g) and ethanol at room temperature for 20 h to facilitate an ion exchange reaction, converting the chloride salt to a BF_4 salt. The reaction mixture was evaporated, and the residual salt was purified through multiple recrystallization steps using ethanol. Following recrystallization, the product was thoroughly dried under vacuum at 90 °C to afford the ionic liquid additive [SBP]BF_4 in a suitable white crystalline form with a purity of 99.99%.

2.3. The Chemical Structure Characterization of [SBP]BF_4

The synthesized [SBP]BF_4 sample was dissolved in a dimethyl sulfoxide (DMSO) solution, and the anions and cations of the sample were analyzed using nuclear magnetic resonance spectroscopy (NMR) (Figures S2 and S3). The ^1H NMR (400 MHz, DMSO-d_6) spectrum is depicted in the figure, with an integral ratio of δ 3.41 (s, ^1H) to δ 2.03 (s, ^1H) protons at 1:1. The ^{19}F NMR (376 MHz, DMSO-d_6) spectrum showed an integral of δ-148.62 (d, J = 20.4 Hz) proton at 1. The presence of DMSO solvent and residual water was also confirmed at dw2.49 and dw3.33, respectively. In conclusion, the NMR data confirm that the synthesized sample was [SBP]BF_4.

2.4. Preparation of V_2O_5 Cathode

The synthesis of the V_2O_5 cathode was conducted following methods described in the existing literature [18]. Initially, a mixture of 70% V_2O_5 (0.21 g), 20% acetylene black (0.06 g), and 10% PVDF (0.03 g) was combined in a mortar. Then, 5 mL of NMP solvent was added and stirred manually for 3 h to obtain a slurry with a viscosity of 6000 mPa·S. Subsequently, the electrode slurry was uniformly coated onto graphite paper. The coated electrode was then subjected to vacuum drying at 50 °C for 12 h to facilitate solvent removal.

2.5. Materials Characterization

The crystal structure and phase of the material were ascertained using X-ray diffraction (XRD, GBC MMA diffractometer, Super-Dimensional Technology Co., Ltd., Taiwan, China). Fourier transform infrared spectroscopy (FTIR, Nexus 870, Thermo Fisher Nicolet, Waltham, MA, USA) and Raman spectroscopy (DXR3xi, Thermo Scientific, Waltham, MA, USA) were employed to investigate changes in functional groups and the sample's molecular structure. The elemental composition and atomic valence were analyzed using X-ray photoelectron spectroscopy (XPS, ESCALAB 250, Thermal Power Corporation, Waltham, MA, USA). Field emission scanning electron microscopy (FE-SEM, Hitachi Regulus 8100, Hitachi, Tokyo, Japan) and transmission electron microscopy (TEM, JEM-2100FEG, Nippon Electronics Co., Tokyo, Japan) were utilized to study the material's morphology.

2.6. Electrochemical Tests

The electrochemical properties of the Zn | | Zn symmetric battery, Zn | | Cu asymmetric battery, and Zn | | V_2O_5 full battery were studied by assembling the CR2032 button battery. Zinc–zinc symmetric batteries were prepared by using $ZnSO_4$ aqueous solution with and without [SBP]BF_4 additive as an electrolyte, glass fiber as a separator, and two identical zinc foils as positive and negative electrodes. The stability of the zinc anode was evaluated at a current density of 1 mA cm^{-2} using the LAND battery test system (CT3002A Wuhan, China). The Zn | | Cu battery, which employs Zn foil as the negative electrode and Cu foil as the positive electrode, underwent a prolonged cycle test at a capacity of 0.5 mAh cm^{-2} at 2 mA cm^{-2}. The CHI660E electrochemical workstation conducted cyclic voltammetry (CV) testing at a scan rate of 1 m Vs^{-1}, chronoamperometry (CA) testing at an overpotential of 10 mV for 600 s, and electrochemical impedance spectroscopy (EIS) testing within the frequency range of 0.01~105 Hz. The Zn | | V_2O_5 full cell, assembled with a Zn anode and V_2O_5 cathode, was subjected to CV, cycle, and rate tests at different current densities of 0.2~1.6 V and 0.5~1.6 V. These evaluations confirmed the practical applicability of the [SBP]BF_4 additive.

3. Result and Discussion

3.1. Structural Characterization Analysis

It is clear from Figure 1a that the initial stage of Zn deposition is usually uneven and is mainly caused by the recognized "tip effect", and finally, obvious Zn dendrites are formed [35]. Simultaneously, the Zn anode/electrolyte contact experiences the development of H_2 and the formation of by-products, both of which invariably result in fast capacity loss [30]. To solve these problems, we proposed using an IL named [SBP]BF_4 as a functional electrolyte additive, which not only ensures the uniform deposition of Zn but also avoids the occurrence of adverse reactions. To verify these functions, AZIBs were constructed by incorporating different amounts of [SBP]BF_4 into 22wt% $ZnSO_4$ aqueous solution.

As shown in Figure 2a, by using the X-ray diffraction (XRD) technique, the composition of chemicals on the surface of zinc anodes after 70 cycles was examined. Apart from significant zinc deposition on the zinc anode surface during cycling in the absence of the [SBP]BF_4 additive, diffraction peaks equivalent to the $Zn_4SO_4(OH)_6{\cdot}4H_2O$ by-product were also seen at 8.3° and 12.1° [31]. This is because the hydrogen evolution side reaction produced a large amount of OH^-, which formed these by-products by combining with Zn^{2+}, $SO_4{}^{2-}$, and H_2O in the electrolyte. However, no by-product peak was observed on the surface of the zinc electrode after cycling in the electrolyte with $[SBP]BF_4$ additives, indicating that the additive inhibited the generation of by-products. Furthermore, we found that the addition of $[SBP]BF_4$ was advantageous for weakening the solvation between Zn^{2+} and water. The FTIR spectra of pure $ZnSO_4$ electrolytes and electrolytes containing varying concentrations (0.5wt%, 1wt%, and 2wt%) of $[SBP]BF_4$ were compared in Figure 2b–d. The O-H stretching vibration peak of water at 3000–4000 cm^{-1} gradually moved to higher wavenumbers with the increase in additive content, which indicated that SBP^+ cations were beneficial for weakening the interaction between water molecules and Zn^{2+} cations. The vibration stretching peak representing SO_4^{2-} at 900–1200 cm^{-1} also shows a blue shift [32], indicating that the addition of additives damaged the electrostatic coupling between Zn^{2+} and SO_4^{2-}, thereby weakening the binding around SO_4^{2-} [33]. In addition, by comparing Raman spectra of pure $ZnSO_4$ electrolyte and electrolytes with different $[SBP]BF_4$ content, it was thoroughly shown that the addition of $[SBP]BF_4$ can lower the activity of the water molecules in the electrolyte. As illustrated in Figure 2e,f, when the additive content increases, the shoulder peak at 3396 cm^{-1} in the Raman spectra gradually weakened and shifted to a higher wavenumber; this demonstrates how the interaction between $[SBP]BF_4$ and water molecules damaged the initial hydrogen bond network. In summary, Zn^{2+} solvation in water can be lessened by $[SBP]BF_4$ addition, which also accelerates the desolvation of hydrated zinc ions, achieves the homogenization of zinc ion flux, and inhibits the decomposition of water molecules, thereby suppressing dendrite growth and mitigating the occurrence of side reactions.

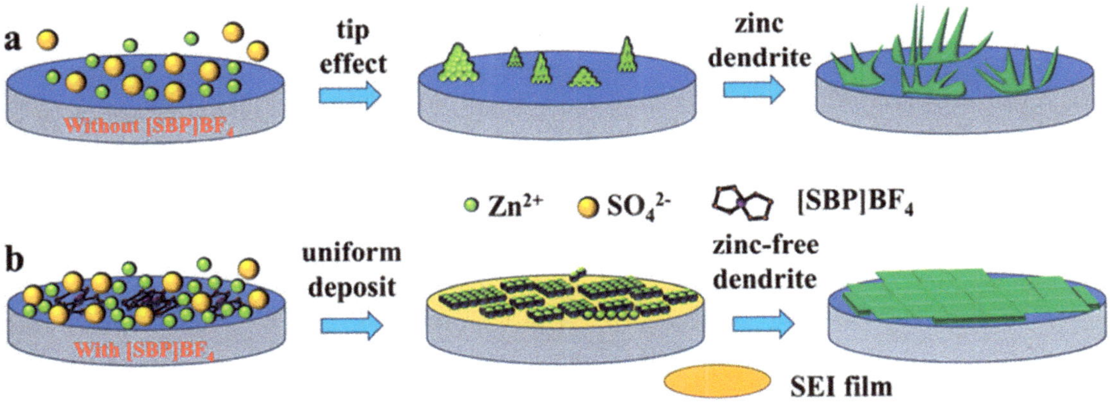

Figure 1. The schematic diagram of the Zn deposition process in the electrolyte (**a**) without and (**b**) with $[SBP]BF_4$ additive.

After 70 cycles, X-ray photoelectron spectroscopy (XPS) was used to characterize the Zn electrode both with and without Ar^+ deep sputtering; the results are displayed in Figure 3a,b. In the absence of $[SBP]BF_4$, C-C and O-C=O can be detected in the C1s spectra of both the zinc electrode surface (no Ar^+ sputtering) and the inner side (Ar^+ sputtering for 5 min), which may mainly be caused by the strong bonded glass fiber separator or CO_2 dissolved in the electrolyte from air [34]. After introducing $[SBP]BF_4$, in addition to the detection of O-C=O peaks related to inorganic $ZnCO_3$, organic components such as C-N, C-H, and C-C were also detected on the surface and inside of the zinc anode, indicating that $[SBP]BF_4$ molecules were adsorbed or decomposed. Analyzing the Zn2p spectra, in the absence of additives, only the loose structure formed by the unevenly distributed ZnO was detected on the surface and inside of the zinc. However, when $[SBP]BF_4$ is present, a large amount of $Zn(OH)_2$ and ZnO inorganic substances was detected both on the surface and inside of the zinc [35]. In addition, for the high-resolution F1 spectra, no F-containing inorganic zinc salt was detected regardless of the sputtering time when there was no [SBP]BF_4. In stark contrast, when [SBP]BF_4 was added, inorganic ZnF_2 was found on the zinc anode surface, and the peak intensity of ZnF_2 escalated with an increase in Ar^+ sputtering depth, suggesting that ZnF_2 not only resided on the surface but also intermingled with organic components embedded within the SEI layer. The existing literature confirms that inorganic constituents like $Zn(OH)_2$, ZnO, and ZnF_2 within the SEI layer have been proven to possess good Zn^{2+} conductivity and hydrophobicity. In summary, adding [SBP]BF_4 additives caused a thick material to develop, as well as a homogeneous SEI layer via organic/inorganic hybridization on the zinc electrode surface. This not only gives it robust toughness but also provides ample transport channels for electrolyte ions [36]. The SEI film not only keeps the aqueous solution and electrode from coming into direct contact, which reduces side reactions related to hydrogen evolution and corrosion, but it also makes it easier for Zn^{2+} to deposit evenly and inhibits the swift proliferation of zinc dendrites.

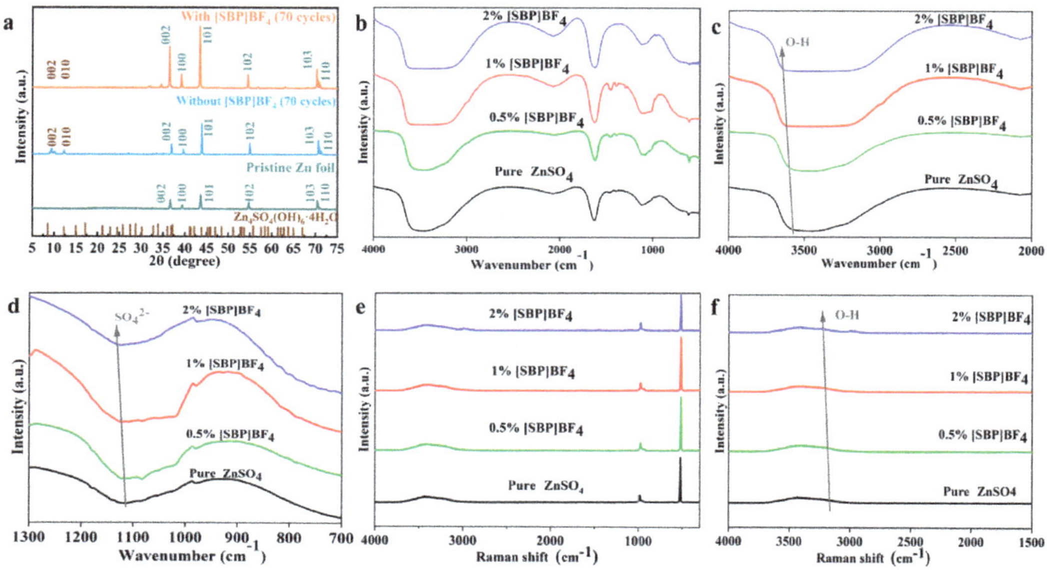

Figure 2. (**a**) XRD patterns of the original zinc foil and the zinc foil after 70 cycles in electrolytes with [SBP]BF_4 and without [SBP]BF_4. FTIR spectra (**b–d**) and Raman spectra (**e**,**f**) of electrolytes containing different proportions of [SBP]BF_4.

To see how the [SBP]BF_4 addition affected the formation of zinc dendrites, scanning electron microscopy (SEM) pictures (Figure 4a,b) were utilized. Without the addition of [SBP]BF_4, disordered and complex dendritic dendrites appeared on the surface of zinc foil, and there were many glass fibers around the dendrites. However, with the addition of additives, zinc foil's surface morphology underwent a dramatic alteration, dendritic development was suppressed, and a dense, homogenous planar structure developed [36]. The optical microscope images further verified that the consistent deposition of zinc ions on the zinc anode's surface was positively impacted by the addition of [SBP]BF_4. Figure 4c,d demonstrate this; many network dendrites appeared on the surface of the zinc sheet without the [SBP]BF_4 additive, resulting in an irregular and rough appearance. After adding [SBP]BF_4, there were no obvious dendrites on the surface of the zinc sheet, and the number of protrusions and dents was also reduced. This showed that the addition of [SBP]BF_4 effectively prevented the formation of unwanted side reactions and the development of zinc dendrites, producing a uniform and smooth zinc sheet surface. Atomic force microscopy (AFM) was used to characterize the three-dimensional profile and roughness of the zinc anode surface in different states (Figure 4e–g). As can be seen from Figure 4g, the polished original zinc foil has a smooth surface with a height of about 94.7 nm. In the $ZnSO_4$ electrolyte without additives, Zn^{2+} primarily deposited on the protrusions on the surface of the zinc anode due to the uneven electric field [37]. Therefore, following 140 h of cycling in pure $ZnSO_4$ electrolyte, the surface of the zinc electrode becomes highly rough, with many zinc dendrites with a height of about 929.4 nm. In contrast, after cycling in the electrolyte containing [SBP]BF_4 for 140 h, the tip effect was weakened, the surface of the zinc foil remained relatively smooth and uniform, and the height was dropped to 537.5 nm. It can be seen that the [SBP]BF_4 additive can successfully stop zinc dendritic growth and has a positive effect on achieving a long cycle of the battery.

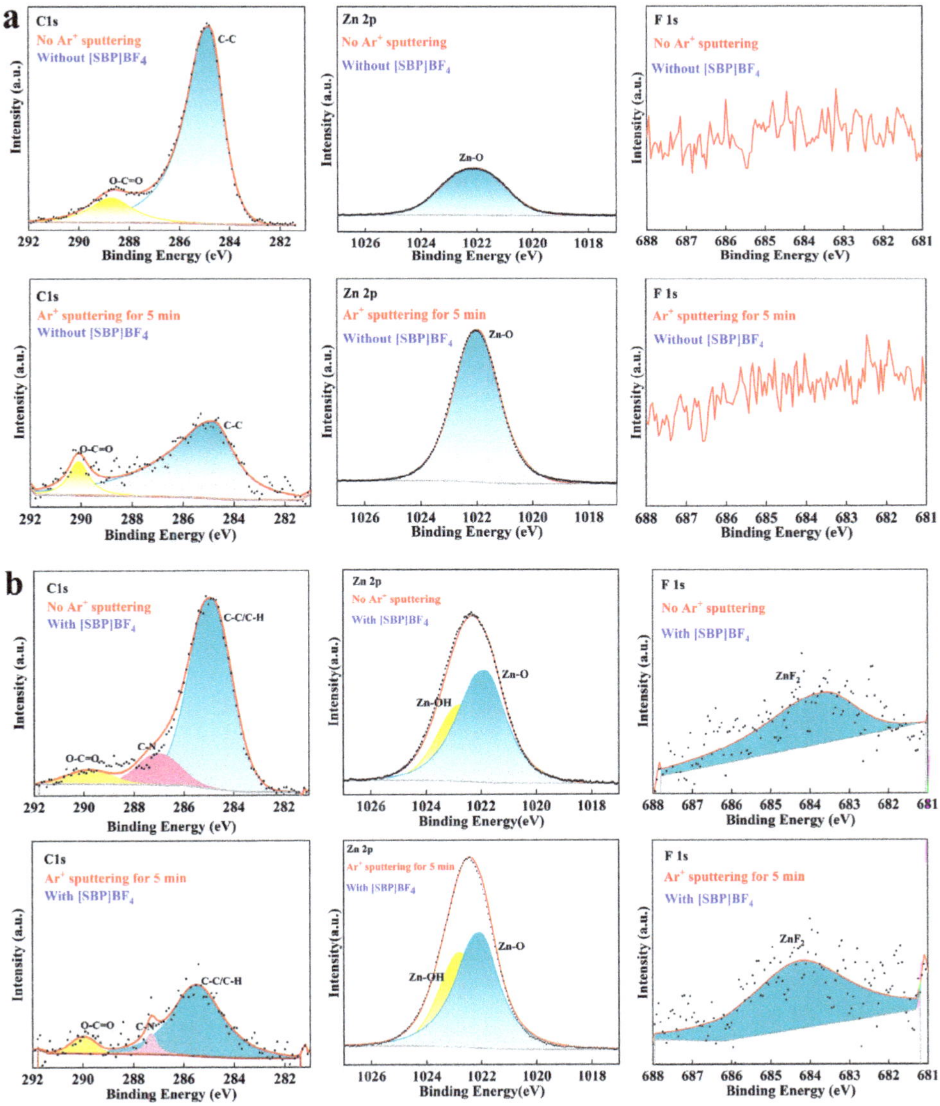

Figure 3. The XPS spectra of the Zn foils after 140 h cycles in (**a**) pure $ZnSO_4$ and (**b**) $ZnSO_4$ + 1wt% [SBP]BF_4 electrolytes.

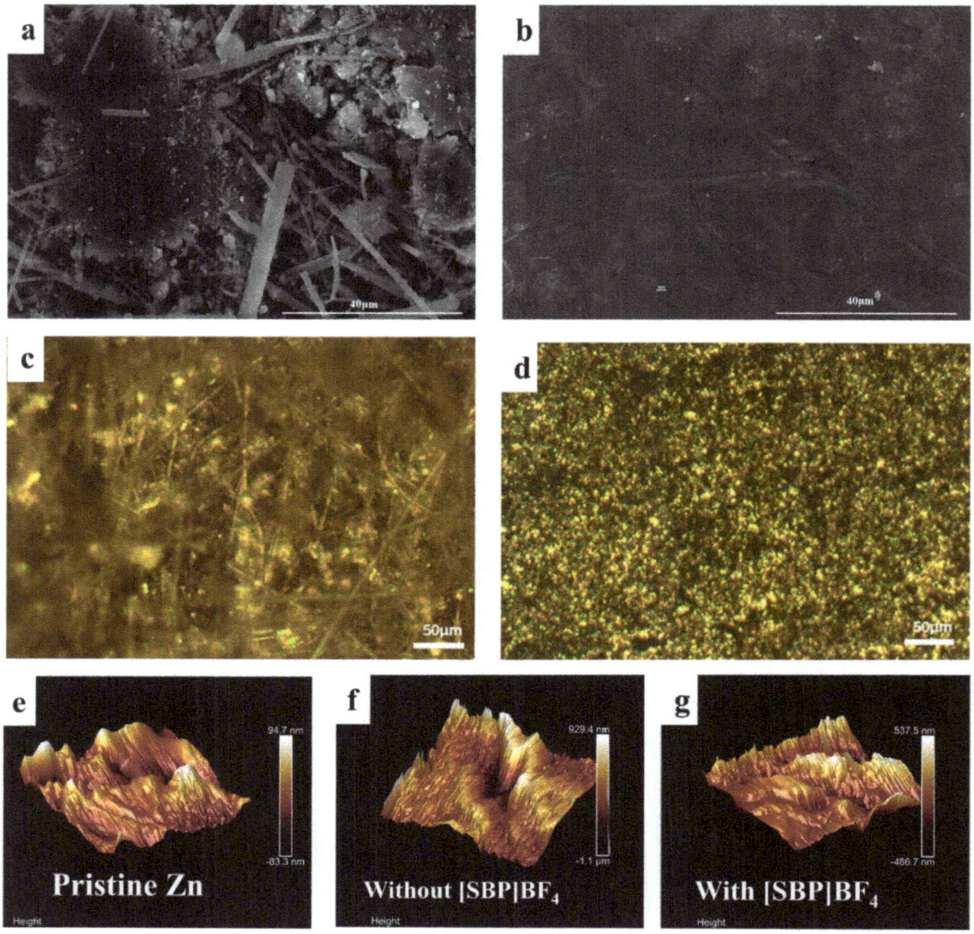

Figure 4. SEM images and optical microscope images of zinc electrode after 140 h cycling in (**a**,**c**) pure $ZnSO_4$ electrolyte and (**b**,**d**) electrolyte containing $[SBP]BF_4$. AFM images of (**e**) the original Zn foil and (**f**,**g**) Zn foil after 140 h cycling in different electrolytes.

3.2. Electrochemical Measurement

To ascertain the impact of the $[SBP]BF_4$ additive on the electrochemical stability of the zinc anode, we assembled a three-electrode system (the working electrode and counter electrode are zinc, and the reference electrode is Ag/AgCl) with Zn | | Zn symmetrical batteries and tested their electrochemical properties in detail. The corrosion activity of zinc anodes in batteries with and without additives was characterized by linear voltammetry, and the result is shown in Figure 5a. It can be observed that the Zn corrosion potential of the zinc anode in batteries with 1wt% $[SBP]BF_4$ and without the $[SBP]BF_4$ additive was −914 mV and −929 mV, respectively, indicating that the $[SBP]BF_4$ additive can prevent the corrosion of the Zn electrode to a certain extent by inhibiting hydrogen evolution reaction [38]. In addition, as shown in Figure S1, the cyclic voltammetry (CV) curves of Zn | | SS (stainless steel) asymmetric batteries with and without the $[SBP]BF_4$ additive were obtained. The introduction of the $[SBP]BF_4$ additive increased the nucleation sites of the cell from 63 to 96 mV, which can reduce the radius of the Zn crystal nucleus and facilitate the formation of a compact, flat, and uniform layer of Zn [39]. Consequently, this prevented the growth of zinc dendrites.

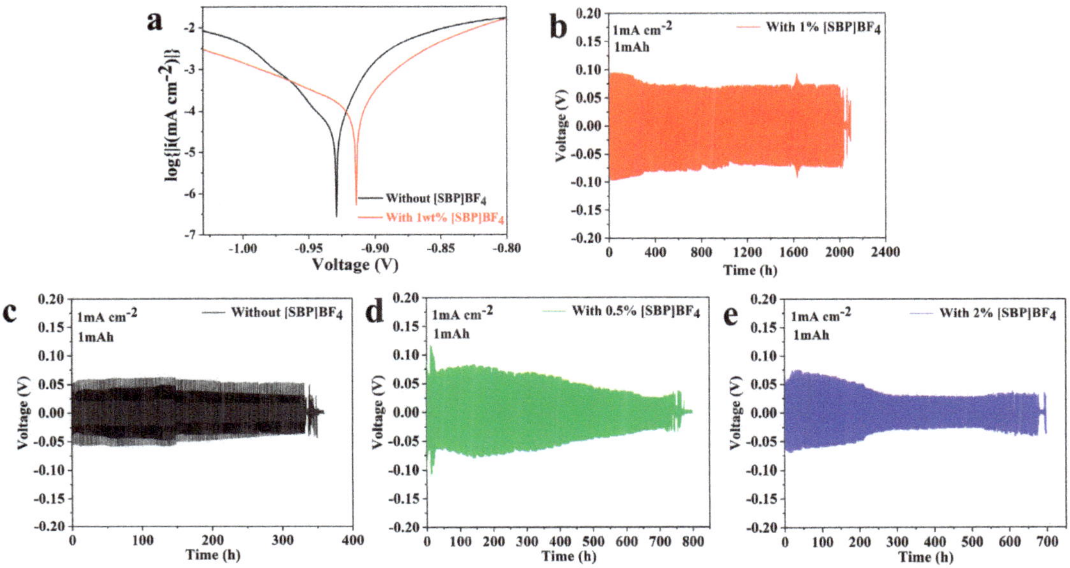

Figure 5. (**a**) The potentiodynamic polarization curves of the zinc anode in three-electrode system in electrolyte with 1wt% $[SBP]BF_4$ and without $[SBP]BF_4$. (**b–e**) The constant current charge and discharge cycle performance of Zn | | Zn symmetrical battery using electrolytes with different proportions of $[SBP]BF_4$ additive at 1 mA cm^{-2}.

By evaluating the Zn | | Zn symmetric battery's constant current charge–discharge curve, the Zn plating/stripping process on the Zn anode during long-term cycling may be examined. The results show that the batteries with different electrolytes exhibit different cycle life at a constant current density of 1 mA cm^{-2} and a cut-off capacity of 1 mAh cm^{-2}. As shown in Figure 5b–e, the battery using pure $ZnSO_4$ electrolyte was damaged after 330 h of cycling and therefore has a short life. After the addition of $[SBP]BF_4$ into the electrolyte at ratios of 0.5% and 2%, it was found that the battery life was prolonged, but the polarization voltage was unstable, and they failed after 670 h and 680 h, respectively. It was worth noting that when the content of $[SBP]BF_4$ is 1%, the resulting Zn | | Zn battery exhibits a steady overpotential, and the cycle lifespan was up to 2000 h, which was significantly better than other batteries. It was proven that the $[SBP]BF_4$ additive can affect the nucleation site of Zn^{2+} and encourage consistent and steady zinc plating or stripping on the zinc anode, which may successfully avoid the occurrence of zinc dendrites [40].

To further demonstrate that the $[SBP]BF_4$ additive can improve the cycle stability of the zinc anode, Zn | | Cu batteries were assembled and their electrochemical performance was tested. In the discharge process, the zinc stripped from the zinc anode in the battery will be deposited on the copper foil. When the battery is charged, the zinc metal on the copper foil will be stripped before the cut-off voltage. Thus, one crucial metric to confirm the reversibility and cycle stability of the zinc anode in zinc/copper batteries was the coulomb efficiency (CE) [40]. As shown in Figure 6a, the Zn | | Cu battery using pure $ZnSO_4$ electrolyte maintained a stable CE for 590 cycles, but then the CE value fluctuated and decreased due to the formation of zinc dendrites/by-products or other side reactions. After the $[SBP]BF_4$ additive was added, the initial cycles of the battery belonged to the remodeling zinc ion coordination stage, so the CE was low. However, the CE value soon reached more than 99% and remained in more than 2200 cycles. These results indicate that the zinc anode can achieve higher cycle stability by using the $[SBP]BF_4$ + $ZnSO_4$ electrolyte. At the same time, the voltage–time curves of Zn | | Cu batteries were also tested at a constant temperature of 25 °C. As shown in Figure 6b, the substrate metal

(Cu) was used to create a zinc storage cell with a capacity of 2.26 mAh cm^{-2} (Qt), and a fixed current density of 0.565 mA cm^{-2} was used during the cycle for stripping and electroplating 0.565 mAh cm^{-2} (Qc) of Zn. In the final stage, the recoverable capacity Qs were stripped through the charging process to achieve a specific voltage of 1 V. According to the above conditions, the average CE value was calculated using Equation (1) [41]. The Qs value of the battery without additives was 1.236 mAh cm^{-2}, and the average CE was calculated to be 87.1%. The Qs value of the battery with additives was 1.7932 mAh cm^{-2}, and the average CE was calculated to be 94.1%. Therefore, it has been established that the reversibility and utilization rate of the zinc anode may be significantly boosted by adding the $[SBP]BF_4$ additive to the $ZnSO_4$ electrolyte.

$$\mathrm{CE} = \frac{9\mathrm{Qc} + \mathrm{Qs}}{9\mathrm{Qc} + \mathrm{Qt}} \tag{1}$$

$$\sigma = \frac{\mathrm{L}}{\mathrm{R_b S}} \tag{2}$$

In order to further prove the role of the $[SBP]BF_4$ additive, we used graphite as the positive and negative electrodes and glass fiber membrane as the diaphragm, assembled graphite | | graphite batteries, tested their electrochemical impedance (EIS) curves (shown in Figure 6c,d), and calculated the ionic conductivity of the two electrolytes in glass fiber separators according to the obtained data and Equation (2) [42]. Where L is the thickness of the glass fiber (0.0675 cm), S is the contact area (1 cm^{-2}), and R_b is the resistance value of the glass fiber obtained by testing, the results showed that the ionic conductivity of the separator with $[SBP]BF_4$ + $ZnSO_4$ electrolyte was 24.19 mS cm^{-1}, which was higher than that of the separator with the pure $ZnSO_4$ electrolyte (12.38 mS cm^{-1}). High ionic conductivity of the glass fiber separator can improve the diffusion efficiency of Zn^{2+}, so these data can also help explain the effect of additives on facilitating uniform deposition of zinc ions.

$$t_{zn^{2+}} = \frac{I_s(\Delta V - I_o R_o)}{I_o(\Delta V - I_s R_s)} \tag{3}$$

In the chronoamperometry experiment (Figure 6e,f), the applied voltage was set to 10 mV, denoted as ΔV. The initial current and resistance were represented by I_o and R_o, respectively, while the steady-state current and resistance were represented by I_s and R_s, respectively. For the cell using electrolyte of 22wt% $ZnSO_4$ + 78wt% H_2O, the initial current (I_o) was 0.158 mA, the initial resistance (R_o) was 218 Ω, the steady-state current (I_s) was 0.099 mA, and the steady-state resistance (R_s) was 474 Ω. Therefore, according to Equation (3) [42], the calculated $t_{zn^{2+}}$ was 0.415. In contrast, for the cell using electrolyte of 22wt% $ZnSO_4$ + 1wt% $[SBP]BF_4$ + 77wt% H_2O, the initial current (I_o) and resistance (R_o) were found to be 7.747 mA and 183 Ω, respectively, while the steady-state current (I_s) and resistance (R_s) were 2.485 mA and 250 Ω, respectively. Therefore, the $t_{zn^{2+}}$ calculated by Equation (3) is 0.739. The result shows that the addition of $[SBP]BF_4$ can improve the migration rate of zinc ions and promote the uniform distribution of ions so as to help uniform galvanizing.

In this work, two full batteries were assembled using V_2O_5 as the cathode, Zn foil as the anode, and $ZnSO_4$ and $[SBP]BF_4$ + $ZnSO_4$ as the electrolyte, respectively. As shown in Figure 7a, the CV curves of both whole cells showed four (two pairs) redox peaks at a scan rate of 1 mV s^{-1}, which linked to the insertion/extraction of protons and Zn^{2+} ions [43]. Remarkably, the voltage difference of the Zn | | $[SBP]BF_4$+$ZnSO_4$ | | V_2O_5 cell is only 6 mV, while the voltage difference of the Zn | | $ZnSO_4$ | | V_2O_5 cell is 32 mV. This indicated that the addition of $[SBP]BF_4$ can alleviate the polarization of the battery and enhance the reversibility, which had a positive effect on preventing the growth of zinc dendrites [44].

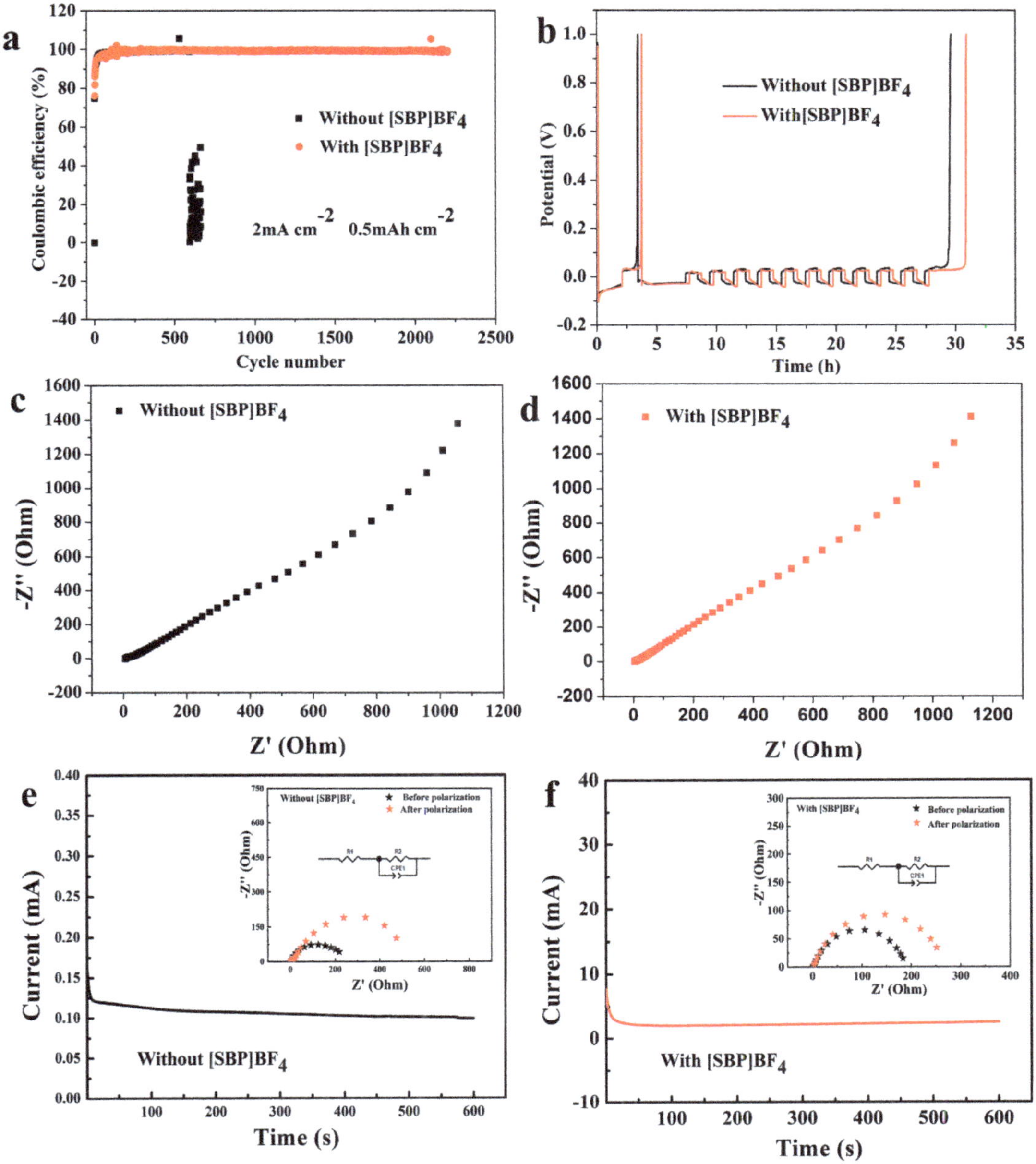

Figure 6. (**a**) The coulombic efficiency and (**b**) the voltage–time curves of Zn | | Cu batteries using electrolytes with and without additives. Nyquist plots tested of the graphite paper | | graphite paper batteries (**c**) without $[SBP]BF_4$ additive and (**d**) with additive in the electrolyte. The chronoamperometry curves of Zn | | Zn cells (**e**) without and (**f**) with 1wt% $[SBP]BF_4$ additives in the electrolyte (insets give the corresponding Nyquist plots at initial and steady states after fitting).

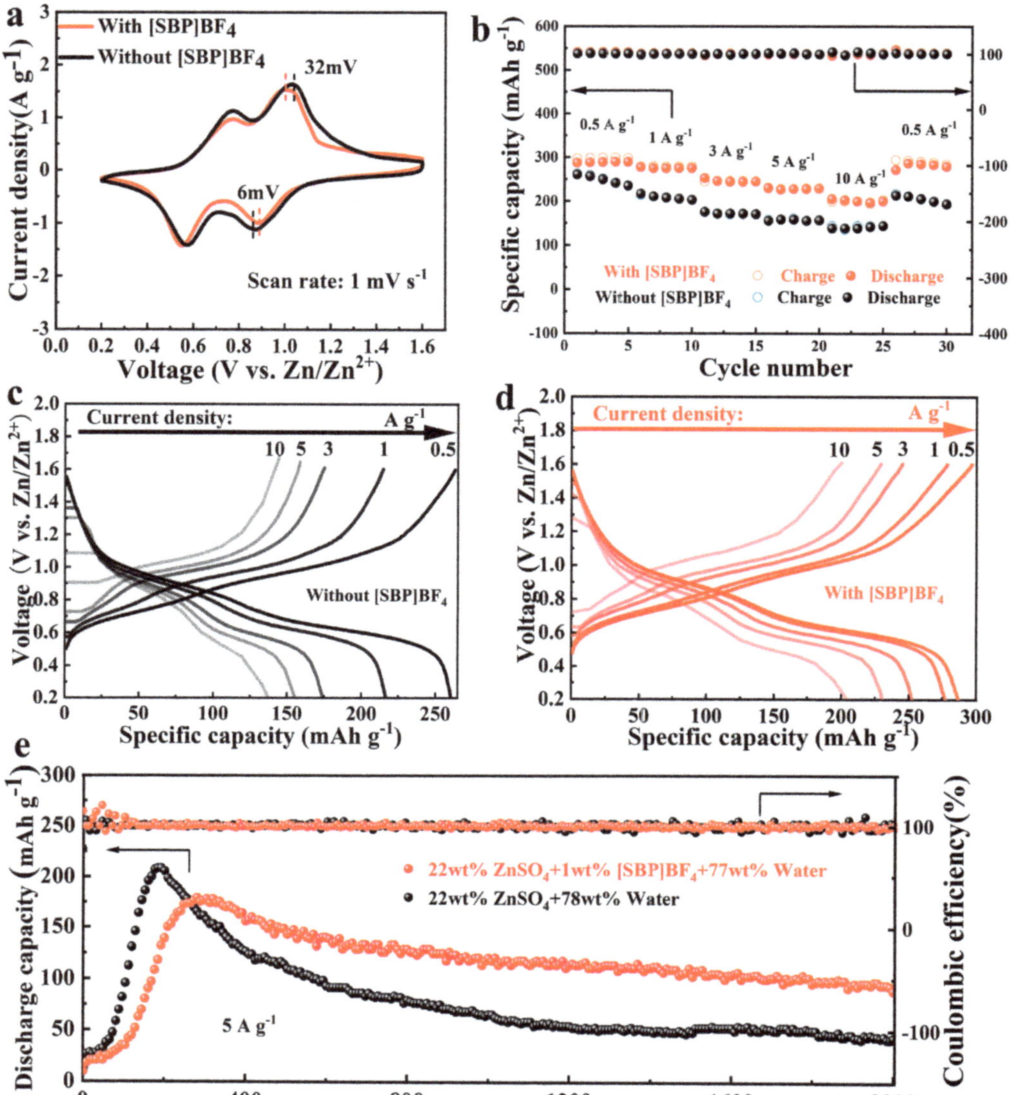

Figure 7. In the study of the electrochemical performance of Zn | | V_2O_5 fuel cells, it was determined that the active material loading on the positive electrode sheet is between 1 and 2 mg cm^{-2}: (**a**) the CV cures at a scan rate of 1 mV s^{-1} in electrolytes with and without [SBP]BF_4 additives; (**b**) rate performance in electrolytes with and without [SBP]BF_4 additives; charge/discharge profiles in electrolytes of (**c**) without and (**d**) with [SBP]BF_4 additive; (**e**) cyclic stabilities and efficiencies in electrolytes with and without [SBP]BF_4 additives at 5 A g^{-1}.

Figure 7b displays the rate performance of the two batteries at various current densities (0.5 A g^{-1}, 1 A g^{-1}, 3 A g^{-1}, 5 A g^{-1}, 10 A g^{-1}, and 0.5 A g^{-1}). The discharge specific capacity of the zinc battery with [SBP]BF_4 added to the electrolyte was 296.6 mAh g^{-1} at a current density of 0.5 A g^{-1}, which was higher than that of the battery without the [SBP]BF_4 additive (260.4 mAh g^{-1}). Surprisingly, even at a high current density of

10 A g^{-1}, the battery with the [SBP]BF_4 additive can still maintain a high specific capacity of 166 mAh g^{-1}, which is significantly higher than that of the battery without additives. This revealed that the rate performance of zinc batteries was much enhanced by the addition of additive [SBP]BF_4. Upon evaluating the rate performance, it was noteworthy that the battery's capacity (294.1 mAh g^{-1}) could still be reached when the current density dropped to 0.5 A g^{-1}, indicating strong cycle stability. Figure 7c,d shows the charge/discharge curves of two cells with different current densities, and there are two discharge plateaus near 0.68 and 1.08 V, corresponding to reversible Zn^{2+} insertion and extraction reaction processes, respectively, which are consistent with CV curves. Furthermore, the two batteries' long-term cycle performance was assessed at a current density of 5 A g^{-1}; the outcome is displayed in Figure 7e. Following 2000 cycles, the Zn | | [SBP]BF_4 + $ZnSO_4$ | | V_2O_5 cell demonstrates a notable 97 mAh g^{-1} specific capacity and a high average CE of 99.7%. In contrast, under the same conditions, the specific capacity of the battery without [SBP]BF_4 decreased to 46 mAh g^{-1}, which was majorly attributed to the electrochemical hydrogen evolution corrosion and zinc dendrite growth [45]. The cycling stability of Zn | | V_2O_5 batteries, both with and without additives, was evaluated at a current density of 20 A g^{-1}, as depicted in Figure S4. After 3000 cycles, the battery equipped with additives maintained a capacity of 66.7 mAh g^{-1}, outperforming its counterpart devoid of additives, which retained a capacity of merely 33.3 mAh g^{-1} after the same number of cycles. Following cycling, the zinc foil negative electrode and V_2O_5 positive electrode were detached from the Zn | | V_2O_5 battery. Photographic evidence revealed that in the absence of additives (Figure S4a), severe corrosion had occurred on the surface of the zinc foil, leading to the proliferation of zinc dendrites. Consequently, there was an abundance of glass fiber separator residue on the zinc foil's surface. Concurrently, the V_2O_5 positive electrode exhibited significant corrosion accompanied by numerous cracks. In stark contrast, the application of the [SBP]BF_4 additive (Figure S4b) mitigated the corrosion and zinc dendrites on the Zn surface while also enhancing the density and regularity of the V_2O_5 pole piece surface. As shown in Table S1 [35,45–52], we also made a comparison with other electrolyte additives to further illustrate the role of [SBP]BF_4 in inhibiting zinc dendrite formation and corrosion. Therefore, it can be concluded that the introduction of additive [SBP]BF_4 has obvious advantages in improving the rate performance and long-term cycle stability of zinc batteries, and this result provides valuable supporting data for promoting the practical application of zinc-ion batteries.

4. Conclusions

AZIBs exhibit promising potential for use in the domains of portable electronics, electric cars, and large-scale energy storage. However, the metal zinc anode faces problems involving zinc dendrite growth and zinc metal corrosion, as well as anode surface passivation, which affect the electrochemical performance of AZIBs. In order to solve these problems, 1wt% [SBP]BF_4 + 22wt% $ZnSO_4$ electrolyte was used in this work to improve the stability of the zinc anode. We found that [SBP]BF_4 can preferentially adsorb on the surface of the zinc anode to regulate the uniform nucleation of zinc ions and induce their epitaxial deposition, can improve the solvation sheath of hydrated Zn^{2+} to promote the transport and charge transfer of Zn^{2+}, and can produce adaptive SEI film on the surface of the zinc anode to further prevent the occurrence of side reactions between the electrolyte and the electrode. This means that the Zn | | Zn battery including the [SBP]BF_4 additive has a cycle life of 2000 h, which was six times longer than the battery lacking [SBP]BF_4. It is worth noting that the Zn | | Cu battery with the [SBP]BF_4 additive can also be stably cycled more than 2200 cycles, during which the CE value has been maintained above 99%, proving the good reversibility of the zinc anode. In addition, the Zn | | [SBP]BF_4+$ZnSO_4$ | | V_2O_5 full cell has a large specific capacity (288 mAh g^{-1} at 0.5 A g^{-1}), good rate performance (205 mAh g^{-1} at 10 A g^{-1}), and a stable long life of 2000 cycles at a current density of 5 A g^{-1}. In summary, the addition of [SBP]BF_4 can increase the rate performance and cycle life of AZIBs in addition to improving the stability and reversibility of the zinc anode.

Supplementary Materials: The following supporting information can be downloaded at: https://www.mdpi.com/article/10.3390/batteries10030102/s1.

Author Contributions: Conceptualization, X.Z.; methodology, K.S.; validation, Y.H.; formal analysis, K.X.; investigation, Y.W.; resources, M.H.; data curation, J.L. All authors have read and agreed to the published version of the manuscript.

Funding: This research was funded by the Taishan Scholars Program (NO. tsqnz20230629); the Western Young Scholars Foundations of Chinese Academy of Sciences; the Science Fund of Shandong Laboratory of Yantai Advanced Materials and Green Manufacturing [AMGM2022A02, AMGM2023A04]; the Collaborative Innovation Alliance Fund for Young Science and Technology Worker (HZJJ23-5).

Data Availability Statement: The raw data supporting the conclusions of this article will be made available by the authors on request.

Conflicts of Interest: The authors declare no conflict of interest.

References

1. Chayambuka, K.; Mulder, G.; Danilov, D.L.; Notten, P.H.L. From Li-Ion Batteries toward Na-Ion Chemistries: Challenges and Opportunities. *Adv. Energy Mater.* **2020**, *10*, 2001310. [CrossRef]
2. Ma, L.; Schroeder, M.A.; Borodin, O.; Pollard, T.P.; Ding, M.S.; Wang, C.; Xu, K. Realizing High Zinc Reversibility in Rechargeable Batteries. *Nat. Energy* **2020**, *5*, 743–749. [CrossRef]
3. Raza, H.; Cheng, J.; Lin, C.; Majumder, S.; Zheng, G.; Chen, G. High-Entropy Stabilized Oxides Derived via a Low-Temperature Template Route for High-Performance Lithium-Sulfur Batteries. *EcoMat* **2023**, *5*, e12324. [CrossRef]
4. Wang, H.; Zhou, A.; Hu, X.; Hu, Z.; Zhang, F.; Huang, Y.; Li, L.; Wu, F.; Chen, R. Bifunctional Dynamic Adaptive Interphase Reconfiguration for Zinc Deposition Modulation and Side Reaction Suppression in Aqueous Zinc Ion Batteries. *ACS Nano* **2023**, *17*, 11946–11956. [CrossRef]
5. Zhou, T.; Xie, L.; Han, Q.; Qiu, X.; Xiao, Y.; Yang, X.; Liu, X.; Yang, S.; Zhu, L.; Cao, X. Progress and Prospect of Vanadates as Aqueous Zn-Ion Batteries Cathodes. *Coord. Chem. Rev.* **2024**, *498*, 215461. [CrossRef]
6. Deng, W.; Xu, Z.; Wang, X. High-Donor Electrolyte Additive Enabling Stable Aqueous Zinc-Ion Batteries. *Energy Storage Mater.* **2022**, *52*, 52–60. [CrossRef]
7. Deng, R.; He, Z.; Chu, F.; Lei, J.; Cheng, Y.; Zhou, Y.; Wu, F. An Aqueous Electrolyte Densified by Perovskite $SrTiO_3$ Enabling High-Voltage Zinc-Ion Batteries. *Nat. Commun.* **2023**, *14*, 4981. [CrossRef] [PubMed]
8. Kuai, X.; Li, K.; Chen, J.; Wang, H.; Yao, J.; Chiang, C.-L.; Liu, T.; Ye, H.; Zhao, J.; Lin, Y.-G.; et al. Interfacial Engineered Vanadium Oxide Nanoheterostructures Synchronizing High-Energy and Long-Term Potassium-Ion Storage. *ACS Nano* **2022**, *16*, 1502–1510. [CrossRef] [PubMed]
9. Xu, X.; Zhu, X.; Li, S.; Xu, Y.; Sun, L.; Shi, L.; Song, M. In Situ Constructing Solid Electrolyte Interphase and Optimizing Solvation Shell for a Stable Zn Anode. *J. Electron. Mater.* **2024**, *53*, 288–297. [CrossRef]
10. Shi, X.; Wang, J.; Yang, F.; Liu, X.; Yu, Y.; Lu, X. Metallic Zinc Anode Working at 50 and 50 mAh cm^{-2} with High Depth of Discharge via Electrical Double Layer Reconstruction. *Adv. Funct. Mater.* **2023**, *33*, 2211917. [CrossRef]
11. Jia, X.; Liu, C.; Neale, Z.G.; Yang, J.; Cao, G. Active Materials for Aqueous Zinc Ion Batteries: Synthesis, Crystal Structure, Morphology, and Electrochemistry. *Chem. Rev.* **2020**, *120*, 7795–7866. [CrossRef]
12. Wang, M.; Wu, X.; Yang, D.; Zhao, H.; He, L.; Su, J.; Zhang, X.; Yin, X.; Zhao, K.; Wang, Y.; et al. A Colloidal Aqueous Electrolyte Modulated by Oleic Acid for Durable Zinc Metal Anode. *Chem. Eng. J.* **2023**, *451*, 138589. [CrossRef]
13. Jia, H.; Wang, Z.; Tawiah, B.; Wang, Y.; Chan, C.-Y.; Fei, B.; Pan, F. Recent Advances in Zinc Anodes for High-Performance Aqueous Zn-Ion Batteries. *Nano Energy* **2020**, *70*, 104523. [CrossRef]
14. Raza, H.; Bai, S.; Cheng, J.; Majumder, S.; Zhu, H.; Liu, Q.; Zheng, G.; Li, X.; Chen, G. Li-S Batteries: Challenges, Achievements and Opportunities. Electrochem. *Energy Rev.* **2023**, *6*, 29. [CrossRef]
15. Chao, D.; Zhou, W.; Xie, F.; Ye, C.; Li, H.; Jaroniec, M.; Qiao, S.-Z. Roadmap for Advanced Aqueous Batteries: From Design of Materials to Applications. *Sci. Adv.* **2020**, *6*, eaba4098. [CrossRef] [PubMed]
16. Zhang, X.; Jia, C.; Zhang, J.; Zhang, L.; Liu, X. Smart Aqueous Zinc Ion Battery: Operation Principles and Design Strategy. *Adv. Sci.* **2024**, *11*, 2305201. [CrossRef] [PubMed]
17. Yoo, G.; Lee, Y.-G.; Im, B.; Kim, D.G.; Jo, Y.-R.; An, G. Integrated Solution for a Stable and High-Performance Zinc-Ion Battery Using an Electrolyte Additive. *Energy Storage Mater.* **2023**, *61*, 102845. [CrossRef]
18. Fang, T.; Liu, Q.; Hu, A.; Meng, J.; Fu, Y.; Shi, Z. Dendrite-Free and Stable Zinc-Ion Batteries Enabled by a Cation-Anion Synergistic Regulation Additive. *J. Power Sources* **2023**, *581*, 233521. [CrossRef]
19. Yin, J.; Li, M.; Feng, X.; Cui, T.; Chen, J.; Li, F.; Wang, M.; Cheng, Y.; Ding, S.; Xu, X.; et al. Unveiling the Multifunctional Regulation Effect of a Glutamine Additive for Highly Reversible Zn Metal Anodes. *J. Mater. Chem. A* **2024**, *12*, 1543–1550. [CrossRef]
20. Abdulla, J.; Cao, J.; Zhang, D.; Zhang, X.; Sriprachuabwong, C.; Kheawhom, S.; Wangyao, P.; Qin, J. Elimination of Zinc Dendrites by Graphene Oxide Electrolyte Additive for Zinc-Ion Batteries. *ACS Appl. Energ. Mater.* **2021**, *4*, 4602–4609. [CrossRef]

21. Lin, Y.; Mai, Z.; Liang, H.; Li, Y.; Yang, G.; Wang, C. Dendrite-Free Zn Anode Enabled by Anionic Surfactant-Induced Horizontal Growth for Highly-Stable Aqueous Zn-Ion Pouch Cells. *Energy Environ. Sci.* **2023**, *16*, 687–697. [CrossRef]
22. Su, K.; Chen, J.; Zhang, X.; Feng, J.; Xu, Y.; Pu, Y.; Wang, C.; Ma, P.; Wang, Y.; Lang, J. Inhibition of Zinc Dendrites by Dopamine Modified Hexagonal Boron Nitride Electrolyte Additive for Zinc-Ion Batteries. *J. Power Sources* **2022**, *548*, 232074. [CrossRef]
23. Wei, T.; Zhang, X.; Ren, Y.; Wang, Y.; Li, Z.; Zhang, H.; Hu, L. Reconstructing Anode/Electrolyte Interface and Solvation Structure towards High Stable Zinc Anode. *Chem. Eng. J.* **2023**, *457*, 141272. [CrossRef]
24. Xu, M.; Ivey, D.G.; Qu, W.; Xie, Z. Study of the Mechanism for Electrodeposition of Dendrite-Free Zinc in an Alkaline Electrolyte Modified with 1-Ethyl-3-Methylimidazolium Dicyanamide. *J. Power Sources* **2015**, *274*, 1249–1253. [CrossRef]
25. Bayer, M.; Overhoff, G.M.; Gui, A.L.; Winter, M.; Bieker, P.; Schulze, S. Influence of Water Content on the Surface Morphology of Zinc Deposited from EMImOTf/Water Mixtures. *J. Electrochem. Soc.* **2019**, *166*, A909–A914. [CrossRef]
26. Ma, Z.; Kan, J. Study of Cylindrical Zn/PANI Secondary Batteries with the Electrolyte Containing Alkylimidazolium Ionic Liquid. *Synth. Met.* **2013**, *174*, 58–62. [CrossRef]
27. Lahiri, A.; Yang, L.; Li, G.; Endres, F. Mechanism of Zn-Ion Intercalation/Deintercalation in a Zn-Polypyrrole Secondary Battery in Aqueous and Bio-Ionic Liquid Electrolytes. *ACS Appl. Mater. Interfaces* **2019**, *11*, 45098–45107. [CrossRef] [PubMed]
28. Cao, L.; Li, D.; Pollard, T.; Deng, T.; Zhang, B.; Yang, C.; Chen, L.; Vatamanu, J.; Hu, E.; Hourwitz, M.J.; et al. Fluorinated Interphase Enables Reversible Aqueous Zinc Battery Chemistries. *Nat. Nanotechnol.* **2021**, *16*, 902–910. [CrossRef]
29. Liu, Z.; El Abedin, S.Z.; Endres, F. Electrodeposition of Zinc Films from Ionic Liquids and Ionic Liquid/Water Mixtures. *Electrochim. Acta* **2013**, *89*, 635–643. [CrossRef]
30. Wang, R.; Fang, C.; Yang, L.; Li, K.; Zhu, K.; Liu, G.; Chen, J. The Novel Ionic Liquid and Its Related Self-Assembly in the Areas of Energy Storage and Conversion. *Small Sci.* **2022**, *2*, 2200048. [CrossRef]
31. Lebedeva, O.; Kultin, D.; Kustov, L. Advanced Research and Prospects on Polymer Ionic Liquids: Trends, Potential and Application. *Green Chem.* **2023**, *25*, 9001–9019. [CrossRef]
32. Zhao, P.; Yang, B.; Chen, J.; Lang, J.; Zhang, T.; Yan, X. A Safe, High-Performance, and Long-Cycle Life Zinc-Ion Hybrid Capacitor Based on Three-Dimensional Porous Activated Carbon. *Acta Phys.-Chim. Sin.* **2020**, *36*, 1904050. [CrossRef]
33. Huo, H.; Li, X.; Chen, Y.; Liang, J.; Deng, S.; Gao, X.; Doyle-Davis, K.; Li, R.; Guo, X.; Shen, Y.; et al. Bifunctional Composite Separator with a Solid-State-Battery Strategy for Dendrite-Free Lithium Metal Batteries. *Energy Storage Mater.* **2020**, *29*, 361–366. [CrossRef]
34. Chen, R.; Zhang, W.; Huang, Q.; Guan, C.; Zong, W.; Dai, Y.; Du, Z.; Zhang, Z.; Li, J.; Guo, F.; et al. Trace Amounts of Triple-Functional Additives Enable Reversible Aqueous Zinc-Ion Batteries from a Comprehensive Perspective. *Nano-Micro Lett.* **2023**, *15*, 81. [CrossRef]
35. Chen, J.; Zhou, W.; Quan, Y.; Liu, B.; Yang, M.; Chen, M.; Han, X.; Xu, X.; Zhang, P.; Shi, S. Ionic Liquid Additive Enabling Anti-Freezing Aqueous Electrolyte and Dendrite-Free Zn Metal Electrode with Organic/Inorganic Hybrid Solid Electrolyte Interphase Layer. *Energy Storage Mater.* **2022**, *53*, 629–637. [CrossRef]
36. Chen, X.; Liu, T.; Ding, Y.; Sun, X.; Huang, J.; Qiao, J.; Peng, S. Controlled Nucleation and Growth for the Dendrite-Free Zinc Anode in Aqueous Zinc-Ion Battery. *J. Alloys Compd.* **2024**, *970*, 172584. [CrossRef]
37. Chen, R.; Liu, Q.; Xu, L.; Zuo, X.; Liu, F.; Zhang, J.; Zhou, X.; Mai, L. Zwitterionic Bifunctional Layer for Reversible Zn Anode. *ACS Energy Lett.* **2022**, *7*, 1719–1727. [CrossRef]
38. Zheng, S.; Wei, L.; Zhang, Z.; Pan, J.; He, J.; Gao, L.; Li, C.C. In Situ Polymerization of Ionic Liquid with Tunable Phase Separation for Highly Reversible and Ultralong Cycle Life Zn-Ion Battery. *Nano Lett.* **2022**, *22*, 9062–9070. [CrossRef] [PubMed]
39. Jang, J.; Jae-Sun, S.; Ko, S.; Park, H.; Song, W.-J.; Park, C.B.; Kang, J. Self-Assembled Protective Layer by Symmetric Ionic Liquid for Long-Cycling Lithium-Metal Batteries. *Adv. Energy Mater.* **2022**, *12*, 2103955. [CrossRef]
40. Gou, Q.; Luo, H.; Zhang, Q.; Deng, J.; Zhao, R.; Odunmbaku, O.; Wang, L.; Li, L.; Zheng, Y.; Li, J.; et al. Electrolyte Regulation of Bio-Inspired Zincophilic Additive toward High-Performance Dendrite-Free Aqueous Zinc-Ion Batteries. *Small* **2023**, *19*, 2207502. [CrossRef] [PubMed]
41. Ma, L.; Schroeder, M.A.; Pollard, T.P.; Borodin, O.; Ding, M.S.; Sun, R.; Cao, L.; Ho, J.; Baker, D.R.; Wang, C.; et al. Critical Factors Dictating Reversibility of the Zinc Metal Anode. *Energy Environ. Mater.* **2020**, *3*, 516–521. [CrossRef]
42. Hao, J.; Li, X.; Zhang, S.; Yang, F.; Zeng, X.; Zhang, S.; Bo, G.; Wang, C.; Guo, Z. Designing Dendrite-Free Zinc Anodes for Advanced Aqueous Zinc Batteries. *Adv. Funct. Mater.* **2020**, *30*, 2001263. [CrossRef]
43. Li, B.; Xue, J.; Lv, X.; Zhang, R.; Ma, K.; Wu, X.; Dai, L.; Wang, L.; He, Z. A Facile Coating Strategy for High Stability Aqueous Zinc Ion Batteries: Porous Rutile Nano-TiO_2 Coating on Zinc Anode. *Surf. Coat. Technol.* **2021**, *421*, 127367. [CrossRef]
44. Yong, B.; Ma, D.; Wang, Y.; Mi, H.; He, C.; Zhang, P. Understanding the Design Principles of Advanced Aqueous Zinc-Ion Battery Cathodes: From Transport Kinetics to Structural Engineering, and Future Perspectives. *Adv. Energy Mater.* **2020**, *10*, 2002354. [CrossRef]
45. Xu, J.; Wang, M.; Asraful Alam, M.; Hoang, T.K.A.; Zhang, Y.; Li, H.; Lv, Y.; Zhao, A.; Xiong, W. Employing Cationic Kraft Lignin as Electrolyte Additive to Enhance the Electrochemical Performance of Rechargeable Aqueous Zinc-Ion Battery. *Fuel* **2023**, *333*, 126450. [CrossRef]
46. Sun, P.; Ma, L.; Zhou, W.; Qiu, M.; Wang, Z.; Chao, D.; Mai, W. Simultaneous Regulation on Solvation Shell and Electrode Interface for Dendrite-Free Zn Ion Batteries Achieved by a Low-Cost Glucose Additive. *Angew. Chem. Int. Ed.* **2021**, *60*, 18247–18255. [CrossRef]

47. Zou, P.; Lin, R.; Pollard, T.P.; Yao, L.; Hu, E.; Zhang, R.; He, Y.; Wang, C.; West, W.C.; Ma, L.; et al. Localized Hydrophobicity in Aqueous Zinc Electrolytes Improves Zinc Metal Reversibility. *Nano Lett.* **2022**, *22*, 7535–7544. [CrossRef] [PubMed]
48. Zhang, Q.; Luan, J.; Fu, L.; Wu, S.; Tang, Y.; Ji, X.; Wang, H. The Three- Dimensional Dendrite-Free Zinc Anode on a Copper Mesh with a Zinc-Oriented Polyacrylamide Electrolyte Additive. *Angew. Chem. Int. Ed.* **2019**, *58*, 15841–15847. [CrossRef]
49. Zhang, Q.; Ma, Y.; Lu, Y.; Zhou, X.; Lin, L.; Li, L.; Yan, Z.; Zhao, Q.; Zhang, K.; Chen, J. Designing Anion-Type Water-Free Zn2+ Solvation Structure for Robust Zn Metal Anode. *Angew. Chem. Int. Ed.* **2021**, *60*, 23357–23364. [CrossRef] [PubMed]
50. Cui, J.; Liu, X.; Xie, Y.; Wu, K.; Wang, Y.; Liu, Y.; Zhang, J.; Yi, J.; Xia, Y. Improved Electrochemical Reversibility of Zn Plating/Stripping: A Promising Approach to Suppress Water-Induced Issues through the Formation of H-Bonding. *Mater. Today Energy* **2020**, *18*, 100563. [CrossRef]
51. Hao, J.; Long, J.; Li, B.; Li, X.; Zhang, S.; Yang, F.; Zeng, X.; Yang, Z.; Pang, W.K.; Guo, Z. Toward High-Performance Hybrid Zn-Based Batteries via Deeply Understanding Their Mechanism and Using Electrolyte Additive. *Adv. Funct. Mater.* **2019**, *29*, 1903605. [CrossRef]
52. Zeng, X.; Mao, J.; Hao, J.; Liu, J.; Liu, S.; Wang, Z.; Wang, Y.; Zhang, S.; Zheng, T.; Liu, J.; et al. Electrolyte Design for In Situ Construction of Highly Zn2+-Conductive Solid Electrolyte Interphase to Enable High-Performance Aqueous Zn-Ion Batteries under Practical Conditions. *Adv. Mater.* **2021**, *33*, 2007416. [CrossRef]

batteries

MDPI

Article

High-Performance Supercapacitors Based on Graphene/Activated Carbon Hybrid Electrodes Prepared via Dry Processing

Shengjun Chen [1,2,*,†], Wenrui Wang [3,†], Xinyue Zhang [3] and Xiaofeng Wang [1]

1 Department of Precision Instrument, Tsinghua University, Beijing 100084, China
2 Ningbo CRRC New Energy Technology Co., Ltd., Ningbo 315112, China
3 School of Mechanical Engineering, Nanjing University of Science and Technology, Nanjing 210094, China
* Correspondence: chenshengjun1974@icloud.com
† Shengjun Chen and Wenrui Wang are co-first authors.

Abstract: Graphene has a high specific surface area and high electrical conductivity, and its addition to activated carbon electrodes should theoretically significantly improve the energy storage performance of supercapacitors. Unfortunately, such an ideal outcome is seldom verified in practical commercial supercapacitor design and production. In this paper, the oxygen-containing functional groups in graphene/activated carbon hybrids, which are prone to induce side reactions, are removed in the material synthesis stage by a special process design, and electrodes with high densities and low internal resistances are prepared by a dry process. On this basis, a carbon-coated aluminum foil collector with a full tab structure is designed and assembled with graphene/activated carbon hybrid electrodes to form a commercial supercapacitor in cylindrical configuration. The experimental tests confirmed that such supercapacitors have high capacity density, power density, low internal resistance (about 0.06 mΩ), good high-current charging/discharging characteristics, and a long lifetime, with more than 80% capacity retention after 10 W cycles.

Keywords: graphene/activated carbon hybrid electrode; dry processing; full tab structure; supercapacitors

1. Introduction

A supercapacitor fills the gap between batteries and traditional capacitors with its simultaneously superior power density and energy density, and it has a wide range of commercial applications in microgrids [1], electrical equipment, consumer electronics [2], and many other fields. For commercial supercapacitors, the most used electrode material is activated carbon because of its large specific surface area and low cost [3,4]. For academic research, graphene has attracted extensive attention in the field of supercapacitors due to its special two-dimensional material properties [5–10]. As an electrode material, graphene has multiple contributions to the energy storage of supercapacitors: firstly, it has a higher specific surface area than traditional activated carbon, providing a richer microscopic interface for energy storage [11]; secondly, it has a good electrochemical activity [12], which can realize a greater energy storage density through the pseudocapacitive effect and other electrochemical reactions [13–17]; and thirdly, it has a good electrical conductivity [18], which reduces the energy loss of the supercapacitor itself. As a result, many researchers have focused on next-generation supercapacitors based on graphene electrodes and successfully fabricated many device prototypes with great performance in the laboratory environment [19–22].

However, until now, supercapacitors based on graphene electrodes did not make ideal progress in the industry and failed to capture a significant market share [23,24]. This is mainly since some active oxygen-containing functional groups of graphene electrode materials constantly have side reactions which produce gases, resulting in an increasing internal resistance and capacity degradation of supercapacitors on longer time scales [25,26]. Although there has been some research devoted to the removal of these potentially unfavorable oxygen-containing functional groups, successful large-scale applications in the industry have not been realized yet.

To solve the above problems, this paper proposes an easy-to-implement manufacturing method for high-performance graphene/activated carbon hybrid electrodes. High-density hybrid electrodes with good mechanical properties are realized by removing oxygen-containing functional groups and humidity in the electrodes via high-temperature treatment and a dry-film-forming process. Based on this hybrid electrode, improvements have been made in both the surface interface design and full tab [27] design of the collector, realizing a commercial-grade, high-performance supercapacitor that can be mass produced. Detailed charge–discharge cycling test data demonstrate the long-term stability of the performance gain of the proposed hybrid electrode supercapacitor.

2. Materials and Methods

To eliminate the oxygen-containing functional groups on the surface of the porous graphene/activated carbon hybrid material, the hybrid was subjected to high-temperature heat treatment at 850 °C for 2 h in a H_2/argon mixed atmosphere (H_2 volume ratio of 10%) within a tube furnace. This treatment effectively removed the oxygen-containing functional groups on the surface of the hybrid material. The surface functional group content of porous graphene/activated carbon hybrid materials is typically characterized using the alkaline Boehm titration method. The surface contains less than 0.1 meq/g of oxygen-containing functional groups, significantly below industry standards. This is expected to enhance the pressure resistance, lifespan, and safety of electrode materials.

Through the above dry method electrode manufacturing technique, we can prepare electrode sheets with a thickness of 140–150 μM. Using a laser slitting machine to cut the electrode plates, the electrode plate size is 75 × 37 mm, and the battery cells are stacked using opposite side electrode tabs. The preparation environment is controlled to a humidity of dew point temperature of −45 °C and baking conditions of 120 °C for 12 h under a vacuum degree of ≤−90 Kpa with nitrogen replenishment. The moisture content in the final electrode is maintained below 200 ppm. The battery cells are stacked in pairs, with an electrolyte injection ratio of 1:1 (electrolyte/cell mass). Meanwhile, the electrolyte used is 1 mol/L tetrafluoroborate spirocyclic quaternary ammonium salt (SBP-BF4)/acetonitrile (AN).

3. Results

3.1. Preparation of Graphene/Activated Carbon Hybrid Electrode

Typically, graphene oxide has a high content of oxygen functional groups on the surface (5–10%), which is prone to both blocking the abundant pores of the activated carbon and inducing gas-producing side reactions at high potentials. Therefore, if graphene oxide and activated carbon hybrids are directly used to prepare electrodes, this will cause problems such as increased internal resistance and the decreased capacity of capacitors.

To fully harness the electrochemical energy storage advantages of graphene materials while mitigating the negative impacts of oxygen-containing functional groups, we have proposed a one-step carbonization activation method for synthesizing porous graphene/activated carbon hybrids, as illustrated in the process flow diagram depicted in Figure 1. The TEM image reveals that the activated products exhibit a wrinkled nanosheet structure with layered porosity, as highlighted by the red circle in Figure 1c. This observation indicates that KOH activation effectively etches and creates pores. The abundant voids formed during the activation process further enhance the specific surface area of the hybrid material, enabling greater energy storage capacity. Leveraging the porous graphene material with a specific surface area exceeding 2000 m^2/g (as a comparison, the results in the literature [28] are only 1861 m^2/g), pore size distribution ranging from 2 to 5 nm, and pore volume of 2.4 mL/g, we homogeneously mixed it with petroleum coke and potassium hydroxide using ultra-high-speed centrifugal dispersion technology. Subsequently, the carbonization and activation of petroleum coke, as well as the activation and reduction of graphene, were accomplished at high temperatures, resulting in the formation of high-specific-surface-area porous activated carbon with a three-dimensional conductive network structure, integrated with activated graphene hybrids.

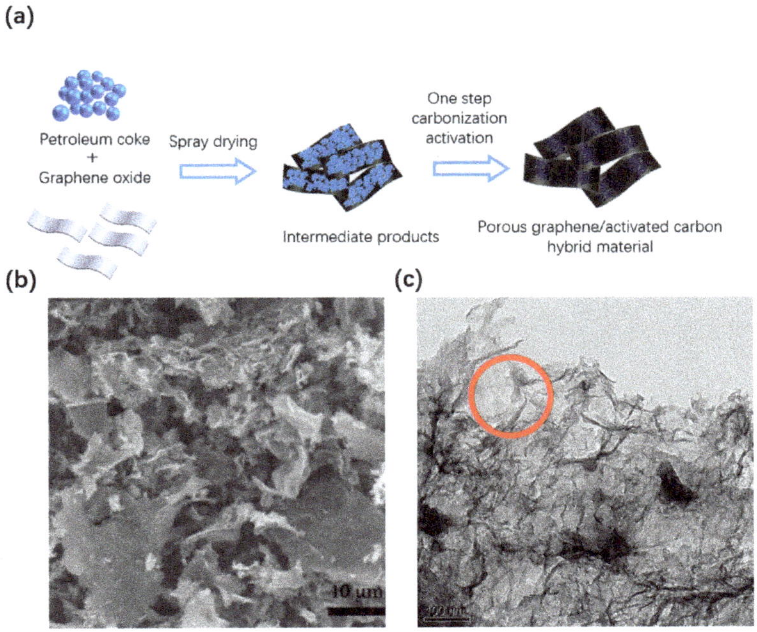

Figure 1. (**a**) Schematic diagram of one-step carbonization activation method for preparing porous graphene/activated carbon hybrid materials; (**b**) SEM image of porous graphene/activated carbon hybrid material; (**c**) TEM image of porous graphene/activated carbon hybrid material.

The morphology of the porous graphene/activated carbon hybrid material is depicted in Figure 1b. As evident from the figure, the product post-one-step carbonization activation retains a wrinkled sheet-like structure akin to the original graphene, featuring a certain degree of agglomeration. The surface is enveloped in an amorphous carbon layer, exhibiting rough and wrinkled features. This graphene/activated carbon hybrid material is theoretically superior in electrochemical energy storage characteristics, yet it proves challenging to produce electrode films using traditional wet processes. This is because capacitors are highly sensitive to moisture during the preparation process. Yet, the traditional wet electrode preparation process necessitates the use of solvents like deionized water to regulate the slurry's viscosity. Even high-temperature and high-vacuum drying processes struggle to eliminate the moisture introduced by the wet electrode. This moisture not only makes the electrode prone to delamination but also leads to increased leakage currents in capacitors, compromising the long-term stability of the product. Additionally, while traditional wet electrode preparation processes boast strong continuous production capabilities and relatively low engineering application difficulties, they result in low electrode densities (usually less than 0.6 g/cm^3), which in turn limit the capacity and voltage endurance of the individual cells (to less than 2.7 V) [29]. Hence, traditional wet processes fail to produce electrode films of graphene/activated carbon hybrids with superior electrochemical performance.

To enhance the voltage resistance and electrode density, this study independently developed a dry preparation technique, as illustrated in Figure 2a.

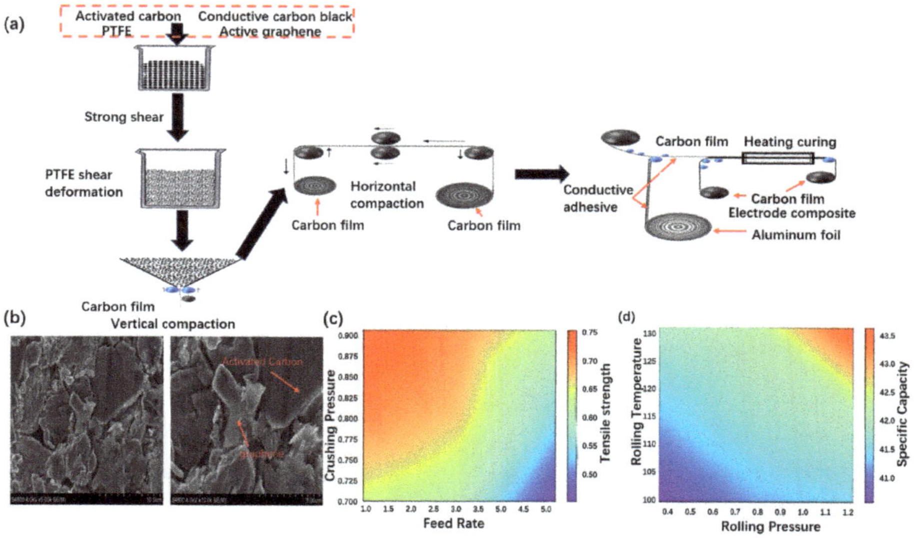

Figure 2. Preparation process and properties of the hybrid electrode film. (**a**) Dry preparation process; (**b**) electron microscope photographs; (**c**) mechanical properties; (**d**) specific capacity.

Initially, we developed ultra-high-speed shear dispersion technology. This process breaks down the agglomeration of micrometer-sized carbon materials, achieving uniform solid-phase mixing. Additionally, ultra-high-speed intense shearing transforms the chain-like polymer structure of the dry powder adhesive polytetrafluoroethylene (PTFE), transforming it from a cluster to a network configuration, evenly distributing it across the surface of the carbon-based hybrid material. This ensures that the electrode does not lose powder, fall off, or rebound during subsequent production steps, thereby enhancing the supercapacitor's long service life. Research on carbon film formation technology utilizes low-temperature vertical rolling to synergistically produce high-density carbon films with electrode hybrids and PTFE. Research on heat-curing rolling technology integrates electrode materials with current collectors by hot-rolling hybrid carbon films and graphene conductive adhesive-coated aluminum foils. Leveraging this dry preparation technique, the electrode density can be increased to 0.65–0.7 g/cm^3 (as a comparison, the results in the literature [30] are only 0.59 g/cm^3), significantly improving the mass of active material per unit volume and significantly enhancing both the overall capacity and specific energy of the monomers.

Furthermore, this process eliminates the liquid-phase steps entirely, avoiding the introduction of moisture that significantly impacts the voltage window, thereby facilitating an increase in the monomer's window voltage.

The microstructure of the hybrid dry electrode sheet is shown in Figure 2b. From the figure, it can be seen that the carbon particles in the electrode sheet are tightly stacked, indicating that the density of the electrode sheet can be increased by the above process, thus increasing the content of active material in the monolithic capacitor. Meanwhile, the size of active graphene is close to the size of active carbon particles, which proves that the above ultra-high-speed shear dispersion technique makes the active graphene well dispersed in the hybrid electrode sheet. Additionally, Figure 2b reveals sheet-like porous graphene (as indicated by the arrow, with a smoother surface) and activated carbon (as indicated by the arrow, in a blocky particle structure), both of which are well dispersed. The high magnification scanning electron microscope (Figure 3b) photo further confirms that the flake active graphene is well dispersed and connected to the active carbon particles, and it also acts as a conductive bridging agent.

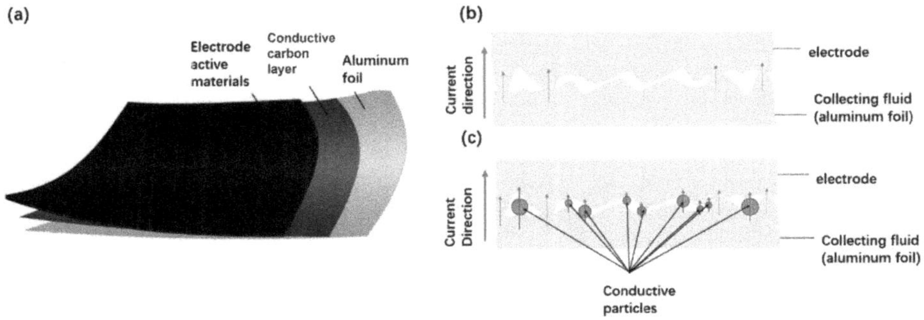

Figure 3. Carbon-coated aluminum foil collector. (**a**) Three-layer structure; (**b**) mechanisms of the conventional conductivity of the interfacial contact; (**c**) enhanced mechanism of interfacial contact conductivity.

The key technology in the dry electrode manufacturing process is the fibrilization of the binder. This is mainly achieved by pulverizing the spherical PTFE binder fibers with a high-pressure air stream. The degree of fibrillation is mainly influenced by factors such as feed rate and comminution pressure. Samples with a length of 9 cm and a width of 1 cm were used for testing. It was experimentally verified that the slower the feed rate and the higher the grinding pressure, the higher the degree of binder fibrillation, the higher the dry electrode film strength, and the better the electrode sheet quality. Figure 2c shows the experimental data for the feed rate, comminution pressure, and tensile strength of the sample with a length of 9 cm and a width of 1 cm. At a feed rate of 1 kg/h and a squeezing pressure of 0.9 MPa, corresponding to the top left point on the data graph, the carbon film has the highest tensile strength along its length.

Dry electrode rolling is the key process from powder to dry electrode sheet. The rolling pressure and temperature directly affect the electrode density and ultimately determine the specific capacitance of the capacitor monolith. It is experimentally verified (the process parameters are shown in Figure 2d) that the higher the rolling pressure and the higher the rolling temperature, the higher the electrode density and the higher the specific capacitance of the monomer.

3.2. Fabrication of Collector-Optimized Supercapacitor

The current collector, which is the carrier material for the electrode material in the composition of capacitors, serves the function of current collection and support. It should possess the attributes of conductivity, corrosion resistance, and overload resistance. Commonly utilized current collectors include stable metal foils or meshes such as Al, Cu, Ni, or Ti, which are known to exhibit high conductivity [31–36]. In specific contexts, rare or precious metals may be employed as current collector materials. In power double-layer capacitors, aluminum foil is typically employed as a current collector due to its advantageous qualities, including high electrical conductivity, low cost, and corrosion resistance. However, due to its high reactivity, the surface of aluminum current collectors tends to form a poorly conductive aluminum oxide film, which increases the resistance between the current collector and the active material and reduces adhesion between them. Therefore, appropriate treatments of aluminum current collectors are necessary.

In the context of industrial applications, the surface treatment of aluminum foil for supercapacitors primarily involves etching. The resulting material, designated "corroded foil", exhibits superior adhesion to electrode materials compared to unetched aluminum foil, yet it exhibits reduced conductivity and mechanical strength and is associated with a higher production cost, largely driven by the reliance on imports of the required raw materials.

The coating of carbon on aluminum foil involves the application of a layer of carbon film (with a thickness of less than 5 micrometers) to the surface of the foil, with the addition of conductive agents such as conductive graphite, carbon black, carbon nanotubes, and graphene. This coating has the effect of significantly reducing the interface resistance between the electrode sheet and the aluminum foil, while also enhancing the adhesion between the active material and the current collector. Furthermore, the coating can also serve to partially suppress the corrosion of the current collector. Consequently, the utilization of this novel type of current collector can facilitate the enhancement of the high-rate charge–discharge capability and cycle life of batteries. The schematic diagram of the structure is depicted in Figure 3a. In the absence of a coating, the contact between interfaces is insufficient, resulting in a reduction in the number of current pathways and a restriction of the flow of the current. The coating facilitates improved contact between the particles, thereby enhancing the current and performance, as illustrated in Figure 3b,c.

Figure 4 presents a comparative analysis of the adhesion properties of various collector electrode coatings. The procedure for testing adhesion can be described as follows: the electrode is placed onto a flat plate of sufficient hardness, the handle of the baguette cutter is held so that the multi-flute cutter is perpendicular to the plane of the test piece, and the cut is made with uniform pressure and at a cutting speed of 20–50 mm/s. The polarizer is then rotated through 90 degrees, the aforementioned operation is repeated at the cutting place to form a grid pattern, and the electrode piece is gently brushed with a soft brush along the two diagonals of the grid pattern, backwards and forwards five times each. Finally, the electrode piece is tested. Upon observation, it was noted that there were small pieces of flaking at the intersection of the cuts. Additionally, the actual damage in the grid area was found to be less severe. The adhesion of the two types of poles, carbon and aluminum foil coated and corroded aluminum foil, is essentially comparable. It has been demonstrated that the carbon aluminum foil coating can enhance the adhesion between the porous carbon material and the collector.

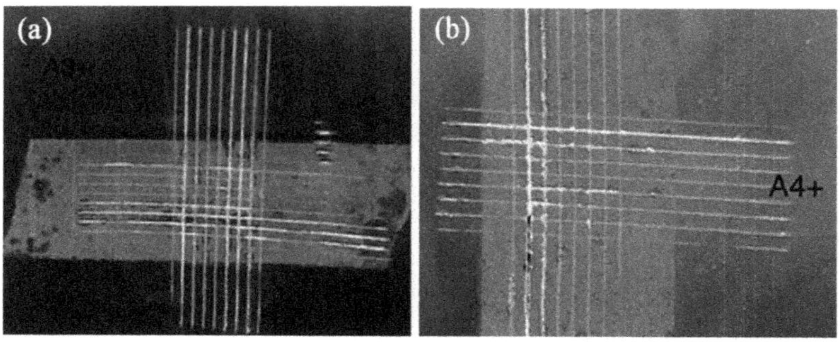

Figure 4. The comparative diagram of film adhesion on different current collector electrodes: (**a**) coating carbon aluminum foil; (**b**) corroded aluminum foil.

Figure 5 illustrates the alterations in the internal resistance and capacity of 100F capacitors assembled with distinct current collector electrodes during high-temperature ageing. From the graph, it can be observed that, during the 500 h high-temperature ageing process, the rate of the increase in the internal resistance of carbon-coated aluminum foil current collector capacitors is relatively gentle, while the rate of capacity retention is higher than that of corroded foil current collector capacitors. This suggests that coated aluminum foil can markedly reduce the interface resistance between the electrode and aluminum foil, thereby enhancing the stability of the electrode.

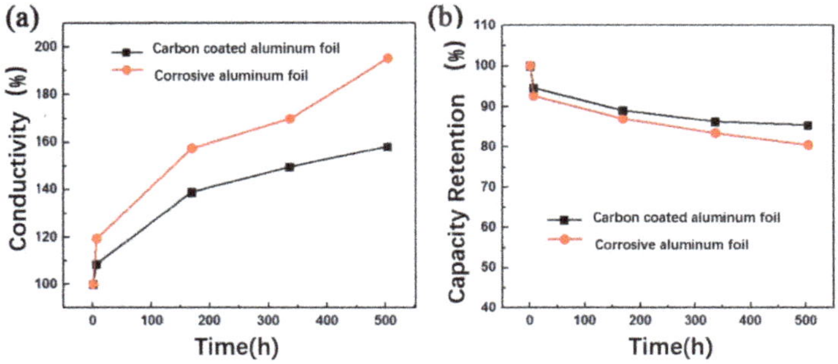

Figure 5. Contribution of carbon-coated aluminum foil to durability. (**a**) Conductivity; (**b**) capacity retention.

In the development of cylindrical devices, the manner in which the tabs are connected and lead out can serve to reduce the internal resistance of the device and increase the power density. In this paper, we have employed a full tab structure in the design of the electrode tab structure with the objective of enhancing the current-carrying capability. The full electrode tab structure is depicted in Figure 6. Initially, the activated carbon was coated on an aluminum foil, with the diaphragm placed between two layers of activated carbon. Thereafter, the activated carbon, aluminum foil, and diaphragm were wound into a compact structure. Following winding, the area of the aluminum foil is marginally larger than that of the toner, with both toner ends encompassing a portion of aluminum foil. The aluminum foil serves as electrode tabs, with the upper end connected to the negative collector as a whole and the lower end connected to the positive collector in the same manner. The negative collector is connected to the top cover, while the positive collector is connected to the bottom case. In this configuration, the cover and the case act as the two ends of the capacitor, with the capacitors able to output voltage to the outside and be insulated from each other. Furthermore, the lead-in terminals employ the direct lead-in method in lieu of the conventional indirect lead-in method. The structure is illustrated in Figure 6, which entails soldering the lead foil directly to the lead-out terminals. The protrusions of the lead-in terminals pass directly through the through-holes of the housing and extend outwardly from the housing. The direct lead-in structure design significantly reduces the internal resistance and enhances the power performance of the supercapacitor.

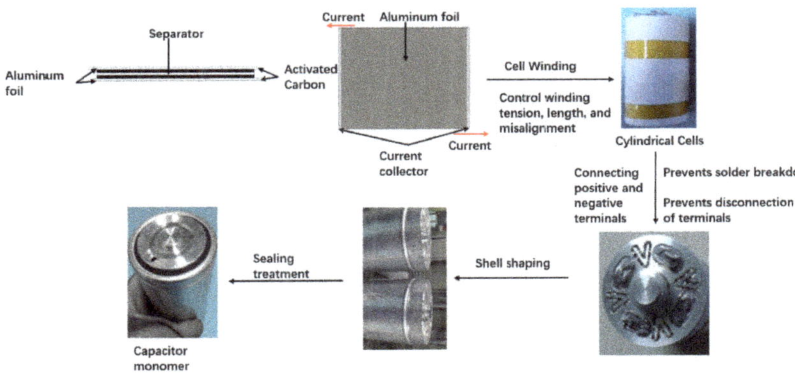

Figure 6. Fabrication process for supercapacitors with full tab structure [36].

A trial production of capacitors based on the aforementioned materials and structures was conducted. In order to verify the withstand voltage characteristics of the capacitor, a cyclic voltammetry (CV) test was first conducted on the capacitor, with a voltage range of

0–3 V and a scanning speed of 5 mV/s. The cyclic voltammetry (CV) curve is presented in Figure 7a. The CV curve approximates a rectangular structure, reflecting favorable capacitance characteristics, particularly at a high voltage (3V). The absence of an evident oxidation curve in the CV indicates that, under high voltage, the electrode is relatively stable and no significant side reactions occur. This indicates that the capacitor has a stable structure at 3V. It should be noted that the current used in the test was I40 = 5 A, I10 = 1.25 A, and I5 = 0.625 A.

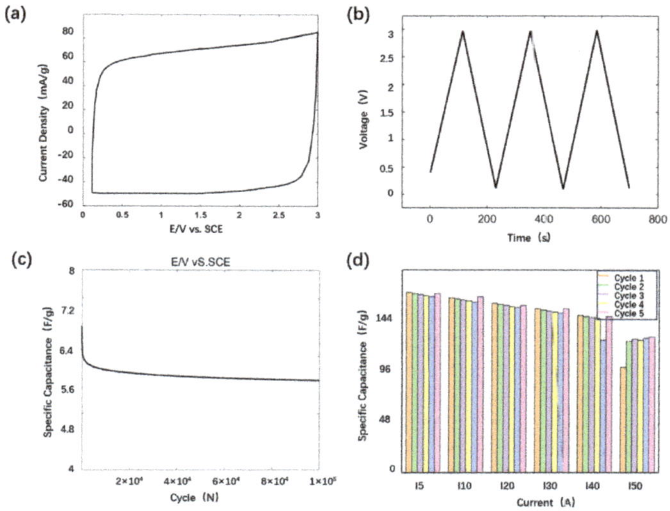

Figure 7. Performance testing of supercapacitors. (**a**) Cyclic voltammetry; (**b**) charging and discharging with constant current; (**c**) cycling performance; (**d**) rate performance.

The capacitor's charging and discharging test range is illustrated in Figure 7b, spanning a voltage range of 0–3 V. This range exhibits a regular triangular sawtooth shape and clear capacitance behavior. To ascertain the capacitor's operational lifetime, we conducted cyclic tests, and the results are presented in Figure 7c. Following the completion of the cycling tests, the observed capacity decay was found to be relatively minor, thereby indicating that the capacitor in question has a longer lifespan. In order to ascertain the rate performance of the capacitor, charge–discharge tests were conducted on individual capacitors at different rates. The voltage range was 0–3 V, and the charge–discharge current was 5–50 I. Each rate of current was subjected to five cycles of charge–discharge. The test results are presented in Figure 7d. The initial capacity is 165 F (under 5 I testing conditions), and as the testing current increases (5 I to 30 I), the capacity does not decay significantly, demonstrating good rate performance. Upon increasing the charging and discharging current to 40 I, a slight decrease in capacity was observed. At a test current of 50 I, the capacity is approximately 125 F, with a capacity retention rate of 75.8%. Furthermore, following repeated cycles of charge and discharge testing under different magnification currents, the capacitor did not demonstrate any notable gas swelling, suggesting that it has excellent voltage resistance. Furthermore, electrochemical impedance spectroscopy (EIS) testing was conducted on supercapacitors, with the results presented in the Supplementary Materials. It should be noted that the internal resistance of the capacitor prepared in this work is about 0.06 mΩ (as a comparison, the results in the literature [28] are only 0.5 Ω). Table 1 shows some specific test data for capacitors, such as weight, internal resistance, power density and energy density.

Table 1. Test data of cylindrical capacitors.

Number	Weight (g)	Capacity (F)	Specific Energy (Wh/Kg)	Internal Resistance (mΩ)	Power Density (kW/kg)	Capacitance Retention (%)
1#	543.5	2960	6.81	0.066	62.72	81.5%
2#	534.5	2902	6.79	0.057	73.85	80.3%

4. Discussion

In this paper, porous graphene/activated carbon hybrids were prepared by a one-step carbonization activation process on the basis of porous graphene materials with a specific surface area of more than 2000 m^2/g, a pore size distribution of 2–5 nm, and a void volume of 2.4 mL/g. An ultra-high-speed centrifugal dispersion technique was employed to achieve a homogeneous slurry comprising graphene oxide, petroleum coke, and potassium hydroxide. At elevated temperatures, petroleum coke undergoes carbonization and activation, while graphene undergoes activation and reduction to form a three-dimensional conductive network structure of high-specific-surface-area porous activated carbon/activated graphene hybrids. The product that resulted from the one-step carbonization and activation process exhibited a folded sheet structure that was similar to that of pristine graphene. However, it also contained some agglomerates and was coated with an amorphous carbon layer that exhibited a rough and folded surface. The development of ultra-high-speed shear dispersion technology was employed to ensure the long-term service life of the supercapacitors. The hybrid electrodes prepared using dry technology exhibited a notable enhancement in total capacitance and specific energy relative to the monomer. The development of an all-pole tab high-power soft-packed capacitor with good voltage resistance, large monomer power density, and excellent electrochemical performance was achieved through the dry preparation of electrode sheets as electrodes. The capacitor's performance was evaluated through cyclic voltammetry, which demonstrated that the electrode exhibited stability at high voltages without significant side reactions. The results of the cyclic testing of the capacitors demonstrated a relatively small reduction in capacitance, indicating that they have a relatively long life. Conversely, the capacitor did not exhibit any discernible gas expansion following multiple charge/discharge tests with varying amplification currents, thereby indicating that the capacitor exhibits robust voltage resistance.

Supplementary Materials: The following supporting information can be downloaded at: https://www.mdpi.com/article/10.3390/batteries10060195/s1, Table S1. Test data of electrodes prepared by two methods, Table S2. Electrical conductivity of several metals, Figure S1: AC Impedance Spectrum of Supercapacitors.

Author Contributions: Conceptualization, X.W.; methodology, S.C.; validation, S.C.; formal analysis, W.W. and X.Z.; investigation, S.C.; data curation, S.C.; writing—original draft preparation, S.C. and W.W.; writing—review and editing, X.W.; visualization, X.Z.; supervision, X.W.; All authors have read and agreed to the published version of the manuscript.

Funding: This research received no external funding.

Data Availability Statement: Data is contained within the article or Supplementary Material.

Conflicts of Interest: The authors declare no conflict of interest.

References

1. Arunkumar, C.R.; Manthati, U.B.; Punna, S. Supercapacitor-based transient power supply for DC microgrid applications. *Electr. Eng.* **2022**, *104*, 463–472. [CrossRef]
2. Fernando, J.; Kularatna, N.; Silva, S.; Thotabaddadurage, S.S. Supercapacitor assisted surge absorber technique: High performance transient surge protectors for consumer electronics. *IEEE Power Electron. Mag.* **2022**, *9*, 48–60. [CrossRef]
3. Luo, L.; Lan, Y.; Zhang, Q.; Deng, J.; Luo, L.; Zeng, Q.; Gao, H.; Zhao, W. A review on biomass-derived activated carbon as electrode materials for energy storage supercapacitors. *J. Energy Storage* **2022**, *55*, 105839. [CrossRef]

4. Jiang, Y.; Li, J.; Jiang, Z.; Shi, M.; Sheng, R.; Liu, Z.; Zhang, S.; Cao, Y.; Wei, T.; Fan, Z. Large-surface-area activated carbon with high density by electrostatic densification for supercapacitor electrodes. *Carbon* **2021**, *175*, 281–288. [CrossRef]
5. Nithya, V.D. A review on holey graphene electrode for supercapacitor. *J. Energy Storage* **2021**, *44*, 103380. [CrossRef]
6. Yang, Z.; Tian, J.; Ye, Z.; Jin, Y.; Cui, C.; Xie, Q.; Wang, J.; Zhang, G.; Dong, Z.; Miao, Y.; et al. High energy and high power density supercapacitor with 3D Al foam-based thick graphene electrode: Fabrication and simulation. *Energy Storage Mater.* **2020**, *33*, 18–25. [CrossRef]
7. Yan, Z.; Gao, Z.; Zhang, Z.; Dai, C.; Wei, W.; Shen, P.K. Graphene nanosphere as advanced electrode material to promote high performance symmetrical supercapacitor. *Small* **2021**, *17*, 2007915. [CrossRef] [PubMed]
8. Jiang, Q.; Liu, D.; Liu, B.; Zhou, T.; Zhou, J. Blotting paper-derived activated porous carbon/reduced graphene oxide composite electrodes for supercapacitor applications. *Molecules* **2019**, *24*, 4625. [CrossRef]
9. Thalji, M.R.; Ali, G.A.; Liu, P.; Zhong, Y.L.; Chong, K.F. W18O49 nanowires-graphene nanocomposite for asymmetric supercapacitors employing AlCl3 aqueous electrolyte. *Chem. Eng. J.* **2021**, *409*, 128216. [CrossRef]
10. Wang, J.; Li, Q.; Peng, C.; Shu, N.; Niu, L.; Zhu, Y. To increase electrochemical performance of electrode material by attaching activated carbon particles on reduced graphene oxide sheets for supercapacitor. *J. Power Sources* **2020**, *450*, 227611. [CrossRef]
11. Iakunkov, A.; Skrypnychuk, V.; Nordenström, A.; Shilayeva, E.A.; Korobov, M.; Prodana, M.; Enachescu, M.; Larsson, S.H.; Talyzin, A.V. Activated graphene as a material for supercapacitor electrodes: Effects of surface area, pore size distribution and hydrophilicity. *Phys. Chem. Chem. Phys.* **2019**, *21*, 17901–17912. [CrossRef] [PubMed]
12. Ramadoss, A.; Kim, S.J. Improved activity of a graphene–TiO_2 hybrid electrode in an electrochemical supercapacitor. *Carbon* **2013**, *63*, 434–445. [CrossRef]
13. Costentin, C.; Savéant, J.M. Energy storage: Pseudocapacitance in prospect. *Chem. Sci.* **2019**, *10*, 5656–5666. [CrossRef] [PubMed]
14. George, J.; Balachandran, M. Extrinsic pseudocapacitance: Tapering the borderline between pseudocapacitive and battery type electrode materials for energy storage applications. *J. Energy Storage* **2023**, *74*, 109292. [CrossRef]
15. Aderyani, S.; Flouda, P.; Shah, S.A.; Green, M.J.; Lutkenhaus, J.L.; Ardebili, H. Simulation of cyclic voltammetry in structural supercapacitors with pseudocapacitance behavior. *Electrochim. Acta* **2021**, *390*, 138822. [CrossRef]
16. Park, H.W.; Roh, K.C. Recent advances in and perspectives on pseudocapacitive materials for supercapacitors—A review. *J. Power Sources* **2023**, *557*, 232558. [CrossRef]
17. Hu, Y.R.; Dong, X.L.; Zhuang, H.K.; Yan, D.; Hou, L.; Li, W.C. Introducing Electrochemically Active Oxygen Species to Boost the Pseudocapacitance of Carbon-based Supercapacitor. *ChemElectroChem* **2021**, *8*, 3073–3079. [CrossRef]
18. Bokhari, S.W.; Siddique, A.H.; Sherrell, P.C.; Yue, X.; Karumbaiah, K.M.; Wei, S.; Ellis, A.V.; Gao, W. Advances in graphene-based supercapacitor electrodes. *Energy Rep.* **2020**, *6*, 2768–2784. [CrossRef]
19. Down, M.P.; Rowley-Neale, S.J.; Smith, G.C.; Banks, C.E. Fabrication of graphene oxide supercapacitor devices. *ACS Appl. Energy Mater.* **2018**, *1*, 707–714. [CrossRef]
20. Zhang, S.; Pan, N. Supercapacitors performance evaluation. *Adv. Energy Mater.* **2015**, *5*, 1401401. [CrossRef]
21. Ke, Q.; Wang, J. Graphene-based materials for supercapacitor electrodes—A review. *J. Mater.* **2016**, *2*, 37–54. [CrossRef]
22. Li, Z.J.; Yang, B.C.; Zhang, S.R.; Zhao, C.M. Graphene oxide with improved electrical conductivity for supercapacitor electrodes. *Appl. Surf. Sci.* **2012**, *258*, 3726–3731. [CrossRef]
23. Zhang, H.; Yang, D.; Lau, A.; Ma, T.; Lin, H.; Jia, B. Hybridized graphene for supercapacitors: Beyond the limitation of pure graphene. *Small* **2021**, *17*, 2007311. [CrossRef] [PubMed]
24. Zequine, C.; Bhoyate, S.; de Souza, F.; Arukula, R.; Kahol, P.K.; Gupta, R.K. Recent advancements and key challenges of graphene for flexible supercapacitors. In *Adapting 2D Nanomaterials for Advanced Applications*, American Chemical Society: Washington, DC, USA, 2020; pp. 49–77.
25. Pameté, E.; Köps, L.; Kreth, F.A.; Pohlmann, S.; Varzi, A.; Brousse, T.; Balducci, A.; Presser, V. The many deaths of supercapacitors: Degradation, aging, and performance fading. *Adv. Energy Mater.* **2023**, *13*, 2301008. [CrossRef]
26. Kurzweil, P.; Schottenbauer, J.; Schell, C. Past, present and future of electrochemical capacitors: Pseudocapacitance, aging mechanisms and service life estimation. *J. Energy Storage* **2021**, *35*, 102311. [CrossRef]
27. Pegel, H.; Wycisk, D.; Scheible, A.; Tendera, L.; Latz, A.; Sauer, D.U. Fast-charging performance and optimal thermal management of large-format full-tab cylindrical lithium-ion cells under varying environmental conditions. *J. Power Sources* **2023**, *556*, 232408. [CrossRef]
28. Zhang, J.; Zhao, Y.; Zang, X.; Wang, B.; Ma, F. Application of three-dimensional porous graphene in high-magnification supercapacitors. *Energy Eng.* **2020**, 58–64. [CrossRef]
29. Balducci, A.; Dugas, R.; Taberna, P.L.; Simon, P.; Plée, D.; Mastragostino, M.; Passerini, S. High temperature carbon-carbon supercapacitor using ionic liquid as electrolyte. *J. Power Sources* **2007**, *165*, 922–927. [CrossRef]
30. Liu, F.; Xue, L. Effect of moulding process on the performance of activated carbon electrodes for supercapacitors. *Electron. Compon. Mater.* **2017**, *36*, 25–28.
31. Dai, X. Research and Application of Modified Aluminum Foil Current Collector and Interface Optimization by Electric Spark Method in Supercapacitors. Master's Thesis, Jilin University, Changchun, China, 2020.
32. Zhang, D.; Wu, Y.; Li, T.; Huang, Y.; Zhang, A.; Miao, M. High performance carbon nanotube yarn supercapacitors with a surface-oxidized copper current collector. *ACS Appl. Mater. Interfaces* **2015**, *7*, 25835–25842. [CrossRef]

33. Chae, C.; Han, J.H.; Lee, S.S.; Choi, Y.; Kim, T.H.; Jeong, S. A printable metallic current collector for all-printed high-voltage micro-supercapacitors: Instantaneous surface passivation by flash-light-sintering reaction. *Adv. Funct. Mater.* **2020**, *30*, 2000715. [CrossRef]
34. An, G.H.; Cha, S.N.; Ahn, H.J. Surface functionalization of the terraced surface-based current collector for a supercapacitor with an improved energy storage performance. *Appl. Surf. Sci.* **2019**, *478*, 435–440. [CrossRef]
35. Yu, J.; Yu, C.; Guo, W.; Wang, Z.; Ding, Y.; Xie, Y.; Liu, K.; Wang, H.; Tan, X.; Huang, H.; et al. Insight into the effects of current collectors and in situ Ni leaching in high-voltage aqueous supercapacitors. *Adv. Funct. Mater.* **2022**, *32*, 2204609. [CrossRef]
36. Ruan, D. *Dynamic Double-Layer Capacitors—Principles, Manufacturing, and Applications*; Science Press: Beijing, China, 2018.

batteries

MDPI

Review

Battery-Type Transition Metal Oxides in Hybrid Supercapacitors: Synthesis and Applications

Bikash Raut [1], Md. Shahriar Ahmed [1], Hae-Yong Kim [2], Mohammad Mizanur Rahman Khan [3], Gazi A. K. M. Rafiqul Bari [3], Mobinul Islam [1,*] and Kyung-Wan Nam [1,*]

[1] Department of Energy & Materials Engineering, Dongguk University, Seoul 04620, Republic of Korea; 2023121093@dgu.ac.kr (B.R.); shahriar.emcl@dgu.ac.kr (M.S.A.)
[2] Department of Advanced Battery Convergence Engineering, Dongguk University, Seoul 04620, Republic of Korea
[3] School of Mechanical Smart and Industrial Engineering, Gachon University, Gyeonggi-do 13120, Republic of Korea; mmrkhan@gachon.ac.kr (M.M.R.K.); grafiqulbari@gachon.ac.kr (G.A.K.M.R.B.)
* Correspondence: mobin85@dongguk.edu (M.I.); knam@dongguk.edu (K.-W.N.); Tel.: +82-2-2260-4978 (K.-W.N.)

Abstract: Hybrid supercapacitors (HSCs) have garnered growing interest for their ability to combine the high energy storage capability of batteries with the rapid charge–discharge characteristics of supercapacitors. This review examines the evolution of HSCs, emphasizing the synergistic mechanisms that integrate both Faradaic and non-Faradaic charge storage processes. Transition metal oxides (TMOs) are highlighted as promising battery-type electrodes owing to their notable energy storage potential and compatibility with various synthesis routes, including hydro/solvothermal methods, electrospinning, electrodeposition, and sol–gel processes. Particular attention is directed toward Ti-, Co-, and V-based TMOs, with a focus on tailoring their properties through morphology control, composite formation, and doping to enhance electrochemical performance. Overall, the discussion underscores the potential of HSCs to meet the growing demand for next-generation energy storage systems by bridging the gap between high energy and high power requirements.

Keywords: hybrid supercapacitor; transition metal oxide; battery; supercapacitor; energy storage device; electrospinning; electrodeposition

1. Introduction

The rapid depletion of non-renewable energy sources, such as fossil fuels, and the alarming rise in environmental pollution caused by their extensive use have spurred the development of renewable energy alternatives in recent years [1,2]. Technologies such as solar energy, electrochemical energy storage (including batteries and supercapacitors), and hydrogen energy have gained significant attention. Among various electrochemical energy storage systems (EES), batteries and supercapacitors stand out as the most promising devices [3–7]. The high specific capacity and energy density of secondary batteries, attributed to bulk redox reactions, position them at the top end of the energy density axis on the Ragone plot; a graphical tool that compares the energy density and power density of various energy storage systems including batteries, supercapacitors and hybrid supercapacitors in terms of both how much energy they can store and how quickly they can deliver that energy [8–10] as shown in Scheme 1A. The high energy density of batteries (150–300 $WhKg^{-1}$) enables long driving range for EVs [11]. But when it comes to power density (below 350 Wkg^{-1}), batteries have inferior power density [12,13]. The power density on the other hand in the case of supercapacitors is high but supercapacitors have very low energy density, almost 10 times lower than that of batteries, so, it is practically not feasible to use supercapacitors where high energy density is required [14–17]. As both high energy density and high power density are critical for advanced energy storage systems, there is a growing need to develop energy materials capable of delivering both simultaneously. In the context of the Ragone plot, which visually maps the trade-off between energy density and power density, and balance between these two-performance metrics. One promising class of energy materials that has attracted extensive research attention is hybrid supercapacitors (HSCs) [18]. HSCs are innovative energy storage devices that combine the advantages of battery-like materials (Faradaic redox reactions) with supercapacitor-like materials (non-Faradaic double-layer capacitance) [19]. This hybridization enables HSCs to bridge the gap between two extremes of the Ragone plot, batteries, which provide high energy density but suffer from low power density, and supercapacitors, which offer high power density but exhibit limited energy density. By leveraging the complementary mechanisms of batteries and supercapacitors, HSCs open up new and previously unexplored regions of the Ragone plot [1,20]. The battery-like component of HSCs contributes to higher energy storage through Faradaic redox reactions, where energy is stored in the bulk of the active material via ion insertion/extraction processes or conversion, and alloying processes. Simultaneously, the supercapacitor-like component facilitates rapid charge/discharge capability, delivering high power density through surface-based, non-Faradaic charge storage mechanisms. This dual nature of HSCs enables a unique synergy that addresses the inherent limitations of both batteries and supercapacitors, providing a balance between fast energy delivery and large energy storage capacity. Due to these advantages, hybrid supercapacitors are emerging as a key technology for next-generation energy storage systems, garnering significant research focus globally. Over the past decade, the field has witnessed exponential growth, with a substantial increase in the number of publications each year, reflecting the increasing interest and progress in this area as shown in Scheme 1B. This trend underscores the potential of HSCs to revolutionize energy storage by simultaneously enhancing both energy and power performance. Ongoing innovations in material design, electrode architecture, and device engineering are driving the next generation of HSCs by enhancing both energy and power densities, thereby offering more efficient and versatile energy storage solutions.

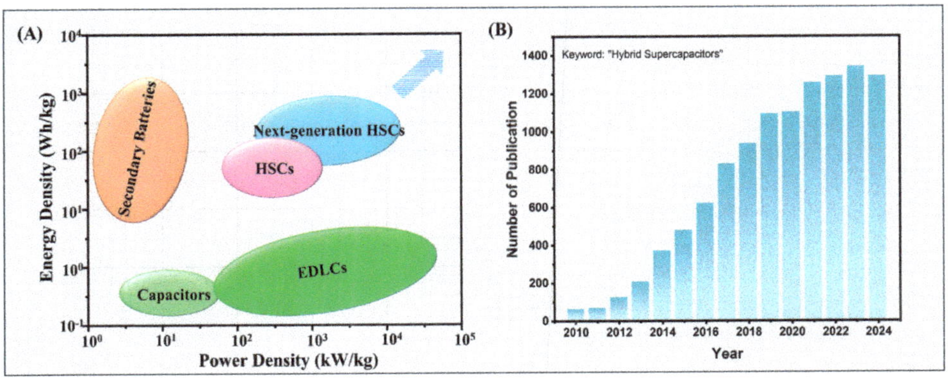

Scheme 1. (**A**) Ragone plots for secondary batteries, supercapacitors, and HSCs. (**B**) Publication numbers over time.

Transition metal oxides (TMOs) have emerged as a pivotal class of materials in the development of hybrid supercapacitors (HSCs) due to their remarkable redox activity, high theoretical capacitance, and tunable electronic properties [21–24]. Their ability to facilitate

Faradaic charge storage mechanisms enables them to deliver energy densities comparable to batteries while retaining the power capabilities of supercapacitors. This dual functionality makes them indispensable for next-generation energy storage technologies aimed at addressing the growing demand for efficient, sustainable, and versatile energy solutions. Previous reviews have extensively covered TMO fundamentals and their electrochemical properties [1,18,20,25]. In contrast, this review provides a comprehensive overview of fundamentals and history of HSCs, detailed analysis of the diverse TMOs used as battery-type electrodes, including oxides of vanadium, cobalt, and titanium. Special attention is given to the influence of their crystalline structure, oxidation states, and defect chemistry on their electrochemical behavior. These factors play a critical role in determining the charge storage capacity, rate capability, and long-term stability of HSC systems. Furthermore, this article delves into state-of-the-art synthesis techniques such as sol–gel, hydrothermal, electrospinning, and electrodeposition methods. These methods enable precise control over the morphology, particle size, and surface area of TMOs, which are crucial for optimizing their interaction with electrolytes. Recent advancements in nanostructuring strategies, including the development of hierarchical architectures and composite materials, are discussed for their ability to enhance ion diffusion and electronic conductivity. The review also addresses key challenges in the application of TMOs in HSCs, such as cyclic stability, and conductivity limitations. Emerging solutions, including doping strategies, and hybridization with carbon-based materials are critically evaluated for their effectiveness in overcoming these challenges. Finally, this review outlines future directions for research and development in this field, emphasizing the need for scalable synthesis techniques, environmentally friendly production methods, and the integration of TMOs into flexible and wearable energy storage devices. By bridging the performance gap between batteries and supercapacitors, TMOs hold the potential to revolutionize energy storage systems and pave the way for sustainable technological advancements.

2. History

The early history of HSCs, where two electrodes combine to form a new hybridized device, can be traced back to the mid-1990s. In 1995, Aleksei Stepanov et al. reported the successful hybridization of a nickel oxide electrode and a fibrous carbonic material to form a hybrid device with enhanced capacity compared to the bare double-layer capacitor [26]. This hybridization of two different types of materials to create a new device with enhanced electrochemical properties opened up a new door for the exploration of hybrid devices. The possibility of hybridization of two materials either similar or different (symmetric and asymmetric) allows researchers to explore various combinations to make hybrid devices with different working mechanisms on either side of the electrodes. A timeline of other types of hybrid devices is presented in Scheme 2. In 2006, Aida et al. reported a Li-based hybrid supercapacitor using carbon-based electrodes and Li-based non-aqueous electrolytes. The non-aqueous Li-based electrolyte coupling with carbon-based electrodes enhanced the energy output of the hybrid system by increasing the operational potential window [27]. The hybridization of composite-type negative electrodes with carbon-based positive electrodes with a non-aqueous electrolyte was reported by Naoi et al. in 2009. A nanocomposite of lithium titanate ($Li_4Ti_5O_{12}$; LTO) with carbon nanofiber (CNF) was used as a negative electrode and hybridized with an activated carbon electrode to obtain a hybrid supercapacitor, which was termed a nano-hybrid supercapacitor (NHSC) or nano-hybrid capacitor (NHC). The zero-strain ion insertion capability of LTO allows an insertion-type redox mechanism in LTO/CNF composite electrodes without significant volume change during the process. Also, the operational potential window of LTO lies in between the decomposition limit of traditional non-aqueous-based electrolytes which eliminates the possibility of electrolyte decomposition and unwanted side reaction-based non-reversible capacity contribution [28,29]. In 2016, Yoo et al. reported an HSC with "rocking-chair"-type mechanism. The HSC was comprised of di-valent metal negative electrode, activated carbon positive electrode and a non-aqueous electrolyte. This new "rocking-chair" type mechanism enhances the energy density of the HSC and lowers the volume of required electrolytes in the HSC system [30].

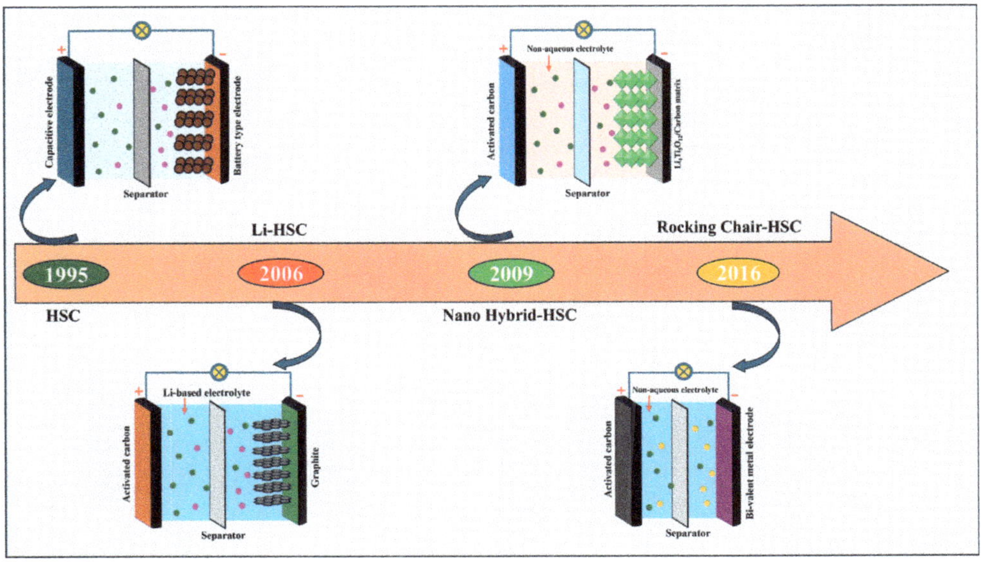

Scheme 2. Timeline of different types of HSCs.

3. Fundamental Concepts of HSC

Before delving into the concepts behind the HSCs, it is essential to understand their two components, i.e., batteries and supercapacitors. Batteries and supercapacitors both are charge storage systems with different electrochemical charge storage mechanisms, the former stores charge based on bulk Faradaic processes, and the latter is based on near-surface non-Faradaic processes [31–34]. The bulk redox process of batteries results in high energy density, but low power density stems from the sluggish ionic diffusion (diffusion limitation). On the other hand, supercapacitors possess low energy density due to the absence of redox reactions and high power density resulting from fast ionic transport without significant diffusion limitations. As both have different charge storage mechanisms, they can be distinguished by cyclic voltammograms. For instance, due to the involvement of redox reactions clear well-separated anodic (oxidation) and cathodic (reduction) peaks are visible in cyclic voltammograms of batteries, for supercapacitors, these clear cathodic-anodic peaks are missing, and rectangular-shaped voltammograms can be seen stemming from non-Faradaic processes [35,36]. The mechanism of these two energy storage systems can be obtained from sweep voltammetry, like, whether the process is diffusion controlled (like in batteries) or diffusion independent (supercapacitors). For an electrochemical process, the following equation provides insight into the charge storage mechanism of the electrochemical system [37,38]:

$$i = av^b \tag{1}$$

Here, i is the peak current, v is the sweep rate, and b is a variable, which can be obtained from the slope of the logarithmic plot between i and v. The value of b = 0.5 indicates battery-like diffusion-controlled redox processes and this can be expressed as a square root relationship between peak current and sweep rate by Equation (2):

$$i = a\sqrt{v} \tag{2}$$

For a capacitive process (supercapacitor-like), the peak current is linearly related to the sweep rate, i.e., b = 1, and can be expressed by Equation (3):

$$i = av \tag{3}$$

We can combine these two charge storage mechanisms in a single system to obtain a cumulative effect on electrochemical performance compared to its single components, i.e., hybrid supercapacitors (HSCs). This hybridization can solve the low power-density issue of batteries and the low energy-density issue of supercapacitors by utilizing both Faradaic and non-Faradaic reactions. So, for HSC, we can expect the value of b between 0.5 and 1 (0.5 < b < 1) from Equations (1) and (2), and this can be expressed as the linear combination of both equations as follows [39,40]:

$$i(HSC) = a\sqrt{v} + av \tag{4}$$

The hybridization of battery-like and supercapacitor-like components to form a hybrid supercapacitor system with their cyclic voltammograms is shown schematically in Scheme 3.

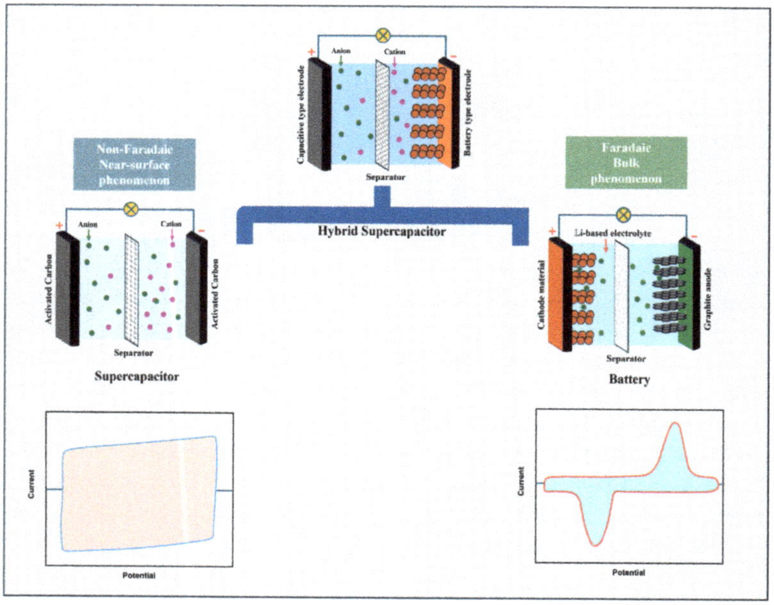

Scheme 3. Hybridization of battery and supercapacitor type electrodes to hybrid supercapacitor.

4. Transition Metal Oxides as Battery-Type Electrodes

Transition metal oxides are emerging as a new paradigm in energy-related applications owing to their unique properties. This review delves into the important role of transition metal oxides (TMO) as battery-type electrodes in hybrid supercapacitor (HSC) systems, highlighting their unique electrochemical properties and potential to bridge the gap between batteries and capacitors. Additionally, it explores the various synthesis techniques employed to tailor their structure, composition, and performance, offering insights into advancements and challenges in this dynamic field.

5. Synthesis of TMOs

5.1. *Hydro/Solvothermal*

Hydro/solvothermal synthesis is one of the most popular, scalable, simple, and cost-effective synthesis techniques. In this technique reaction is carried out in a sealed vessel such as an autoclave, to create a high-pressure and high-temperature environment. High pressure and temperature inside the autoclave facilitate chemical reactions and material formation. Hydrothermal and solvothermal both work on this same principle just with the assistance of different solvents, in hydrothermal, water is used as a solvent, and in solvothermal other solvents like alcohols, amines, or glycerol's are used as solvents for the dissolution of precursors. A schematic of the hydrothermal synthesis process is shown in Figure 1A. In hydrothermal synthesis depending upon reactant ratio, synthesis temperature, and synthesis time, products with different morphologies and functionalities can be obtained, which makes it an important synthesis technique to synthesize different materials including battery-type TMOs. For instance, MnO_2 nanowires [41], $MgCo_2O_4$ nanoflakes [42], $FeCo_2O_4$ nanoflakes [43], and Mn_3O_4 nanorods [44] were synthesized utilizing hydrothermal synthesis. The solvothermal method is also used to synthesize oxides like $MnCo_2O_{4.5}$ with pod-like microstructure [45]. The hydro/solvothermal technique is also used to synthesize composite materials. Recently, Yu and colleagues synthesized $ZnMoO_4/MoO_3$ composite materials by hydrothermal method [46]. Other composites like Cu_2O-CuO-reduced graphene oxide (RGO) [47] and ZnO@Fe_2O_3 [48] were also synthesized using this synthesis method.

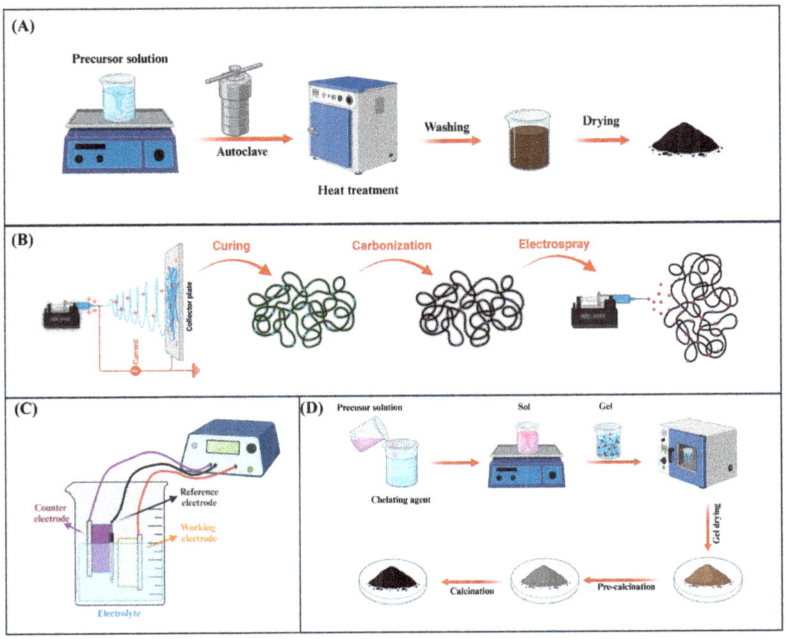

Figure 1. Schematic of different synthesis techniques. (**A**) Hydro/solvothermal. (**B**) Electrospinning. (**C**) Electrodeposition. (**D**) Sol–gel. Images created by BioRender.com.

5.2. Electrospinning

Electrospinning is a versatile and efficient method for fabricating ultrafine fibers. It is a simple and effective technique for producing continuous fibers at the micro- and nanoscale from a polymer solution or melt under the influence of a high-voltage electric field [49]. The method is known for its ability to fabricate fibers with high surface area, fine diameters, and controlled morphology. This electrospinning technique can be utilized to synthesize transition metal oxide nanofibers with several benefits such as self-standing electrodes with no requirement of binders, high surface area, enhanced electron transport, and enhanced stability [50]. These benefits motivated researchers to utilize this technique to synthesize TMOs for several applications including energy applications. For instance, Cao et al. synthesized a series of $La_xSr_{1-x}NiO_{3-\delta}$ ($0.3 \leq x \leq 1$) nanofibers using an electrospinning technique [51]. Several other TMOs, hierarchical Co_3O_4 [52], $MgCo_2O_4$ nanofibers [53], $ZnFe_2O_4$ nanofibers [54], porous-high aspect ratio-Mn_3O_4 nanofibers [55], and coaxial-cable-like Mn_2O_3 nanofibers [56] were also fabricated using electrospinning technique and utilized in energy applications. In 2019, Xu et al. reported the synthesis of a core–shell $NiCo_2O_4$-C@$Ni(OH)_2$ nanofibers using the electrospinning technique, which also proves the versatility of the electrospinning technique to synthesize structurally different TMO nanofibers [57]. A schematic of the electrospinning process is shown in Figure 1B, the versatility of the electrospinning method to couple with techniques like electrospraying to synthesize composite TMOs is also shown schematically.

5.3. Electrodeposition

Electrodeposition or electrochemical deposition is another versatile and cost-effective technique for fabricating TMO-based electrode materials with controlled properties like morphology, thickness, and composition. In this technique, an electric potential is used to deposit the intended material from its precursor solution on a conductive electrode surface like carbon fibers, stainless steel, and nickel foam. Depending on the charges of the substrate on which electrodeposition was performed, it can be divided into cathodic (negatively charged) and anodic (positively charged) electrodeposition. A wide range of TMOs are fabricated with this technique, with controlled composition, morphology, and thickness, and proper optimization of solution pH, applied potential, and precursor ratio [58]. For instance, Joo and co-workers electrodeposited porous and amorphous CuO film on a stainless-steel substrate by cathodic electrodeposition using copper sulfate solution with an optimized pH of 9.0 and an applied potential of 1.05 V vs. saturated calomel electrode (SCE) [59]. Several other TMOs were also fabricated using this electrodeposition technique, like α-MnO_2 [60], MnO_2 nanowires [61], V_2O_5 [62], NiO [63], $MnCo_2O_4$ nanosheets [64], 3D mesoporous Fe_2O_3 [65] and Co_3O_4 nanoflakes [66] were electrodeposited on different substrates. A schematic of the electrodeposition process is shown in Figure 1C.

5.4. Sol–Gel

The sol–gel method is a widely used wet-chemical synthesis technique for synthesizing inorganic materials such as transition metal oxides. Sol–gel is a bottom-up synthesis technique that involves transforming a liquid "sol" of precursors into a solid "gel" network, which is then processed to obtain the final material [67]. Figure 1D schematically shows the detailed synthesis process. The sol–gel approach offers several notable advantages, including uniform mixing at the molecular level (ensuring high homogeneity), broad versatility in material design, and precise control over structural features [68]. This technique has been used to synthesize various TMOs, such as spinel Co_3O_4 [69], $NiCo_2O_4$ [70,71], $FeCo_2O_4$ [72], MnO_2 nanowires [73], and composite $LaNiO_3$/NiO [74].

6. TMOs as Battery-Type Electrodes for HSC

6.1. Ti-Based TMOs

Relying on high natural abundance, chemical stability, cost-effectiveness, and excellent safety, Ti-based oxide materials such as TiO_2 and $Li_4Ti_5O_{12}$ are widely used as electrode materials for supercapacitors and secondary battery applications, such as anodes in Li and Na-ion batteries [75,76].

6.1.1. TiO_2

The most abundant oxide of titanium, TiO_2 exists in four major polymorphs in nature, including rutile, anatase, brookite, and TiO_2 (B) as shown in Figure 2A [77]. Most of these polymorphs exhibit intrinsically low electrical conductivity except TiO_2 (B), which possesses higher electrical conductivity compared to other polymorphs. Because of its good electrical conductivity, TiO_2 (B) is used as an anode for Li-ion batteries (LIBs) and Li-ion capacitors (LICs) without further modification like carbon coating or nanocomposites with other conducting materials [78,79]. Wang et al. reported the successful fabrication of an HSC using TiO_2 (B) nanowires as an anode and carbon nanotube (CNT) as a cathode with enhanced energy density, power density, and cycling stability, higher than CNT-CNT symmetric supercapacitors [80]. However, while TiO_2(B) resolves the conductivity issue compared to other TiO_2 polymorphs, it is more structurally complex and can be challenging to synthesize in large quantities with uniform morphologies. In addition, despite its relatively higher conductivity, TiO_2(B) still falls short of ideal metallic or graphitic levels, sometimes necessitating further optimization through doping or carbon integration for applications requiring ultrahigh power densities. Scale-up and cost also remain considerations: bulk production of phase-pure TiO_2(B) may require carefully controlled synthesis routes that can drive up manufacturing expenses. On the other hand, due to intrinsic low electrical conductivity, other polymorphs of TiO_2 are mostly used as composites with carbon or other functional materials. For instance, Kang and co-workers reported the fabrication of an HSC utilizing an anatase TiO_2-RGO composite anode paired with an activated carbon (AC) cathode. The HSC device with TiO_2-RGO anode exhibited enhanced electrochemical performance compared to the HSC device having a synthesized TiO_2-without RGO and a commercial TiO_2 [81]. A critical drawback here is that integrating carbon (e.g., reduced graphene oxide, carbon nanotubes, or other conductive phases adds to the complexity of material synthesis and can introduce additional costs for large-scale production. Moreover, achieving a uniform dispersion of carbon within TiO_2 is essential for stable long-term performance, and any agglomeration could lead to localized conductivity issues. Reproducibility is also a key concern, especially for industrial-scale applications, where large batch sizes must maintain consistent phase composition and particle size distribution. To address the intrinsic issues of TiO_2, Ramasubbu et al. explored morphologically different TiO_2. They reported the successful synthesis of 3D porous TiO_2 aerogel/Cobalt-metal-organic framework (TiO_2 aerogel/Co-MOF) composites using sol–gel method and used as a battery-type electrode to fabricate an HSC device [82]. The TiO_2 aerogel and Co-MOF synergistically enhanced the contact between electrolyte and electrode interface, increased the electroactive sites for Faradaic reactions, decreased the internal resistance. TiO_2 aerogel/Co-MOF composite also enhanced the ionic transportation by shortening the diffusion path and improved the mechanical stability of the composite electrode. The HSC fabricated using TiO_2 aerogel/Co-MOF as battery-type electrode and activated carbon (AC) as capacitive electrode (as shown in Figure 2B) shows enhanced electrochemical performance. The fabricated HSC shows an enhanced specific capacity of 111.2 $C \cdot g^{-1}$ at a current density of 0.8 $A \cdot g^{-1}$ and a high specific power density of 1875 $W \cdot kg^{-1}$, which is greater than bare TiO_2 aerogel and other TiO_2/carbon composites without Co-MOF. The AC and

TiO_2 aerogel/Co-MOF electrodes operate at different potential windows and therefore for the HSC device, combination of these two potential windows can be expected, which leads to the enhancement of workable potential window of 1.5 V (as shown in Figure 2C(a)). The battery and supercapacitor-like behavior of the HSC device arising from different energy storage mechanisms (Faradaic and non-Faradaic) is evident from the broad peaks and rectangular shape of the CV curves as can be seen in Figure 2C(b). The distorted triangular shape and small plateau in the GCD curve also confirm the HSC-like behavior (as shown in Figure 2C(c)). The fabricated HSC device shows long-term cycling stability. When cycling the HSC at a current density of 2 $A \cdot g^{-1}$ for 5000 cycles, it shows 93% of capacity retention (as shown in Figure 2C(d)). Other TiO_2 composites were also reported as an anode for HSC applications.

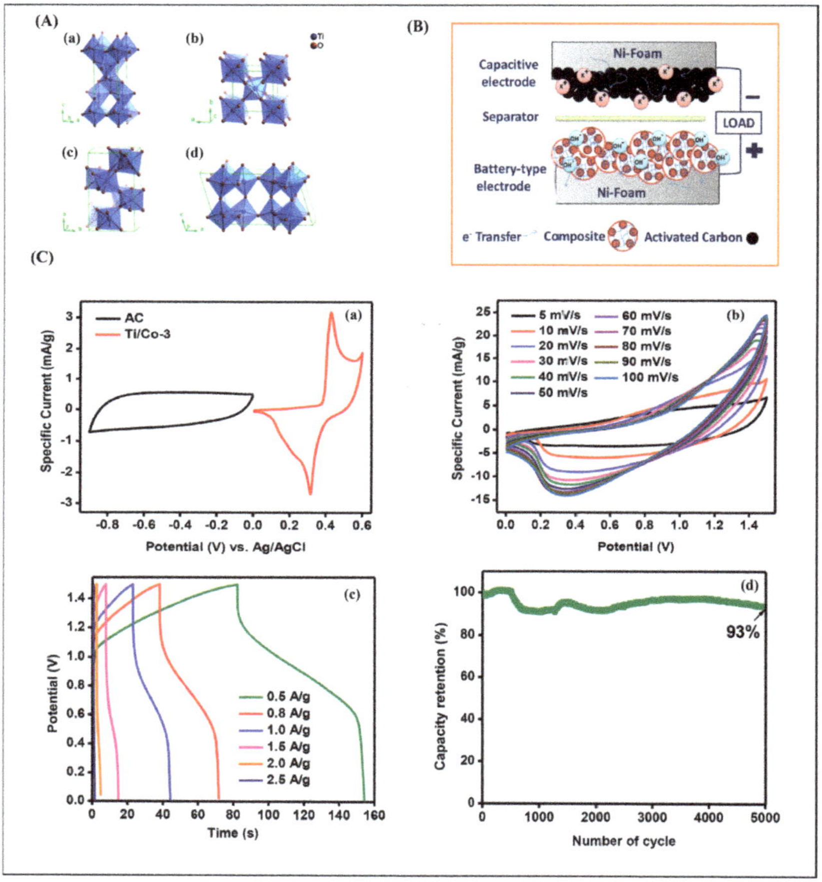

Figure 2. (**A**) Different crystalline phases of TiO_2: (**a**) anatase, (**b**) rutile, (**c**) brookite, and (**d**) TiO_2-B, reproduced with permission from ref. [77], American Chemical Society. (**B**) Schematic diagram of the HSC. (**C**) (**a**) Comparative CV of AC and TiO_2 aerogel/Co-MOF composite electrode, (**b**) CV of the HSC at different scan rates, and (**c**) GCD of the HSCs at various current rates, (**d**) Cycle stability of the AC and TiO_2 aerogel/Co-MOF electrodes, reproduced with permission from ref. [82], Elsevier.

Nonetheless, while TiO_2–MOF and other composite approaches successfully enhance conductivity, specific capacity, and mechanical stability, they introduce added complexity in synthetic steps and may face issues in large-scale reproducibility. MOFs, for example, can be expensive or sensitive to environmental conditions (e.g., humidity), potentially limiting their viability for mass production. The integration of multiple components (TiO_2 aerogel + Co-MOF + carbon) must also ensure strong interfacial contact and consistent distribution across the electrode to avoid inhomogeneous regions that could degrade electrochemical performance.

6.1.2. $Li_4Ti_5O_{12}$

$Li_4Ti_5O_{12}$ (LTO) is a well-studied material as an intercalation type anode for Li-ion batteries due to its several advantages including low cost, long cycle life, and safety. LTO can accommodate three Li-ions per formula unit during cycling with two phase reactions and a negligible volume change of 0.2% (zero-strain) [83]. LTO is also a promising material for HSC but inherent issues like poor electronic conductivity (2.7×10^{-7} Scm^{-1}), poor ionic conductivity (<10^{-12} cm^2s^{-1}), and poor rate capability inhibit the overall electrochemical performance of the HSC device [84,85]. Several methods are utilized to overcome these shortcomings and to enhance the performance of the HSC, such as nanosizing LTO to decrease the diffusion length, composite design with conducting materials like carbon, and doping with other elements like Al^{3+}, Cr^{3+}, or Mg^{2+} [86,87]. Naoi et al. coupled the nanosizing of LTO and composite design with carbon nanofibers (CNF) to attain high-rate capability and enhanced electrochemical performance [88]. They reported the successful synthesis of a high-rate-capable nanocrystalline LTO anode grafted on CNF (nc-LTO/CNF). The schematic of the formation process of nc-LTO/CNF is shown in Figure 3A, and the successful formation of nc-LTO/CNF is evident from the high-resolution transmission electron microscopy (HR-TEM) image of nc-LTO/CNF (as shown in Figure 3B). The enhanced rate capability of the nc-LTO/CNF-based electrochemical cell can be seen in Figure 3C. The HSC device comprising nc-LTO/CNF anode and AC cathode exhibits improved energy and power density compared to the conventional AC/AC symmetric supercapacitor. Also, the nc-LTO/CNF anode with a higher LTO weight percentage than CNF (70:30) shows better performance than the equal weight ratio between LTO and CNF (50:50) (as shown in Figure 3D). Kim and co-workers reported the synthesis of a composite of LTO and spherical holey graphene (HG) to enhance the electronic conductivity of LTO which arises from the high conductivity of graphene and high ionic transport arising from open pores in graphene sheets of HG [89]. The HSC device fabricated using spherical LTO/HG composite anode shows improved electrochemical performance compared to the HSC devices fabricated with LTO/RGO composite. The HSC device with LTO/HG composite anode delivered an energy density as high as 117.3 $Wh.kg^{-1}$, which arises from the synergistic effect of both components. Several other HSC devices with anode of modified LTO by carbon materials were also reported with enhanced electrochemical performances [90–92].

Nevertheless, despite the clear advantages of LTO-based anodes (particularly their near-zero strain and commendable safety features), several critical limitations need further scrutiny. First, while carbon or graphene-based modifications do effectively address the low conductivity, they increase the complexity and cost of synthesis—factors that can hinder large-scale commercial implementation. Second, the reduced volume change helps maintain mechanical integrity, but LTO's somewhat restricted operating voltage window and moderate theoretical capacity may limit the ultimate energy density achievable in HSC configurations. Third, doping with various cations (e.g., Al^{3+} and Mg^{2+}) can improve certain properties but may introduce trade-offs in terms of crystal stability or can demand precise processing conditions that complicate manufacturing. Finally, strict control over particle size and morphology is crucial for consistent, high-performance devices; batch-to-batch variability remains a challenge, particularly when scaling up nanosized LTO materials.

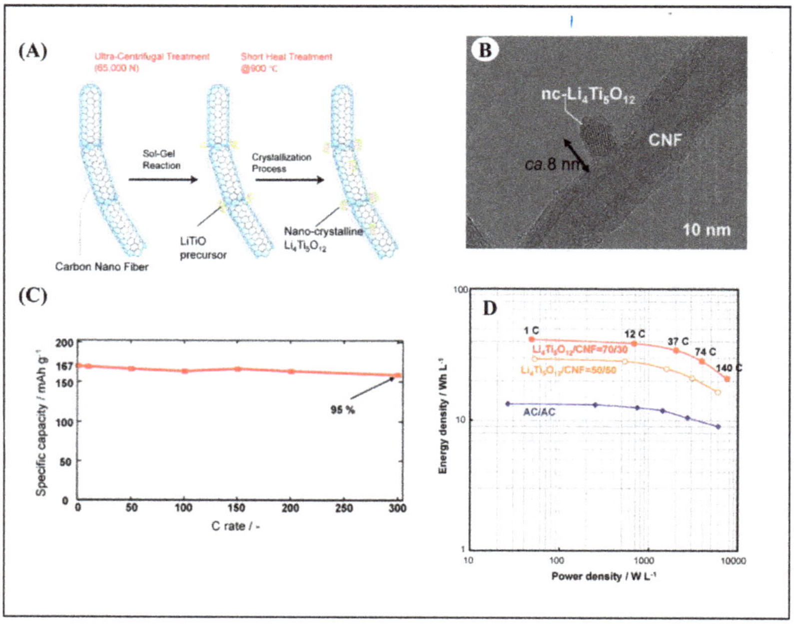

Figure 3. (**A**) schematic illustration for the two-step formation mechanism of the nc-LTO/CNF composite. (**B**) HR-TEM image of the nc-LTO/CNF composite. (**C**) rate capability of nc-LTO/CNF composite at different C-rates. (**D**) Ragone plots of LTO/CNF-based hybrid supercapacitors and AC/AC symmetric supercapacitor. Reproduced with permission from ref. [88], Elsevier.

6.2. Co-Based TMOs

Cobalt as an element plays a key role in battery cathode materials due to its good thermal stability and reversible redox capabilities [93]. In HSC, different Co-based materials such as spinel Co_3O_4 and other bi-metallic cobalt oxides are extensively explored as battery-type electrode materials [94].

6.2.1. Co_3O_4

In 2019, Devi et al. successfully synthesized a spinel Co_3O_4 using the sol–gel technique. They fabricated an HSC device using synthesized Co_3O_4 as a positive electrode, RGO as a negative electrode, and an aqueous-based electrolyte, with an enhanced energy density of 40 $Wh{\cdot}kg^{-1}$ at a power density of 742 $W{\cdot}kg^{-1}$ and Columbic efficiency of 98% [69], which signifies the applicability of Co_3O_4 as a battery-type electrode. Co_3O_4 with different morphologies were also reported to have enhanced electrochemical properties. For instance, Numan et al. reported an HSC with good electrochemical performance (specific capacity: 108 Cg^{-1}, energy density: 23.7 $Wh{\cdot}kg^{-1}$, power density: 307 $W{\cdot}kg^{-1}$) comprising a 2D porous Co_3O_4 nanoflakes (CONF) as positive electrode and an activated carbon as negative electrode in KOH electrolyte [95]. Park and co-workers reported a nano horn-like Co_3O_4 (NHC) using a solvothermal technique, a schematic of the synthesis process is shown in Figure 4A and the TEM image of the NHC is shown in Figure 4B. The authors also presented the influence of calcination temperature on the electrochemical performance of the NHC-containing HSC device [96]. The NHC calcined at 300 °C (NHC@300) shows the best electrochemical performance compared to other NHCs which are calcined at 350, 400, and 450 °C. CV curves of NHC@300 are comprised of well-defined redox peaks stemming from Co^{2+}/Co^{3+} and Co^{3+}/Co^{4+} redox couples (as shown in Figure 4C(a)). The high cathodic and anodic current values of NHC@300 in CV curve compared to NHC@350, 400, and

450 materials confirmed the enhanced electrochemical performance of NHC@300 (as shown in Figure 4C(b)). The enhanced electrochemical performance of NHC@300 stems from the enhanced diffusion kinetics of NHC@300 which is confirmed by the lowered resistance value in the Nyquist plot. The lowest equivalent series resistance (ESR) for NHC@300, which is evident from the intercepts in the high-frequency region of the X-axis and the steepness of the Warburg slope in the low-frequency region confirms the enhanced kinetics of NHC@300 compared to other NHCs (Figure 4C(c)). The plateaus in the galvanostatic charge–discharge (GCD) curves confirmed the battery-like behavior of synthesized NHCs as shown in Figure 4C(d). This plateau is more pronounced in NHC@300 than in other NHCs calcined at different temperatures as evident from Figure 4C(e). The enhanced specific capacitance of different of NHC@300 (2751 $F \cdot g^{-1}$ at 1 $A \cdot g^{-1}$) compared to other NHCs confirmed the enhanced electrochemical performance of NHC@300 (as shown in Figure 4C(f)). The authors also evaluated the cycling stability of an HSC with NHC@300 as a positive electrode and AC as a negative electrode, they found 91.37% capacity retention even after 350,000 GCD cycles at 20 $A \cdot g^{-1}$ of current density as can be seen in Figure 4D.

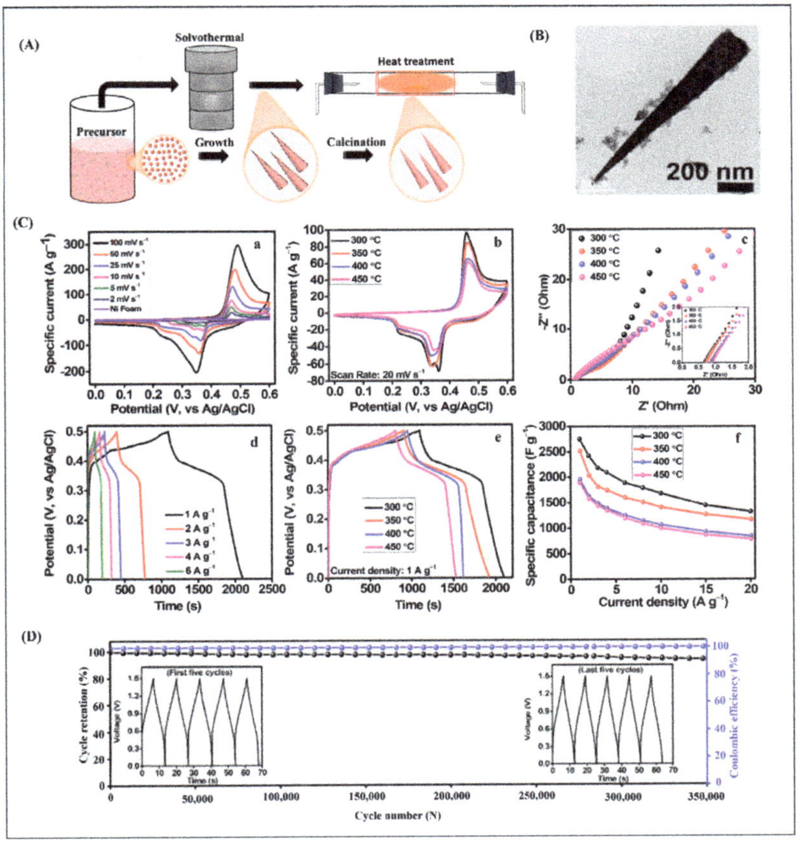

Figure 4. (**A**) schematic illustration of synthetic procedures and proposed formation mechanism of the Co_3O_4 nano horn. (**B**) TEM image of the synthesized nano horn. (**C**) (**a**) CV curve of NHC@300 at different scan rates, (**b**) CV curves of NHCs synthesized at different temperatures, (**c**) Nyquist plots of EIS for different NHCs, (**d**) GCD curves of NHC@300, (**e**) GCD curves of NHCs synthesized at different temperatures, and (**f**) specific capacitance of the NHCs at different current densities. (**D**) cycling stabilities and Coulombic efficiency of the NHC@300-based HSC. Reproduced with permission from ref. [96], Elsevier.

Apart from the morphologically different Co_3O_4, Co_3O_4 coupled with different other materials were also reported and utilized in HSC applications. Vattikuti et al. reported the successful fabrication of an HSC device using carbon nitride/Co_3O_4 material with improved electrochemical performance [97]. Arjunan et al. reported the synthesis of Co_3O_4 nanoparticles embedded in nitrogen-doped porous carbon spheres and utilized it in the fabrication of an HSC device. The fabricated HSC device shows an enhanced specific capacitance of 464 $F{\cdot}g^{-1}$ at a current density of 1 $A{\cdot}g^{-1}$ [98]. A composite of Co_3O_4 and carbon (Co_3O_4/C), derived from a cobalt-containing metal-organic framework (MOF), was reported by Kazemi and co-workers, they utilized this synthesized Co_3O_4/C as a positive electrode to fabricate an HSC with improved electrochemical performance [99]. Different other Co_3O_4-based materials were reported by researchers and their efficacy as a battery-type electrode in an HSC device was studied [100–102]

6.2.2. Co-Based Bi-Metallic Oxides

Bi-metallic oxides where other metals apart from Co are also present are extensively studied as an electrode material for HSC applications. For instance, Chen et al. reported the successful fabrication of an HSC device using bi-metallic manganese-cobalt oxide ($MnCo_2O_{4.5}$) as a cathode and AC as an anode with KOH electrolyte [103]. The $MnCo_2O_{4.5}$-based HSC device shows enhanced electrochemical performance. The authors conducted electrochemical tests on the HSC device using different potential windows and different current densities. Figure 5A(a) shows the CV of the HSC device at a potential window ranging from 0 to 1.2 to 1.8 V at a scan rate of 10 mVs^{-1}. The potential window from 0–1.75 V was found suitable with negligible polarization phenomenon. Battery-like features can be observed in the GCD curves with low voltage drop (as shown in Figure 5A(b)). With the increase in operating potential, an increase in both specific capacity (77.40 $C.g^{-1}$ at 1.2 V to 126.29 $C.g^{-1}$ at 1.75 V) and specific energy density (15.02 $Wh.kg^{-1}$ at 1.2 V to 33.36 $Wh.kg^{-1}$ at 1.75 V) is observed as evident from Figure 5A(c,d). Battery-like redox behavior (peaks arise from Faradaic reactions) and capacitive behavior (distorted rectangular shape) are evident from the CV profiles of the HSC device at different scan rates as can be seen in Figure 5B(a). The appearance of the weak plateau region along with the distorted triangular shape in the GCD curve confirms the capacity contribution from both battery-like capacitive mechanisms arising from battery-type $MnCo_2O_{4.5}$ cathode and capacitive-type AC anode (as shown in Figure 5B(b)). The enhanced rate performance of the bimetallic cathode-based HSC device at different current densities is evident from Figure 5B(c). The $MnCo_2O_{4.5}$-based HSC device showed enhanced cycling stability. No significant capacity decay was observed even after 5000 cycles (100.04% capacity retention), cycling at a current density as high as 6 $A{\cdot}g^{-1}$, which can be seen in Figure 5B(d). Nearly 100% Coulombic efficiency even after long-term cycling suggested good cycling stability of the $MnCo_2O_{4.5}$-containing HSC device. $MnCo_2O_{4.5}$ with different morphologies were also reported and utilized as a battery-type electrode in HSC applications. For instance, urchin-like and micro flower-like $MnCo_2O_{4.5}$ bi-metallic oxides were synthesized by different research groups and used as battery-type electrodes in HSC devices with improved electrochemical performances [104,105]. Apart from $MnCo_2O_{4.5}$, other Mn, Co-based bi-metallic were also reported as a battery-type electrode for HSC [106–108].

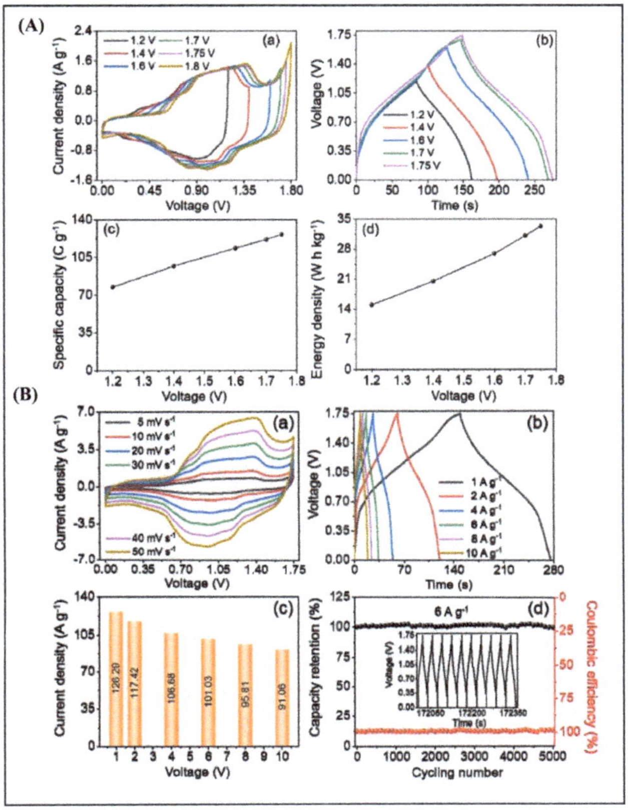

Figure 5. (**A**) (**a**) CV curves of $MnCo_2O_{4.5}$-based HSC at different potential windows, (**b**) GCD curves of $MnCo_2O_{4.5}$-based HSC at different potential windows, (**c**) The plot of specific capacitance as a function of voltage window for $MnCo_2O_{4.5}$-based HSC, and (**d**) The plot of energy density as a function of voltage window for $MnCo_2O_{4.5}$-based HSC. (**B**) (**a**) CV curves, (**b**) GCD curves, (**c**) rate performance, and (**d**) cycling performance and Coulombic efficiency of $MnCo_2O_{4.5}$-based HSC. Reproduced with permission from ref. [103], Elsevier.

The efficacy of Co-based bi-metallic oxides with other metals towards battery-type electrodes in HSC devices were also explored extensively. Recently, Ahmad et al. reported an improvement in the electrochemical performance of an HSC using chromium–cobalt-based bi-metallic oxide ($CrCo_2O_4$) synthesized by hydrothermal method [109]. Both Cr and Co play synergistic role to enhance the electrochemical performance compared to their individual oxide counterparts (Co_3O_4 and Cr_2O_3). Figure 6A(a) shows the CV curves of bi-metallic $CrCo_2O_4$, Cr_2O_3, and Co_3O_4, in the potential window of 0.0 to 0.55 V at a scan rate of 5 mVs^{-1}. Between these three oxides, the current response of $CrCo_2O_4$ is highest with the highest area under the CV curve which signifies the capacitance of the material. Figure 6A(b) shows the specific capacitance of the oxides at different current densities with 69.4% capacity retention for bimetallic $CrCo_2O_4$, the highest compared to the other two individual metal oxides Cr_2O_3 (31.9%) and Co_3O_4 (60.7%). Enhanced diffusion-kinetics is one of the reasons behind the improved performance of the bi-metallic oxide. The bi-metallic $CrCo_2O_4$ possesses the lowest values for charge-transfer resistance (R_{ct}) and equivalent series resistance (R_s) with the value of R_{ct} = 0.135 Ω, and R_s = 0.64 Ω compared to Cr_2O_3 (R_{ct}: 0.176 Ω, R_s: 0.84 Ω), and Co_3O_4 (R_{ct}: 0.156 Ω, R_s: 0.87 Ω) (as shown in Figure 6A(c)). The authors also conducted a long-term cycling test of the oxides with a current density of 10 A·g^{-1}. The $CrCo_2O_4$ shows the highest capacity retention of 98.74% even after 10,000 cycles, which is highest among the oxides with 97.34% retention for Co_3O_4 and 96.4% retention for Cr_2O_3 as evident from Figure 6B. To check the efficacy of the bi-metallic oxide as an electrode for an HSC device, authors fabricated an aqueous HSC using a $CrCo_2O_4$-based positive electrode and AC as a negative electrode. Figure 6C shows the rate-performance of the HSC device at different current densities ranging from 1 A·g^{-1} to 20 A·g^{-1} with the number of cycles up to 400 cycles. Capacity retention as high as 72% of the initial capacity was obtained even after reaching a high current density of 20 A·g^{-1} with 98% Coulombic efficiency. The $CrCo_2O_4$-based HSC device also shows a capacity retention of 97% after 10,000 cycles even at a current density of 10 A·g^{-1}, which encompasses good cycling stability (as shown in Figure 6D). These enhanced performances of the $CrCo_2O_4$-based device stem from the synergism between Cr and Co metals in the bi-metallic oxide—these cumulative effects are missing in individual oxides, and are therefore inferior to performance than their bi-metallic counterpart. Different other bi-metallic oxides such as $CoWO_4$ [110], $FeCo_2O_4$ [111], Cu_2CoO_3 [112], $NiCo_2O_4$ [113] and $ZnCo_2O_4$ [114] were also reported and utilized as battery-type electrode in an HSC device.

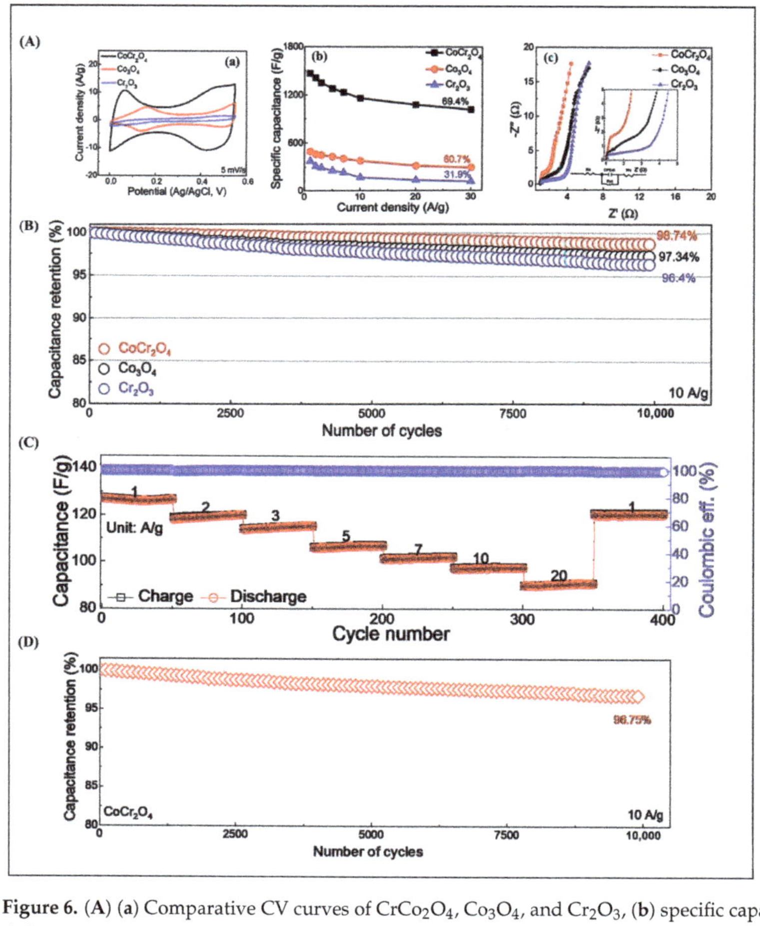

Figure 6. (**A**) (**a**) Comparative CV curves of $CrCo_2O_4$, Co_3O_4, and Cr_2O_3, (**b**) specific capacitance at different current densities, and (**c**) EIS spectra of $CrCo_2O_4$, Co_3O_4, and Cr_2O_3. (**B**) Cycling performance of $CrCo_2O_4$, Co_3O_4, and Cr_2O_3. (**C**) Rate capability and Coulombic efficiency of $CrCo_2O_4$-based HSC. (**D**) Long-term cycling stability of $CrCo_2O_4$-based HSC at 10 A·g^{-1}. Reproduced with permission from ref. [109], American Chemical Society.

Despite these promising outcomes, several challenges and limitations persist for Co-based TMOs. First, the intrinsic electronic conductivity of Co-based TMOs can remain suboptimal, necessitating complex morphologies or carbon integration to achieve high-rate capabilities. Second, engineering unique architectures (e.g., horn-like shapes, nanoflakes, or MOF-derived composites) often involves multiple synthesis steps that can be costly and difficult to reproduce consistently at large scales. Additionally, while Co-based TMO materials exhibit stable cycling in many lab-scale demonstrations, stability can still degrade under prolonged high-rate operations or in certain aqueous electrolytes. Finally, cobalt-based compounds raise concerns regarding cost fluctuations and potential environmental impacts related to cobalt extraction and processing. Future endeavors should thus focus on scalable, eco-friendly routes to form robust architecture and composites, as well as advanced characterization methods to unravel failure mechanisms in real time. Addressing these issues will be crucial for accelerating the commercial deployment of cobalt-based TMOs in HSCs with both high energy and power densities.

6.3. V-Based TMOs

Vanadium (V) is another attractive element used in energy storage applications like Li-ion, Na-ion batteries, multivalent batteries, and supercapacitors owing to its diverse oxidation states (from +2 to +5) [115–117].

V_2O_5

Among different vanadium-based compounds vanadium pentoxide (V_2O_5) stands out as a promising electrode material owing to its low cost, layered structure, high capacity, and straightforward synthesis process [115,118]. Although V_2O_5 is considered a promising material as a battery-type electrode, it is plagued with inherent issues like low electronic conductivity and low ionic diffusion kinetics. To counter these issues strategies like morphology tuning (acquiring high surface area or nanostructuring to reduce diffusion length), composite formation with conductive materials such as carbon materials, and pre-intercalation with hetero ions such as Li^+ to increase the interlayer spacing were explored [117,119,120]. Recently, Xu et al. reported the enhancement in ionic and electronic conductivity of V_2O_5 by incorporating Co^{2+} ions and polyaniline (PANI) between the layers of V_2O_5 (CoVO-PANI) and utilized it as a cathode material in Zn-ion hybrid supercapacitor (ZHSC) [121]. The fabricated ZHSC device with CoVO-PANI@CC as cathode, AC@CC as anode, and using Zn-based electrolyte shows enhanced electrochemical performance stemming from the synergistic effects of Co^{2+} and PANI. Figure 7A shows the fabrication process of CoVO-PANI. The higher capacitance of CoVO-PANI-based ZHSC is evident from the area of the CV curves compared to VO-PANI, CoVO, and VO-based ZHSC device, which also confirms the synergism between Co^{2+} and PANI in enhancing the performance of ZHSC (as shown in Figure 7B(a)). The enhanced specific capacitance of CoVO-PANI-based ZHSC is evident from GCD curves as can be seen in Figure 7B(b). The lowered charge-transfer resistance of CoVO-PANI-based ZHSC (R_{ct}: 0.45 Ω) shows the enhanced diffusion rate of Zn^{2+} ions. The ZHSC devices based on VO-PANI, CoVO, and VO exhibit much higher R_{ct} values signifying sluggish Zn^{2+} ion diffusion (as shown in Figure 7B(c)). The synergism of Co^{2+} and PANI is also evident from the enhanced areal capacitance and enhanced long-term cycling stability of CoVO-PANI-based ZHSC compared to its other counterparts. The CoVO-PANI-based ZHSC retained 81.37% of its initial capacitance even after 20,000 cycles at current density of 50 mA.cm^{-2} (as shown in Figure 7C(a) and 7C(b)). Owing to the enhanced performance of CoVO-PANI electrode, authors fabricated a flexible ZHSC device using PVA-Zn(OTF)$_2$ as gel electrolyte and utilized the fabricated flexible

ZHSC device to power a digital watch as can be seen in Figure 7D(a). The schematic structure of the flexible ZHSC device is shown in Figure 7D(b).

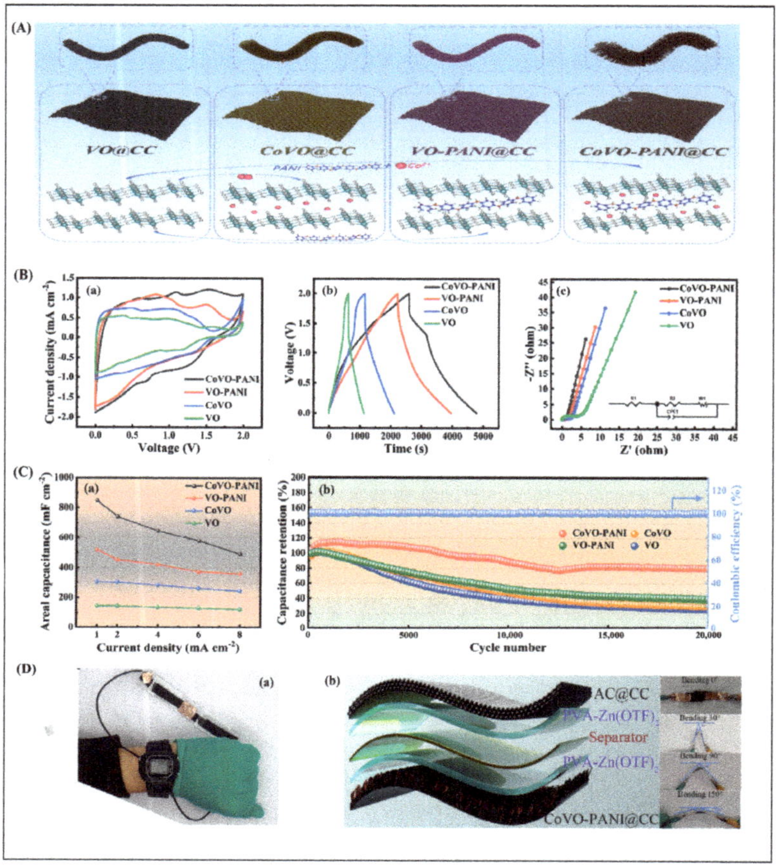

Figure 7. (**A**) preparation schematics of VO@CC, CoVO@CC, VO-PANI@CC, and CoVO-PANI@CC. (**B**) (**a**) CV curves of CoVO-PANI, VO-PANI, CoVO and VO at 1 mVs^{-1}, (**b**) GCD curves of CoVO-PANI, VO-PANI, CoVO and VO, and (**c**) EIS curves of CoVO-PANI, VO-PANI, CoVO and VO. (**C**) (**a**) areal capacitance curves of CoVO-PANI, VO-PANI, CoVO and VO at different current densities; (**b**) long-term cycling stability tests of CoVO-PANI, VO-PANI, CoVO and VO. (**D**) (**a**) Image of digital watch powered by flexible ZHSC devices; (**b**) schematic diagram of the fabrication flexible ZHSC device. Reproduced with permission from ref. [121].

Different other V_2O_5, such as pre-lithiated V_2O_5 [122], hollow-structured V_2O_5 [123], V_2O_5 nanofibers [124], V_2O_5/graphene oxide nanocomposites [125] were also reported as an electrode material for HSC applications, demonstrating significant improvements in performance compared to their unmodified counterparts. These structural and compositional modifications address the inherent limitations of pristine V_2O_5, such as low electrical conductivity, poor cycling stability, and limited rate capability. Pre-lithiated V_2O_5, for instance, benefits from a pre-intercalation strategy that introduces lithium ions into the V_2O_5 lattice, resulting in enhanced electrical conductivity and reduced lattice strain during the charge/discharge cycles. This modification not only stabilizes the crystal structure but also boosts the overall capacity and cycling life of the electrode. Hollow-structured V_2O_5 takes advantage of its unique architecture, which provides a high surface area, reduced ion diffusion pathways, and improved electrolyte penetration. These features significantly enhance the redox reaction kinetics and increase the electrode's ability to store and deliver charge efficiently. V_2O_5 nanofibers, synthesized through electrospinning and other advanced fabrication methods, offer a one-dimensional structure with excellent electron transport properties and a high aspect ratio. The interconnected network of nanofibers facilitates rapid ion and electron transfer, making them particularly suitable for high power applications. V_2O_5/graphene oxide nanocomposites represent another promising approach, where the synergistic combination of V_2O_5's redox activity and graphene oxide's high conductivity and mechanical stability results in superior electrochemical performance. The graphene oxide not only enhances the electrical conductivity of the composite but also prevents the agglomeration of V_2O_5 particles, ensuring a more stable cycling performance.

To provide a clear comparison of the electrochemical performance of transition metal oxide (TMO) battery-type electrodes in hybrid supercapacitors, a tabulated summary of specific power and energy densities is presented in Table 1. This comprehensive overview will help readers quickly assess the relative merits of various TMO-based systems, including cycling performance, energy density and power density.

Table 1. The performance matrices of different HSCs based on TMOs.

TMO-Based HSC	Synthesis	Energy Density ($Whkg^{-1}$)	Power Density (Wkg^{-1})	Cycling Stability	Reference
Co_3O_4@Mn-Ni$(OH)_2$/CC	Hydrothermal	65.5	800	93.0%, 10,000 cycles	[126]
Co_3O_4/Ni//N@g-C_3N_4	Hydrothermal	22.26	4000	82.4% 5000 cycles,	[127]
Co_3O_4 MBs//AC	Hydrothermal	38.5	962	126.4%, 5000 cycles	[128]
Co_3O_4//rGO	Sol–gel	40	742	-	[69]
$MnCo_2O_4$//AC	Aqueous chemical synthesis	36	4274	91.1%, 5000 cycles	[129]
$CoFe_2O_4$/Cr_2CTx//AC	Hydrothermal	125	15,264.4	99%, 2500 cycles	[130]
MWCNT/$LiCo_2O_4$//AC	Co-precipitation	54.84	775	94%, 5000 cycles	[131]
Nio@Co_3O_4/C//AC	Precipitation method	32.6	750	87.1%, 5000 cycles	[132]
$CoFe_2O_4$//NPC	Reflux condensation	56.2	1091.5	97.91%, 5000 cycles	[133]
Fe_3O_4 NPs	Solvothermal	152.06	1822	89.71%, 5000 cycles	[134]
$Ni_xFe_yO_z$@rGO//AC	Microwave synthesis	43	2500	86%, 1000 cycles	[135]
Li_3VO_4//AC	Hydrothermal	136.4	532	87%, 1500 cycles	[136]
Li_3VO_4@C//AC	Micro-emulsion-based method	190	18,500	–	[137]
3DG/$Ni_xCo_{1-x}O$//3DG	Spontaneous assembly process	27.2	14,500	86%, 10,000 cycles	[138]
NiCo-P/$Ni_3V_2O_8$//AC	Hydrothermal	65	750	84.21%, 10,000 cycles	[139]
$ZnCo_2O_4$ NSs//AC	Hydrothermal	43.88	1077.45	~100%, 8000 cycles	[140]
$CrCo_2O_4$/AC/PPY//MnO_2/AC	Hydrothermal	97.77	1600	76.75%, 10,000 cycles	[141]
Cu-Mn-Zn Oxide//AC	Hydrothermal	44.26	9432.69	86.65%, 3000 cycles	[142]
$ZnCr_2O_4$@CC//AC	Hydrothermal	26.2	800.6	96.43%, 10,000 cycles	[143]

7. Conclusions and Future Direction

Hybrid supercapacitors (HSCs) represent a transformative advancement in energy storage technologies, effectively bridging the gap between batteries and supercapacitors by combining high energy density with excellent power density. This review has highlighted the critical role of transition metal oxides (TMOs) as battery-type electrodes, underscoring their capacity for Faradaic redox reactions that enable robust charge storage while retaining formidable power performance. Across the broad spectrum of synthesis techniques, spanning hydro/solvothermal processes, electrospinning, electrodeposition, and sol–gel methods, researchers have demonstrated how careful control of TMO structures and morphologies can mitigate key limitations such as low electrical conductivity and restricted ionic diffusion. The spotlight on Ti-, Co-, and V-based TMOs, in particular, illustrates how strategies like nanostructuring, composite integration, and doping can boost electrochemical outputs.

Nonetheless, certain challenges must be tackled for TMO-based HSCs to achieve commercial readiness such as (i) cycling stability: while striving for higher energy density, TMOs undergo substantial volumetric changes during ion insertion/extraction, which can lead to microcracks and structural deformation. This underscores the delicate balance between boosting capacity and preserving electrode integrity over extended cycles. (ii) Scalability: many of the advanced nanostructuring and composite methods that deliver excellent lab-scale performance can be difficult or costly to implement at industrial scales, necessitating more streamlined, cost-effective manufacturing processes. (iii) Cost-effectiveness: high-performance materials and specialized synthesis conditions can drive up production expenses. Achieving a favorable balance of performance and affordability is paramount for real-world adoption.

Looking ahead, the future outlook for TMO-based HSCs hinges on integrating innovative research avenues and practical engineering solutions. Emerging strategies for developing advanced TMO electrodes concentrate on simultaneously boosting electrochemical performance and maintaining structural resilience, including (i) defect engineering: creating and tailoring oxygen vacancies or interstitials to regulate ion transport and accommodating volume expansion [144,145]. (ii) Hierarchical nanostructures: designing multi-level architectures, such as core–shell or hollow structures, that offer larger reactive surface areas while tolerating mechanical strain more effectively [146]. (iii) Functional coatings and protective layers: using thin coatings (e.g., polymeric, metallic, or carbonaceous layers) to stabilize electrode surfaces, mitigate unwanted side reactions, and reduce structural deterioration [147]. (iv) Hybrid composites with 2D materials: incorporating emerging two-dimensional materials (like graphene derivatives or MXenes) to reinforce conductivity, boost mechanical integrity, and provide additional ion diffusion channels [146]. (v) In situ/operando characterization tools: developing advanced analytical techniques to monitor structural evolution and reaction kinetics in real time, enabling more targeted material and device optimization [147].

Beyond these emerging strategies, there are several additional avenues with potential to advance TMO-based HSCs further:

1. Advanced Material Design

Innovations in defect chemistry, nanostructured morphologies, and composite interfaces can significantly improve both energy/power densities and mechanical resilience.

2. Sustainable Synthesis Methods

Green chemistry principles, eco-friendly solvents, and the recycling or reuse of precursors will be essential for minimizing environmental impact and reducing costs associated with large-scale production.

3. Flexible and Wearable HSCs

Architectures capable of withstanding mechanical stress under bending or stretching broaden the scope of HSC usage in consumer electronics and wearable technologies.

4. Machine Learning and Computational Modeling

High-throughput simulations and predictive algorithms can accelerate the discovery of new TMO compositions, dopants, and architectures, helping to efficiently pinpoint the most promising optimization pathways.

By focusing on these interconnected goals, enhancing cycling stability, refining scalability, and ensuring economic viability, TMO-based HSCs can continue to evolve as a cornerstone of next-generation energy storage. As researchers deepen their mechanistic insights and leverage computational tools, we can expect further breakthroughs that balance high energy density, structural robustness, and environmental sustainability, ultimately driving the widespread adoption of HSCs in a variety of commercial and industrial applications. All the challenges of HSCs and future directions for TMO-based HSCs are summarized schematically in Figure 8.

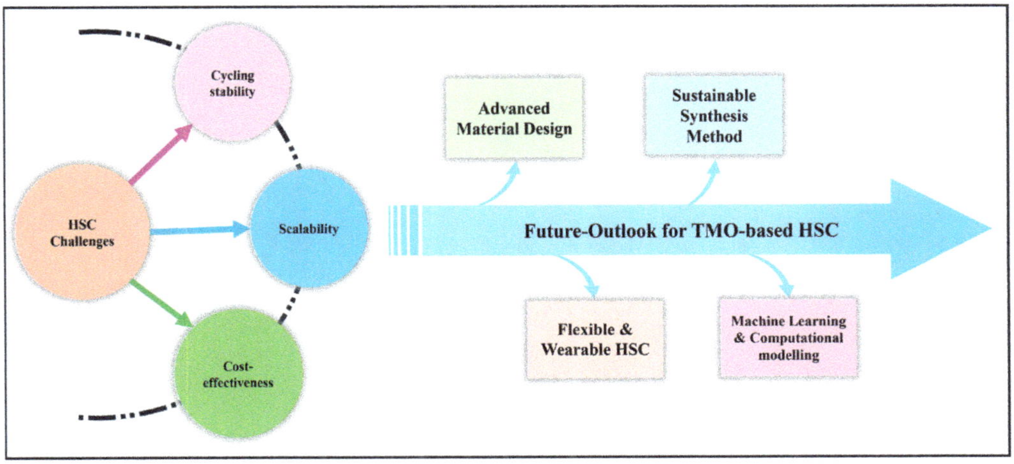

Figure 8. Schematic representation of challenges faced by HSCs and future directions for TMO-based HSCs.

In conclusion, HSCs hold immense potential to revolutionize energy storage by combining the best attributes of batteries and supercapacitors. Continued interdisciplinary research encompassing materials science, electrochemistry, and device engineering will be essential to overcome existing challenges and unlock the full potential of HSCs for a sustainable and energy-efficient future.

Author Contributions: Conceptualization, B.R., M.I., M.M.R.K. and G.A.K.M.R.B.; validation: M.S.A., M.M.R.K. and H.-Y.K.; data curation, G.A.K.M.R.B. and H.-Y.K.; funding acquisition, K.-W.N.; resources, K.-W.N.; supervision, M.I., and K.-W.N.; writing—original draft, B.R.; writing—review and editing, M.I., M.S.A. and K.-W.N. All authors have read and agreed to the published version of the manuscript.

Funding: This work was supported by the Ministry of Trade, Industry & Energy (MOTIE) of the Republic of Korea (Grant No. RS-2022-00155717 and 20224000000020).

Data Availability Statement: Not applicable.

Conflicts of Interest: The authors declare no conflicts of interest.

References

1. Liu, H.; Liu, X.; Wang, S.; Liu, H.K.; Li, L. Transition Metal Based Battery-Type Electrodes in Hybrid Supercapacitors: A Review. *Energy Storage Mater.* **2020**, *28*, 122–145. [CrossRef]
2. Simon, P.; Gogotsi, Y.; Dunn, B. Where Do Batteries End and Supercapacitors Begin? *Science* **2014**, *343*, 1208–1210. [CrossRef] [PubMed]
3. Winter, M.; Brodd, R.J. What Are Batteries, Fuel Cells, and Supercapacitors? *Chem. Rev.* **2004**, *104*, 4245–4269. [CrossRef] [PubMed]

4. Goodenough, J.B. How We Made the Li-Ion Rechargeable Battery. *Nat. Electron.* **2018**, *1*, 204. [CrossRef]
5. Ponrouch, A.; Bitenc, J.; Dominko, R.; Lindahl, N.; Johansson, P.; Palacin, M.R. Multivalent Rechargeable Batteries. *Energy Storage Mater.* **2019**, *20*, 253–262. [CrossRef]
6. Palacín, M.R. Recent Advances in Rechargeable Battery Materials: A Chemist's Perspective. *Chem. Soc. Rev.* **2009**, *38*, 2565–2575. [CrossRef]
7. Gicha, B.B.; Tufa, L.T.; Nwaji, N.; Hu, X.; Lee, J. Advances in All-Solid-State Lithium–Sulfur Batteries for Commercialization. *Nano-Micro Lett.* **2024**, *16*, 172. [CrossRef]
8. Ragone, D.V.; Ragone, D.V. *Review of Battery Systems for Electrically Powered Vehicles Review of Battery Systems for Electrically Powered Vehicles*; SAE: Warrendale, PA, USA, 2018.
9. Beyers, I.; Bensmann, A.; Hanke-Rauschenbach, R. Ragone Plots Revisited: A Review of Methodology and Application across Energy Storage Technologies. *J. Energy Storage* **2023**, *73*, 109097. [CrossRef]
10. Christen, T.; Carlen, M.W. Theory of Ragone Plots. *J. Power Sources* **2000**, *91*, 210–216. [CrossRef]
11. Wang, C.Y.; Liu, T.; Yang, X.G.; Ge, S.; Stanley, N.V.; Rountree, E.S.; Leng, Y.; McCarthy, B.D. Fast Charging of Energy- Dense Lithium-Ion Batteries. *Nature* **2022**, *611*, 485–490. [CrossRef]
12. Kim, U.H.; Lee, S.B.; Park, N.Y.; Kim, S.J.; Yoon, C.S.; Sun, Y.K. High-Energy-Density Li-Ion Battery Reaching Full Charge in 12 Min. *ACS Energy Lett.* **2022**, *7*, 3880–3888. [CrossRef]
13. Frith, J.T.; Lacey, M.J.; Ulissi, U. A Non-Academic Perspective on the Future of Lithium-Based Batteries. *Nat. Commun.* **2023**, *14*, 420. [CrossRef] [PubMed]
14. Satpathy, S.; Das, S.; Bhattacharyya, B.K. How and Where to Use Super-Capacitors Effectively, an Integration of Review of Past and New Characterization Works on Super-Capacitors. *J. Energy Storage* **2020**, *27*, 101044. [CrossRef]
15. Kumar, Y.A.; Roy, N.; Ramachandran, T.; Hussien, M.; Moniruzzaman, M.; Joo, S.W. Shaping the Future of Energy: The Rise of Supercapacitors Progress in the Last Five Years. *J. Energy Storage* **2024**, *98*, 113040. [CrossRef]
16. Zhang, M.; Du, H.; Wei, Z.; Zhang, X.; Wang, R. Ultrafast Microwave Synthesis of Nickel-Cobalt Sulfide/Graphene Hybrid Electrodes for High-Performance Asymmetrical Supercapacitors. *ACS Appl. Energy Mater.* **2021**, *4*, 8262–8274. [CrossRef]
17. Zhang, M.; Nautiyal, A.; Du, H.; Wei, Z.; Zhang, X.; Wang, R. Electropolymerization of Polyaniline as High-Performance Binder Free Electrodes for Flexible Supercapacitor. *Electrochim. Acta* **2021**, *376*, 138037. [CrossRef]
18. Dubal, D.P.; Ayyad, O.; Ruiz, V.; Gómez-Romero, P. Hybrid Energy Storage: The Merging of Battery and Supercapacitor Chemistries. *Chem. Soc. Rev.* **2015**, *44*, 1777–1790. [CrossRef]
19. Ding, J.; Hu, W.; Paek, E.; Mitlin, D. Review of Hybrid Ion Capacitors: From Aqueous to Lithium to Sodium. *Chem. Rev.* **2018**, *118*, 6457–6498. [CrossRef]
20. Zuo, W.; Li, R.; Zhou, C.; Li, Y.; Xia, J.; Liu, J. Battery-Supercapacitor Hybrid Devices: Recent Progress and Future Prospects. *Adv. Sci.* **2017**, *4*, 1600539. [CrossRef]
21. Cakici, M.; Reddy, K.R.; Alonso-Marroquin, F. Advanced Electrochemical Energy Storage Supercapacitors Based on the Flexible Carbon Fiber Fabric-Coated with Uniform Coral-like MnO_2 structured Electrodes. *Chem. Eng. J.* **2017**, *309*, 151–158. [CrossRef]
22. Pramitha, A.; Raviprakash, Y. Recent Developments and Viable Approaches for High-Performance Supercapacitors Using Transition Metal-Based Electrode Materials. *J. Energy Storage* **2022**, *49*, 104120. [CrossRef]
23. Yuan, S.; Duan, X.; Liu, J.; Ye, Y.; Lv, F.; Liu, T.; Wang, Q.; Zhang, X. Recent Progress on Transition Metal Oxides as Advanced Materials for Energy Conversion and Storage. *Energy Storage Mater.* **2021**, *42*, 317–369. [CrossRef]
24. Chen, H.; Du, X.; Liu, X.; Wu, R.; Li, Y.; Xu, C. Facile growth of nickel foam-supported $MnCo_2O_4$. 5 porous nanowires as binder-free electrodes for high-performance hybrid supercapacitors. *J. Energy Storage* **2022**, *50*, 104297. [CrossRef]
25. Subhadarshini, S.; Peyada, N.K.; Goswami, D.K.; Das, N.C. Review of Battery-Type Transition Metal (Cu, Co, and Ni) Oxide Based Electrodes: From Fundamental Science to Fabrication of a Hybrid Supercapacitor Device. *Energy Fuels* **2024**, *38*, 11455–11493. [CrossRef]
26. Stepanov, A.B.; Varakin, I.N.; Menukhov, V.V. Double-Layer Capacitor. U.S. Patent 5,986,876, 16 November 1999.
27. Aida, T.; Yamada, K.; Morita, M. An Advanced Hybrid Electrochemical Capacitor That Uses a Wide Potential Range at the Positive Electrode. *Electrochem. Solid-State Lett.* **2006**, *9*, 534–536. [CrossRef]
28. Naoi, K. "Nanohybrid Capacitor": The next Generation Electrochemical Capacitors. *Fuel Cells* **2010**, *10*, 825–833. [CrossRef]
29. Naoi, K.; Simon, P. New Materials and New Confgurations for Advanced Electrochemical Capacitors. *Electrochem. Soc. Interface* **2008**, *17*, 34–37. [CrossRef]
30. Yoo, H.D.; Han, S.D.; Bayliss, R.D.; Gewirth, A.A.; Genorio, B.; Rajput, N.N.; Persson, K.A.; Burrell, A.K.; Cabana, J. "rocking-Chair"-Type Metal Hybrid Supercapacitors. *ACS Appl. Mater. Interfaces* **2016**, *8*, 30853–30862. [CrossRef]
31. Nitta, N.; Wu, F.; Lee, J.T.; Yushin, G. Li-Ion Battery Materials: Present and Future. *Mater. Today* **2015**, *18*, 252–264. [CrossRef]
32. Schmidt-Rohr, K. How Batteries Store and Release Energy: Explaining Basic Electrochemistry. *J. Chem. Educ.* **2018**, *95*, 1801–1810. [CrossRef]

33. Salanne, M.; Rotenberg, B.; Naoi, K.; Kaneko, K.; Taberna, P.L.; Grey, C.P.; Dunn, B.; Simon, P. Efficient Storage Mechanisms for Building Better Supercapacitors. *Nat. Energy* **2016**, *1*, 16070. [CrossRef]
34. Miller, J.R.; Simon, P. Materials Science: Electrochemical Capacitors for Energy Management. *Science* **2008**, *321*, 651–652. [CrossRef] [PubMed]
35. Huang, J.; Xie, Y.; You, Y.; Yuan, J.; Xu, Q.; Xie, H.; Chen, Y. Rational Design of Electrode Materials for Advanced Supercapacitors: From Lab Research to Commercialization. *Adv. Funct. Mater.* **2023**, *33*, 2213095. [CrossRef]
36. Gogotsi, Y.; Simon, P. True Performance Metrics in Electrochemical Energy Storage. *Science* **2011**, *334*, 917–918. [CrossRef]
37. Lindström, H.; Södergren, S.; Solbrand, A.; Rensmo, H.; Hjelm, J.; Hagfeldt, A.; Lindquist, S.E. Li+ Ion Insertion in TiO_2 (Anatase). 1. Chronoamperometry on CVD Films and Nanoporous Films. *J. Phys. Chem. B* **1997**, *101*, 7710–7716. [CrossRef]
38. Bard, A.j. *Fundamentals and Applications Plasmonics: Fundamentals and Applications*; Springer: Berlin/Heidelberg, Germany, 2004; Volume 677.
39. Dong, Y.; Li, D.; Gao, C.; Liu, Y.; Zhang, J. A Self-Assembled 3D Urchin-like $Ti_{0.8}Sn_{0.2}O_2$-RGO Hybrid Nanostructure as an Anode Material for High-Rate and Long Cycle Life Li-Ion Batteries. *J. Mater. Chem. A* **2017**, *5*, 8087–8094. [CrossRef]
40. Noori, A.; El-Kady, M.F.; Rahmanifar, M.S.; Kaner, R.B.; Mousavi, M.F. Towards Establishing Standard Performance Metrics for Batteries, Supercapacitors and Beyond. *Chem. Soc. Rev.* **2019**, *48*, 1272–1341. [CrossRef]
41. Yang, P.; Ding, Y.; Lin, Z.; Chen, Z.; Li, Y.; Qiang, P.; Ebrahimi, M.; Mai, W.; Ping Wong, C.; Lin Wang, Z. Low-Cost High-Performance Solid-State Asymmetric Supercapacitors Based on MnO_2 Nanowires and Fe_2O_3 Nanotubes. *Nano Lett.* **2014**, *14*, 731–736. [CrossRef]
42. Chen, H.; Du, X.; Wu, R.; Wang, Y.; Sun, J.; Zhang, Y.; Xu, C. Facile Hydrothermal Synthesis of Porous $MgCo_2O_4$ nanoflakes as an Electrode Material for High-Performance Asymmetric Supercapacitors. *Nanoscale Adv.* **2020**, *2*, 3263–3275. [CrossRef]
43. Mohamed, S.G.; Chen, C.J.; Chen, C.K.; Hu, S.F.; Liu, R.S. High-Performance Lithium-Ion Battery and Symmetric Supercapacitors Based on $FeCo_2O_4$ Nanoflakes Electrodes. *ACS Appl. Mater. Interfaces* **2014**, *6*, 22701–22708. [CrossRef]
44. Lee, J.W.; Hall, A.S.; Kim, J.D.; Mallouk, T.E. A Facile and Template-Free Hydrothermal Synthesis of Mn_3O_4 Nanorods on Graphene Sheets for Supercapacitor Electrodes with Long Cycle Stability. *Chem. Mater.* **2012**, *24*, 1158–1164. [CrossRef]
45. Chen, H.; Du, X.; Sun, J.; Wang, Y.; Zhang, Y.; Xu, C. Solvothermal Synthesis of Novel Pod-like $MnCo_2O_{4.5}$ Microstructures as High-Performance Electrode Materials for Supercapacitors. *Int. J. Hydrog. Energy* **2020**, *45*, 3016–3027. [CrossRef]
46. Ankinapalli, O.R.; Krishna, B.N.V.; Ayyaluri, R.R.; Yu, J.S. $ZnMoO_4$/MoO_3 Composite Materials via Facile One-Step Hydrothermal Route for Efficient Hybrid Supercapacitors. *J. Energy Storage* **2024**, *85*, 111043. [CrossRef]
47. Sun, L.; Deng, Q.; Li, Y.; Deng, L.; Wang, Y.; Ren, X.; Zhang, P. Solvothermal Synthesis of Ternary Cu_2O-CuO-RGO Composites as Anode Materials for High Performance Lithium-Ion Batteries. *Electrochim. Acta* **2016**, *222*, 1650–1659. [CrossRef]
48. Kaneti, Y.V.; Zakaria, Q.M.D.; Zhang, Z.; Chen, C.; Yue, J.; Liu, M.; Jiang, X.; Yu, A. Solvothermal Synthesis of ZnO-Decorated α-Fe_2O_3 Nanorods with Highly Enhanced Gas-Sensing Performance toward n-Butanol. *J. Mater. Chem. A* **2014**, *2*, 13283–13292. [CrossRef]
49. Pullanchiyodan, A.; Joy, R.; Sreeram, P.; Raphael, L.R.; Das, A.; Balakrishnan, N.T.M.; Ahn, J.H.; Vlad, A.; Sreejith, S.; Raghavan, P. Recent Advances in Electrospun Fibers Based on Transition Metal Oxides for Supercapacitor Applications: A Review. *Energy Adv.* **2023**, *2*, 922–947. [CrossRef]
50. Kenawy, E.R.; Moharram, Y.I.; Abouharga, F.S.; Elfiky, M. Electrospun Network Based on Polyacrylonitrile-Polyphenyl/Titanium Oxide Nanofibers for High-Performance Supercapacitor Device. *Sci. Rep.* **2024**, *14*, 6683. [CrossRef]
51. Cao, Y.; Lin, B.; Sun, Y.; Yang, H.; Zhang, X. Sr-Doped Lanthanum Nickelate Nanofibers for High Energy Density Supercapacitors. *Electrochim. Acta* **2015**, *174*, 41–50. [CrossRef]
52. Lu, Y.; Liu, Y.; Mo, J.; Deng, B.; Wang, J.; Zhu, Y.; Xiao, X.; Xu, G. Construction of Hierarchical Structure of Co_3O_4 Electrode Based on Electrospinning Technique for Supercapacitor. *J. Alloys Compd.* **2021**, *853*, 157271. [CrossRef]
53. Ghaziani, M.M.; Mazloom, J.; Ghodsi, F.E. Electrospun $MgCo_2O_4$ Nanofibers as an Efficient Electrode Material for Pseudocapacitor Applications: Effect of Calcination Temperature on Electrochemical Performance. *J. Phys. Chem. Solids* **2021**, *152*, 109981. [CrossRef]
54. Yang, S.; Ai, J.; Han, Z.; Zhang, L.; Zhao, D.; Wang, J.; Yang, C.; Cao, B. Electrospun $ZnFe_2O_4$/Carbon Nanofibers as High-Rate Supercapacitor Electrodes. *J. Power Sources* **2020**, *469*, 228416. [CrossRef]
55. Bhagwan, J.; Sahoo, A.; Yadav, K.L.; Sharma, Y. Porous, One Dimensional and High Aspect Ratio Mn_3O_4 Nanofibers: Fabrication and Optimization for Enhanced Supercapacitive Properties. *Electrochim. Acta* **2015**, *174*, 992–1001. [CrossRef]
56. Liang, J.; Bu, L.T.; Cao, W.G.; Chen, T.; Cao, Y.C. Facile Fabrication of Coaxial-Cable like Mn_2O_3 Nanofiber by Electrospinning: Application as Electrode Material for Supercapacitor. *J. Taiwan Inst. Chem. Eng.* **2016**, *65*, 584–590. [CrossRef]
57. Xu, L.; Zhang, L.; Cheng, B.; Yu, J. Rationally Designed Hierarchical $NiCo_2O_4$-C@Ni(OH)$_2$ Core-Shell Nanofibers for High Performance Supercapacitors. *Carbon N. Y.* **2019**, *152*, 652–660. [CrossRef]

58. Abbas, Q.; Khurshid, H.; Yoosuf, R.; Lawrence, J.; Issa, B.A.; Abdelkareem, M.A.; Olabi, A.G. Engineering of Nickel, Cobalt Oxides and Nickel/Cobalt Binary Oxides by Electrodeposition and Application as Binder Free Electrodes in Supercapacitors. *Sci. Rep.* **2023**, *13*, 15654. [CrossRef] [PubMed]
59. Patake, V.D.; Joshi, S.S.; Lokhande, C.D.; Joo, O.S. Electrodeposited Porous and Amorphous Copper Oxide Film for Application in Supercapacitor. *Mater. Chem. Phys.* **2009**, *114*, 6–9. [CrossRef]
60. Chen, Y.; Guan, J.H.; Gan, H.; Chen, B.Z.; Shi, X.C. Electrochemical Growth of α-MnO_2 on Carbon Fibers for High-Performance Binder-Free Electrodes of Supercapacitors. *J. Appl. Electrochem.* **2018**, *48*, 105–113. [CrossRef]
61. Chou, S.L.; Wang, J.Z.; Chew, S.Y.; Liu, H.K.; Dou, S.X. Electrodeposition of MnO_2 Nanowires on Carbon Nanotube Paper as Free-Standing, Flexible Electrode for Supercapacitors. *Electrochem. Commun.* **2008**, *10*, 1724–1727. [CrossRef]
62. Velayutham, R.; Manikandan, R.; Raj, C.J.; Kale, A.M.; Kaya, C.; Palanisamy, K.; Kim, B.C. Electrodeposition of Vanadium Pentoxide on Carbon Fiber Cloth as a Binder-Free Electrode for High-Performance Asymmetric Supercapacitor. *J. Alloys Compd.* **2021**, *863*, 158332. [CrossRef]
63. Zhang, J.; Yi, X.-B.; Wang, X.C.; Ma, J.; Liu, S.; Wang, X.J. Nickel Oxide Grown on Carbon Nanotubes/Carbon Fiber Paper by Electrodeposition as Flexible Electrode for High-Performance Supercapacitors. *J. Mater. Sci. Mater. Electron.* **2015**, *26*, 7901–7908. [CrossRef]
64. Sahoo, S.; Naik, K.K.; Rout, C.S. Electrodeposition of Spinel $MnCo_2O_4$ Nanosheets for Supercapacitor Applications. *Nanotechnology* **2015**, *26*, 455401. [CrossRef] [PubMed]
65. Zhao, P.; Wang, N.; Hu, W.; Komarneni, S. Anode Electrodeposition of 3D Mesoporous Fe_2O_3 Nanosheets on Carbon Fabric for Flexible Solid-State Asymmetric Supercapacitor. *Ceram. Int.* **2019**, *45*, 10420–10428. [CrossRef]
66. Kazemi, S.H.; Asghari, A.; Kiani, M.A. High Performance Supercapacitors Based on the Electrodeposited Co_3O_4 Nanoflakes on Electro-Etched Carbon Fibers. *Electrochim. Acta* **2014**, *138*, 9–14. [CrossRef]
67. Danks, A.E.; Hall, S.R.; Schnepp, Z. The Evolution of "sol-Gel" Chemistry as a Technique for Materials Synthesis. *Mater. Horizons* **2016**, *3*, 91–112. [CrossRef]
68. Parashar, M.; Shukla, V.K.; Singh, R. Metal Oxides Nanoparticles via Sol–Gel Method: A Review on Synthesis, Characterization and Applications. *J. Mater. Sci. Mater. Electron.* **2020**, *31*, 3729–3749. [CrossRef]
69. Sankar Devi, V.; Athika, M.; Duraisamy, E.; Prasath, A.; Selva Sharma, A.; Elumalai, P. Facile Sol-Gel Derived Nanostructured Spinel Co_3O_4 as Electrode Material for High-Performance Supercapattery and Lithium-Ion Storage. *J. Energy Storage* **2019**, *25*, 100815. [CrossRef]
70. Wu, Y.Q.; Chen, X.Y.; Ji, P.T.; Zhou, Q.Q. Sol-Gel Approach for Controllable Synthesis and Electrochemical Properties of $NiCo_2O_4$ Crystals as Electrode Materials for Application in Supercapacitors. *Electrochim. Acta* **2011**, *56*, 7517–7522. [CrossRef]
71. Hu, G.; Tang, C.; Li, C.; Li, H.; Wang, Y.; Gong, H. The Sol-Gel-Derived Nickel-Cobalt Oxides with High Supercapacitor Performances. *J. Electrochem. Soc.* **2011**, *158*, A695–A699. [CrossRef]
72. Lobo, L.S.; Kalainathan, S.; Kumar, A.R. Investigation of Electrical Studies of Spinel $FeCo_2O_4$ Synthesized by Sol-Gel Method. *Superlattices Microstruct.* **2015**, *88*, 116–126. [CrossRef]
73. Tang, W.; Shan, X.; Li, S.; Liu, H.; Wu, X.; Chen, Y. Sol-Gel Process for the Synthesis of Ultrafine MnO_2 Nanowires and Nanorods. *Mater. Lett.* **2014**, *132*, 317–321. [CrossRef]
74. Wang, N.; Zhang, Q.; Zhao, P.; Yao, M.; Hu, W.; Komarneni, S. Highly Mesoporous $LaNiO_3$/NiO Composite with High Specific Surface Area as a Battery-Type Electrode. *Ceram. Int.* **2017**, *43*, 5687–5692. [CrossRef]
75. Thangavel, R.; Ganesan, B.K.; Thangavel, V.; Yoon, W.S.; Lee, Y.S. Emerging Materials for Sodium-Ion Hybrid Capacitors: A Brief Review. *ACS Appl. Energy Mater.* **2021**, *4*, 13376–13394. [CrossRef]
76. Zhu, G.N.; Wang, Y.G.; Xia, Y.Y. Ti-Based Compounds as Anode Materials for Li-Ion Batteries. *Energy Environ. Sci.* **2012**, *5*, 6652–6667. [CrossRef]
77. Ma, Y.; Wang, X.; Jia, Y.; Chen, X.; Han, H.; Li, C. Titanium Dioxide-Based Nanomaterials for Photocatalytic Fuel Generations. *Chem. Rev.* **2014**, *114*, 9987–10043. [CrossRef]
78. Aravindan, V.; Shubha, N.; Chui Ling, W.; Madhavi, S. Constructing High Energy Density Non-Aqueous Li-Ion Capacitors Using Monoclinic TiO_2-B Nanorods as Insertion Host. *J. Mater. Chem. A* **2013**, *1*, 6145–6151. [CrossRef]
79. Liu, H.; Bi, Z.; Sun, X.G.; Unocic, R.R.; Paranthaman, M.P.; Dai, S.; Brown, G.M. Mesoporous TiO_2-B Microspheres with Superior Rate Performance for Lithium Ion Batteries. *Adv. Mater.* **2011**, *23*, 3450–3454. [CrossRef]
80. Wang, Q.; Wen, Z.; Li, J. A Hybrid Supercapacitor Fabricated with a Carbon Nanotube Cathode and a TiO_2-B Nanowire Anode. *Adv. Funct. Mater.* **2006**, *16*, 2141–2146. [CrossRef]
81. Kim, H.; Cho, M.Y.; Kim, M.H.; Park, K.Y.; Gwon, H.; Lee, Y.; Roh, K.C.; Kang, K. A Novel High-Energy Hybrid Supercapacitor with an Anatase TiO_2-Reduced Graphene Oxide Anode and an Activated Carbon Cathode. *Adv. Energy Mater.* **2013**, *3*, 1500–1506. [CrossRef]

82. Ramasubbu, V.; Omar, F.S.; Ramesh, K.; Ramesh, S.; Shajan, X.S. Three-Dimensional Hierarchical Nanostructured Porous TiO_2 Aerogel/Cobalt Based Metal-Organic Framework (MOF) Composite as an Electrode Material for Supercapattery. *J. Energy Storage* **2020**, *32*, 101750. [CrossRef]
83. Ohzuku, T.; Ueda, A.; Yamamoto, N. Zero-Strain Insertion Material of Li $[Li_{1/3}Ti_{5/3}]O_4$ for Rechargeable Lithium Cells. *J. Electrochem. Soc.* **1995**, *142*, 1431–1435. [CrossRef]
84. Li, X.; Qu, M.; Yu, Z. Structural and Electrochemical Performances of $Li_4Ti_{5-X}Zr_XO_{12}$ as Anode Material for Lithium-Ion Batteries. *J. Alloys Compd.* **2009**, *487*, 12–17. [CrossRef]
85. Khairy, M.; Bayoumy, W.A.; Qasim, K.F.; El-Shereafy, E.; Mousa, M.A. Ternary V-Doped $Li_4Ti_5O_{12}$-Polyaniline-Graphene Nanostructure with Enhanced Electrochemical Capacitance Performance. *Mater. Sci. Eng. B* **2021**, *271*, 115312. [CrossRef]
86. Naoi, K.; Naoi, W.; Aoyagi, S.; Miyamoto, J.I.; Kamino, T. New Generation "Nanohybrid Supercapacitor". *Acc. Chem. Res.* **2013**, *46*, 1075–1083. [CrossRef] [PubMed]
87. Lee, B.; Yoon, J.R. Preparation and Characteristics of $Li_4Ti_5O_{12}$ with Various Dopants as Anode Electrode for Hybrid Supercapacitor. *Curr. Appl. Phys.* **2013**, *13*, 1350–1353. [CrossRef]
88. Naoi, K.; Ishimoto, S.; Isobe, Y.; Aoyagi, S. High-Rate Nano-Crystalline $Li_4Ti_5O_{12}$ Attached on Carbon Nano-Fibers for Hybrid Supercapacitors. *J. Power Sources* **2010**, *195*, 6250–6254. [CrossRef]
89. Jeong, J.H.; Lee, G.W.; Kim, Y.H.; Choi, Y.J.; Roh, K.C.; Kim, K.B. A Holey Graphene-Based Hybrid Supercapacitor. *Chem. Eng. J.* **2019**, *378*, 122126. [CrossRef]
90. Choi, H.S.; Im, J.H.; Kim, T.; Park, J.H.; Park, C.R. Advanced Energy Storage Device: A Hybrid BatCap System Consisting of Battery-Supercapacitor Hybrid Electrodes Based on $Li_4Ti_5O_{12}$-Activated-Carbon Hybrid Nanotubes. *J. Mater. Chem.* **2012**, *22*, 16986–16993. [CrossRef]
91. Jung, H.G.; Venugopal, N.; Scrosati, B.; Sun, Y.K. A High Energy and Power Density Hybrid Supercapacitor Based on an Advanced Carbon-Coated $Li_4Ti_5O_{12}$ Electrode. *J. Power Sources* **2013**, *221*, 266–271. [CrossRef]
92. Kim, H.; Park, K.Y.; Cho, M.Y.; Kim, M.H.; Hong, J.; Jung, S.K.; Roh, K.C.; Kang, K. High-Performance Hybrid Supercapacitor Based on Graphene-Wrapped $Li_4Ti_5O_{12}$ and Activated Carbon. *ChemElectroChem* **2014**, *1*, 125–130. [CrossRef]
93. Manthiram, A. A Reflection on Lithium-Ion Battery Cathode Chemistry. *Nat. Commun.* **2020**, *11*, 1–9. [CrossRef]
94. Godillot, G.; Taberna, P.L.; Daffos, B.; Simon, P.; Delmas, C.; Guerlou-Demourgues, L. High Power Density Aqueous Hybrid Supercapacitor Combining Activated Carbon and Highly Conductive Spinel Cobalt Oxide. *J. Power Sources* **2016**, *331*, 277–284. [CrossRef]
95. Numan, A.; Khalid, M.; Ramesh, S.; Ramesh, K.; Shamsudin, E.M.; Zhan, Y.; Jagadesh, P. Facile Sonochemical Synthesis of 2D Porous Co_3O_4 Nanoflake for Supercapattery. *J. Alloys Compd.* **2020**, *819*, 153019. [CrossRef]
96. Sivakumar, P.; Jana, M.; Kota, M.; Jung, M.G.; Gedanken, A.; Park, H.S. Controllable Synthesis of Nanohorn-like Architectured Cobalt Oxide for Hybrid Supercapacitor Application. *J. Power Sources* **2018**, *402*, 147–156. [CrossRef]
97. Vattikuti, S.V.P.; Hoang Ngoc, C.T.; Nguyen, H.; Nguyen Thi, N.H.; Shim, J.; Dang, N.N. Carbon Nitride Coupled Co_3O_4: A Pyrolysis-Based Approach for High-Performance Hybrid Energy Storage. *J. Phys. Chem. Lett.* **2023**, *14*, 9412–9423. [CrossRef]
98. Arjunan, A.; Ramasamy, S.; Kim, J.; Kim, S.K. Co_3O_4 Nanoparticles-Embedded Nitrogen-Doped Porous Carbon Spheres for High-Energy Hybrid Supercapacitor Electrodes. *J. Energy Storage* **2023**, *68*, 107758. [CrossRef]
99. Hosseinzadeh, B.; Nagar, B.; Benages-Vilau, R.; Gomez-Romero, P.; Kazemi, S.H. MOF-Derived Conformal Cobalt Oxide/C Composite Material as High-Performance Electrode in Hybrid Supercapacitors. *Electrochim. Acta* **2021**, *389*, 138657. [CrossRef]
100. Sahoo, S.; Milton, A.; Sood, A.; Kumar, R.; Choi, S.; Maity, C.K.; Han, S.S. Microwave-Assisted Synthesis of Perovskite Hydroxide-Derived Co_3O_4/SnO_2/Reduced Graphene Oxide Nanocomposites for Advanced Hybrid Supercapacitor Devices. *J. Energy Storage* **2024**, *99*, 113321. [CrossRef]
101. Muralee Gopi, C.V.V.; Vinodh, R.; Sambasivam, S.; Obaidat, I.M.; Naidu Kalla, R.M.; Kim, H.J. One-Pot Synthesis of Copper Oxide–Cobalt Oxide Core–Shell Nanocactus-like Heterostructures as Binder-Free Electrode Materials for High-Rate Hybrid Supercapacitors. *Mater. Today Energy* **2019**, *14*, 100358. [CrossRef]
102. Izwan Misnon, I.; Krishnan, S.G.; Jose, R. Thin Chemisorbed Polyaniline Film on Cobalt Oxide as an Electrode for Hybrid Energy Storage Devices. *ChemistrySelect* **2020**, *5*, 7973–7983. [CrossRef]
103. Chen, H.; Bao, E.; Du, X.; Ren, X.; Liu, X.; Li, Y.; Xu, C. Advanced Hybrid Supercapacitors Assembled with High-Performance Porous $MnCo_2O_{4.5}$ Nanosheets as Battery-Type Cathode Materials. *Colloids Surfaces A Physicochem. Eng. Asp.* **2023**, *657*, 130663. [CrossRef]
104. Li, W.; Xu, K.; Song, G.; Zhou, X.; Zou, R.; Yang, J.; Chen, Z.; Hu, J. Facile Synthesis of Porous $MnCo_2O_{4.5}$ Hierarchical Architectures for High-Rate Supercapacitors. *CrystEngComm* **2014**, *16*, 2335–2339. [CrossRef]
105. Liu, Y.; Du, X.; Li, Y.; Bao, E.; Ren, X.; Chen, H.; Tian, X.; Xu, C. Nanosheet-Assembled Porous $MnCo_2O_{4.5}$ Microflowers as Electrode Material for Hybrid Supercapacitors and Lithium-Ion Batteries. *J. Colloid Interface Sci.* **2022**, *627*, 815–826. [CrossRef]
106. Yang, C.H.; Chen, Y.C.; Wu, C.F.; Chung, R.J.; Yougbaré, S.; Lin, L.Y. Novel Synthesis of ZIF67-Derived $MnCo_2O_4$ Nanotubes Using Electrospinning and Hydrothermal Techniques for Supercapacitor. *J. Solid State Chem.* **2022**, *313*, 123351. [CrossRef]

107. Che, H.; Wang, Y.; Mao, Y. Novel Flower-like $MnCo_2O_4$ Microstructure Self-Assembled by Ultrathin Nanoflakes on the Microspheres for High-Performance Supercapacitors. *J. Alloys Compd.* **2016**, *680*, 586–594. [CrossRef]
108. Gao, Y.; Xia, Y.; Wan, H.; Xu, X.; Jiang, S. Enhanced Cycle Performance of Hierarchical Porous Sphere $MnCo_2O_4$ for Asymmetric Supercapacitors. *Electrochim. Acta* **2019**, *301*, 294–303. [CrossRef]
109. Ahmad, A.; Khan, S.; Javed, M.S.; Osman, S.; Li, H.; Majeed, S.; Luque, R. Improved Electrochemical Performance of Aqueous Hybrid Supercapacitors Using $CrCo_2O_4$ Mesoporous Nanowires: An Innovative Strategy toward Sustainable Energy Devices. *ACS Appl. Mater. Interfaces* **2024**, *16*, 6920–6930. [CrossRef]
110. Sivakumar, P.; Kulandaivel, L.; Park, J.W.; Raj, C.J.; Savariraj, A.D.; Manikandan, R.; Rajendran, R.; Jung, H. Electroactive Site Enriched Battery-Type Worm-like Cobalt Tungstate Nanoarchitecture Electrode Material for Performance-Enhanced Hybrid Supercapacitor. *Surf. Interfaces* **2023**, *40*, 103111. [CrossRef]
111. Zhao, Y.; Zheng, J.; Yuan, M.; Wang, Y.; Liu, W.; Yang, S.; Li, G.; Lian, J.; Bu, Y. Boosting the Energy Density of Iron-Cobalt Oxide Based Hybrid Supercapacitors by Redox-Additive Electrolytes. *J. Alloys Compd.* **2021**, *885*, 160886. [CrossRef]
112. Nagaraju, M.; Ramulu, B.; Girija Shankar, E.; Su Yu, J. Rational Design of Hierarchical Zeolitic Imidazolate Framework-67@Cu_2CoO_3 Core-Shell Architectures for Hybrid Supercapacitor Applications. *Appl. Surf. Sci.* **2023**, *640*, 158339. [CrossRef]
113. Srivastav, S.; Paliwal, M.K.; Meher, S.K. Ribbon-like Nickel Cobaltite with Layer-by-Layer-Assembled Ordered Nanocrystallites for Next-Generation All-Solid-State Hybrid Supercapatteries. *Langmuir* **2022**, *38*, 3969–3983. [CrossRef]
114. Ganesan, M.; Alagar, S.; Bagchi, V.; Piraman, S. Surface Oxygen Engineered $ZnCo_2O_4$ Planar Hybrid Supercapacitor Electrode for High Energy Applications. *J. Energy Storage* **2024**, *98*, 112954. [CrossRef]
115. Zhang, S.; Tan, H.; Rui, X.; Yu, Y. Vanadium-Based Materials: Next Generation Electrodes Powering the Battery Revolution? *Acc. Chem. Res.* **2020**, *53*, 1660–1671. [CrossRef] [PubMed]
116. Tang, H.; Peng, Z.; Wu, L.; Xiong, F.; Pei, C.; An, Q.; Mai, L. Vanadium-Based Cathode Materials for Rechargeable Multivalent Batteries: Challenges and Opportunities. *Electrochem. Energy Rev.* **2018**, *1*, 169–199. [CrossRef]
117. Yan, Y.; Li, B.; Guo, W.; Pang, H.; Xue, H. Vanadium Based Materials as Electrode Materials for High-Performance Supercapacitors. *J. Power Sources* **2016**, *329*, 148–169. [CrossRef]
118. Yue, Y.; Liang, H. Micro- and Nano-Structured Vanadium Pentoxide (V_2O_5) for Electrodes of Lithium-Ion Batteries. *Adv. Energy Mater.* **2017**, *7*, 1–32. [CrossRef]
119. Yang, Y.; Tang, Y.; Fang, G.; Shan, L.; Guo, J.; Zhang, W.; Wang, C.; Wang, L.; Zhou, J.; Liang, S. Li^+ Intercalated V_2O_5: N H_2O with Enlarged Layer Spacing and Fast Ion Diffusion as an Aqueous Zinc-Ion Battery Cathode. *Energy Environ. Sci.* **2018**, *11*, 3157–3162. [CrossRef]
120. Liu, M.; Su, B.; Tang, Y.; Jiang, X.; Yu, A. Recent Advances in Nanostructured Vanadium Oxides and Composites for Energy Conversion. *Adv. Energy Mater.* **2017**, *7*, 1700885. [CrossRef]
121. Xu, Y.; Yang, X.; Li, X.; Gao, Y.; Wang, L.; Lü, W. Flexible Zinc-Ion Hybrid Supercapacitor Based on Co^{2+}-Doped Polyaniline/V_2O_5 Electrode. *J. Power Sources* **2024**, *623*, 235399. [CrossRef]
122. Keum, K.; Park, D.; Park, M.; Lee, Y.; Lee, H.; Jeong, H.; Kim, J.W.; Kim, D.-w.; Ha, J.S. All Vanadium-Based Li-Ion Hybrid Supercapacitor with Enhanced Electrochemical Performance via Prelithiation. *J. Alloys Compd.* **2022**, *914*, 165288. [CrossRef]
123. Jing, X.; Zhang, Y.; Jiang, H.; Cheng, Y.; Xing, N.; Meng, C. Facile Template-Free Fabrication of Hierarchical V_2O_5 Hollow Spheres with Excellent Charge Storage Performance for Symmetric and Hybrid Supercapacitor Devices. *J. Alloys Compd.* **2018**, *763*, 180–191. [CrossRef]
124. Aravindan, V.; Cheah, Y.L.; Mak, W.F.; Wee, G.; Chowdari, B.V.R.; Madhavi, S. Fabrication of High Energy-Density Hybrid Supercapacitors Using Electrospun V_2O_5 Nanofibers with a Self-Supported Carbon Nanotube Network. *Chempluschem* **2012**, *77*, 570–575. [CrossRef]
125. Vishwakarma, N.; Mashangva, T.T.; Rajput, S.; Pham, T.D.; Kumar, M.; Sharma, A. Unlocking the Potential for Revolutionary Energy Storage Capabilities of V_2O_5/GO Nanocomposites Synthesized under Microwave Irradiation. *J. Energy Storage* **2024**, *103*, 114405. [CrossRef]
126. Wang, G.; Ding, Y.; Xu, Z.; Wang, G.; Li, Z.; Yan, Z. Co_3O_4@Mn-Ni$(OH)_2$ Core–Shell Heterostructure for Hybrid Supercapacitor Electrode with High Utilization. *Chem. Eng. J.* **2023**, *469*, 143984. [CrossRef]
127. Arun Kumar, S.; Sarasamreen, I.; Balaji, C.; Gowdhaman, A.; Ramesh, R.; Anbarasan, P.M. Elevates the Electrochemical Stability Performance of Hydrothermally Synthesized Co_3O_4 Nanowires/NF for Hybrid Supercapacitors. *Inorg. Chem. Commun.* **2023**, *158*, 111506. [CrossRef]
128. Chen, H.; Du, X.; Sun, J.; Wu, R.; Wang, Y.; Xu, C. Template-Free Synthesis of Novel Co_3O_4 Micro-Bundles Assembled with Flakes for High-Performance Hybrid Supercapacitors. *Ceram. Int.* **2021**, *47*, 716–724. [CrossRef]
129. Iqbal, M.Z.; Shaheen, M.; Aftab, U.; Ahmad, Z.; Solangi, M.Y.; Abro, M.I.; Wabaidur, S.M. Faradically Dominant Pseudocapacitive Manganese Cobalt Oxide Electrode Materials for Hybrid Supercapacitors and Electrochemical Water Splitting. *Energy Fuels* **2024**, *38*, 2416–2425. [CrossRef]

130. Reghunath, B.S.; Sunaja Devi, K.R.; Rajasekaran, S.; Saravanakumar, B.; William, J.J.; Pinheiro, D. $CoFe_2O_4$ Nanoparticles Embedded 2D Cr_2CTx MXene: A New Material for Battery like Hybrid Supercapacitors and Oxygen Evolution Reaction. *J. Energy Storage* **2024**, *84*, 110775. [CrossRef]
131. Bhagwan, J.; Han, J.I. Formation of MWCNT/$LiCo_2O_4$ Nanoplates and Their Application for Hybrid Supercapacitor. *Ceram. Int.* **2024**, *50*, 10676–10687. [CrossRef]
132. Zhao, L.; Zhang, H.; Ma, B. Formation of Carbon-Incorporated NiO@Co_3O_4 Nanostructures via a Direct Calcination Method and Their Application as Battery-Type Electrodes for Hybrid Supercapacitors. *ACS Omega* **2023**, *8*, 10503–10511. [CrossRef]
133. Bhosale, R.; Bhosale, S.; Chavan, V.; Jambhale, C.; Kim, D.K.; Kolekar, S. Hybrid Supercapacitors Based on Nanoporous Carbon and $CoFe_2O_4$ Derived from a Bimetallic Organic Framework. *ACS Appl. Nano Mater.* **2024**, *7*, 2244–2257. [CrossRef]
134. Govindasamy, T.; Mathew, N.K.; Asapu, V.K.; Subramanian, V.; Subramanian, B. Modulating the Structural and Magnetic Properties of Fe_3O_4 NPs for High-Performance Supercapattery and EMI Shielding Applications. *J. Energy Storage* **2024**, *79*, 110243. [CrossRef]
135. Huang, H.; Wang, X.; Tervoort, E.; Zeng, G.; Liu, T.; Chen, X.; Sologubenko, A.; Niederberger, M. Nano-Sized Structurally Disordered Metal Oxide Composite Aerogels as High-Power Anodes in Hybrid Supercapacitors. *ACS Nano* **2018**, *12*, 2753–2763. [CrossRef] [PubMed]
136. Shen, L.; Lv, H.; Chen, S.; Kopold, P.; van Aken, P.A.; Wu, X.; Maier, J.; Yu, Y. Peapod-like Li_3VO_4/N-Doped Carbon Nanowires with Pseudocapacitive Properties as Advanced Materials for High-Energy Lithium-Ion Capacitors. *Adv. Mater.* **2017**, *29*, 1–8. [CrossRef] [PubMed]
137. Lim, E.; Lim, W.G.; Jo, C.; Chun, J.; Kim, M.H.; Roh, K.C.; Lee, J. Rational Design of Li_3VO_4@carbon Core-Shell Nanoparticles as Li-Ion Hybrid Supercapacitor Anode Materials. *J. Mater. Chem. A* **2017**, *5*, 20969–20977. [CrossRef]
138. Zhou, Y.; Wen, L.L.; Zhan, K.; Yan, Y.; Zhao, B. Three-Dimensional Porous Graphene/Nickel Cobalt Mixed Oxide Composites for High-Performance Hybrid Supercapacitor. *Ceram. Int.* **2018**, *44*, 21848–21854. [CrossRef]
139. Hao, Y.; Guo, H.; Ren, H.; Yang, Z.; Peng, L.; Liu, Y.; Yang, W. Highly Dispersive Nickel Vanadium Oxide Nanoparticles Anchored on Nickel Cobalt Phosphate Micron-Sheets as Cathodes for High-Energy Hybrid Supercapacitor Devices. *J. Alloys Compd.* **2025**, *1010*, 177893. [CrossRef]
140. Sun, H.; Miao, Y.; Wang, G.; Han, X.; Xu, C.; Zhu, J.; Chen, H. Battery-Type $ZnCo_2O_4$ Nanosheets and Nanowires as Advanced Cathode Materials for Hybrid Supercapacitors with Ultra-Long Cycling Stability. *J. Energy Storage* **2024**, *92*, 112189. [CrossRef]
141. Kour, S.; Kour, P.; Sharma, A.L. A Chromium Cobaltite Based Ternary Composite as an Efficient Electrode Material for Hybrid Supercapacitors with Theoretical Investigation. *Nanoscale* **2024**, *16*, 21456–21470. [CrossRef]
142. Suganya, S.; Kousi, F.; Sambasivam, S.; Tighezza, A.M.; Velsankar, K.; Sudhahar, S. Investigations of Ternary Cu-Mn-Zn Oxide Nanocomposites as Potential Electrode for Hybrid Supercapacitors by One-Pot Hydrothermal Method. *J. Energy Storage* **2025**, *109*, 115181. [CrossRef]
143. Fei, T.; Ahmad, T.; Usman, M.; Ahmad, A.; Saleem, A.; Hanif, M.B.; Karami, A.M.; Javed, M.S.; Akkinepally, B.; Xia, C. Zn-Doped Cr_2O_3 Oxides Boosted the Electrochemical Performance of Aqueous Hybrid Supercapacitor. *Electrochim. Acta* **2024**, *476*, 143673. [CrossRef]
144. Zheng, J.; Meng, D.; Guo, J.; Liu, X.; Zhou, L.; Wang, Z. Defect Engineering for Enhanced Electrocatalytic Oxygen Reaction on Transition Metal Oxides: The Role of Metal Defects. *Adv. Mater.* **2024**, *36*, e2405129. [CrossRef] [PubMed]
145. Sun, Y.; Wu, T.; Bao, Z.; Moon, J.; Huang, Z.; Chen, Z.; Chen, H.; Li, M.; Yang, Z.; Chi, M.; et al. Defect Engineering of Ceria Nanocrystals for Enhanced Catalysis via a High-Entropy Oxide Strategy. *ACS Cent. Sci.* **2022**, *8*, 1081–1090. [CrossRef] [PubMed]
146. Choi, J.; Im, S.; Choi, J.; Surendran, S.; Moon, D.J.; Kim, J.Y.; Kim, J.K.; Sim, U. Recent Advances in 2D Structured Materials with Defect-Exploiting Design Strategies for Electrocatalysis of Nitrate to Ammonia. *Energy Mater.* **2024**, *4*, 400020. [CrossRef]
147. Lu, W.; Yan, L.; Ye, W.; Ning, J.; Zhong, Y.; Hu, Y. Defect Engineering of Electrode Materials towards Superior Reaction Kinetics for High-Performance Supercapacitors. *J. Mater. Chem. A* **2022**, *10*, 15267–15296. [CrossRef]

batteries

MDPI

Article

Improvement of $Co_3V_2O_8$ Nanowire Driven by Morphology for Supercapacitor and Water Splitting Applications

Manesh Ashok Yewale and Dong Kil Shin *

School of Mechanical Engineering, Yeungnam University, Gyeongsan 38541, Republic of Korea; maneshphd@gmail.com
* Correspondence: dkshin@yu.ac.kr

Abstract: Supercapacitors have a better power density than batteries; however, there is room for improvement in energy density. $Co_3V_2O_8$ nanoparticles were synthesized using the hydrothermal approach, with the reaction duration tuned to enhance energy density. At a 10 h hydrothermal reaction time, bundles of nanowires with void spaces were obtained, demonstrating excellent areal capacitance of 4.67 F/cm^2, energy density of 94 µWh/cm^2, and power density of 573 µW/cm^2 at a current density of 3 mA/cm^2. With activated carbon (AC) and $Co_3V_2O_8$ nanoparticles prepared over a 10-h hydrothermal reaction period, an asymmetric supercapacitor (ASC) was assembled. The device performed admirably in terms of energy storage capacity, with an areal capacitance of 781 mF/cm^2 and a volumetric capacitance of 1.43 F/cm^3. The ASC's cyclic stability demonstrated capacity retention of 83.40% after 5000 cycles. The powering of red LEDs was used to show practical applications. In a 2M KOH electrolyte, the optimized $Co_3V_2O_8$ electrode demonstrated good electrocatalytic performance for the hydrogen evolution process, with an overpotential of 259 mV at a current density of 10 mA/cm^2. Overall, water splitting studies revealed a potential of 1.78 V with little potential enhancement after 8 h of Chrono potentiometric stability. As a result, $Co_3V_2O_8$ nanoparticles prepared at a 10 h hydrothermal reaction time offer excellent electrode materials for energy storage in supercapacitors and electrocatalytic applications for total water splitting.

Keywords: $Co_3V_2O_8$ nanowires; supercapacitor; asymmetric supercapacitor (ASC); overall water splitting; XPS; HRTEM

1. Introduction

Electronic gadgets, rapidly increasing in usage nowadays, operate on energy storage devices. Supercapacitors have emerged as an excellent choice for energy storage in advanced electronic equipment due to their outstanding characteristics such as fast charging time, high energy and power density, light weight, and longevity [1–6]. Their popularity is growing day by day. Nowadays, supercapacitors are used in various electronic devices including portable electronics, sensing devices, and hybrid electric vehicles [7–10]. Supercapacitors are classified into two subclasses based on their energy storage mechanism: electric double-layered capacitors (EDLCs) and pesudocapacitors (PCrs). Pseudocapacitors store charge through Faradic reactions on the electrode material's surface, whereas EDLCs store charge through the adsorption of ions on the electrode material's surface. Materials with a high surface area typically exhibit excellent EDLC characteristics. Mostly, carbon-based materials such as graphene oxide [11–13] and carbon nanotubes [14], as well as some metal oxides [15], demonstrate EDLC properties [16]. On the other hand, conducting polymers [17], transition metal oxides, and metal chalcogenides exhibit charge storage mechanisms typical of pesudocapacitors. Pseudocapacitors have gained popularity as electrochemical energy storage devices due to their large theoretical capacitance, different valences, and diverse electronic architectures. Because of their superior theoretical capacitance, numerous oxidation states, synergetic effects, cost-effectiveness, abundance, and huge potential window, bimetallic metal oxides are now being researched for electrochemical energy storage applications. There have been several study papers on bimetallic oxides such as $NiCo_2O_4$ [18–20], $CuCo_2O_4$ [21–23], $Ni_3V_2O_8$ [24,25], and $Co_3V_2O_8$ [26,27]. Bimetallic oxides have a higher energy storage capacity than monometal oxides such as NiO [28,29], Co_3O_4 [23,30], and V_2O_5 [31] due to their synergetic effect and numerous oxidation states. Researchers are particularly interested in $Co_3V_2O_8$. Cobalt-based bimetallic oxides have gained significant interest because of their inherent properties such as strong electrical conductivity, simple synthesis procedures, and great electrical stability [32,33]. Cobalt also has a large theoretical capacitance and a great reversible redox ability and corrosion resistance [34]. The majority of cobalt-based metal oxides have a densely packed cubic spinel structure [35]. Recently, the hydrogen evolution reaction (HER) via electrocatalysis and supercapacitors has piqued the interest of many researchers in the field of environmentally benign and highly effective electrode materials for energy storage and conversion. Metal oxides with bifunctional applications are receiving more attention for economic reasons [36]. Noble metals with high electrocatalytic activity include palladium and platinum. However, their widespread adoption is hampered by high costs. As a result, it is critical to develop electrode materials that are cost-effective, high-efficiency, and environmentally benign for electrocatalytic and supercapacitor applications [37–40]. Porous microwires have potential uses in electrocatalysis and supercapacitor technology. For multifunctional applications, several cobalt–vanadium oxide microstructures have been described. H. Wang et al. reported on the hydrothermal synthesis of hexagonal nanosheets of $Co_2V_2O_7$ for hybrid supercapacitor applications. They have a high energy storage capacity [41]. H. Sun, on the other hand, developed 3D microflowers with exceptional cycle stability for asymmetric supercapacitor electrodes. These were produced using a chemical technique at precise temperatures and precursor molar concentrations. At 1 A/g, the electrode demonstrated a remarkable energy storage capacity of 351 F/g. The energy storage capacity of an asymmetric supercapacitor (ASC) developed with the optimized electrode and reduced graphene oxide (rGO) was 19 Wh/kg. It had a power density of 375.8 W/kg, having a potential range of 0–1.5 V [42]. G. P. Sharma employed hydrothermal synthesis for synthesizing $Co_3V_2O_8$ and S-$Co_3V_2O_8$ nanosheets for supercapacitor applications. The $Co_3V_2O_8$ nanosheets demonstrated a remarkable storage capacity of 337.8 mAh/g at 2 A/g [43]. There have been few reports on the use of cobalt-mixed metal oxide for catalysis. G. M. Thorat reported on octahedral cobalt vanadate synthesized for an oxygen evolution reaction (OER) employing a deep eutectic solvent method [44]. D. K. Singh addressed cobalt-based vanadium mixed oxide nanocrystals coupled with polymer electrodes for electrochemical water oxidation, demonstrating their electrocatalysis applications [45]. J. Han also reported on cobalt vanadium spinel synthesized at various Co/V molar ratios utilizing the hydrothermal process for electrocatalysis applications [46]. The usage of cobalt-based mixed vanadium oxide as a supercapacitor and electrocatalysis electrode has received little attention.

When synthesizing $Co_3V_2O_8$ (CVO) nanowires, the influence of the hydrothermal reaction duration was investigated in order to achieve a homogeneous nanowire with a porous structure. While other parameters remained constant, the hydrothermal reaction time was increased from 4 h to 16 h. The reaction time has an effect on the surface microstructure of $Co_3V_2O_8$. Nanowire bunch production began at lower hydrothermal reaction durations, and the development of these nanowires accelerated as the reaction time increased. Initial little bundles of nanowires began to develop after a hydrothermal reaction period of 10 h. These clusters merged at one end, leaving empty space at the other. This open area is beneficial for ion diffusion, increasing the electrode's capacity. The 10 h hydrothermal reaction time electrode displayed exceptional areal energy storage capacity and electrocatalytic activity for water splitting via HER and OER. The optimized electrode and activated carbon (AC) were used to make an asymmetric supercapacitor. The ASC demonstrated exceptional areal energy storage capacity and power density. A two-electrode device was constructed for practical applications, and it demonstrated its higher energy storage capability by lighting red LEDs. The $Co_3V_2O_8$ electrode formed by a 10 h hydrothermal reaction appears to be an acceptable electrode material for supercapacitor and electrolysis applications based on the findings obtained for the $Co_3V_2O_8$ nanowires.

2. Experimental

2.1. Chemicals

All chemicals, particularly cobalt nitrate hexahydrate ($Co(NO_3)_2 \cdot 6H_2O$ (98%), urea ($CO(NH_2)_2$) (99%), ammonium metavanadate (NH_4VO_3) (99%), acetone (CH_3CO) (99%), ethanol (C_2H_6O) (99.5%), polyvinylidene fluoride (PVDF), N-Methyl-2-pyrrolidone (NMP), and potassium hydroxide (KOH) (85%), were sourced from Sigma-Aldrich (St. Louis, MO, USA) and used without extra purification processing. HCl were provided by Daejung Chemical (Siheung-si, Republic of Korea).

2.2. Preparation of $Co_3V_2O_8$ (CVO) Nanowires

A simple hydrothermal process was used to prepare $Co_3V_2O_8$ nanowires. To begin the synthesis, 40 mL of water was mixed with 4.8 mL of cobalt nitrate, 7.2 mL of ammonium metavanadate, and 2.24 mL of urea to form a homogenous solution. After stirring, the entire solution was transferred and carefully sealed into a 100 mL Teflon liner reactor. The hydrothermal reactor was then placed in a fan-forced furnace for 4 h to maintain a temperature of 140 °C. After the reaction was completed, the reactor was allowed to cool naturally to ambient temperature. After cooling, the produced product was filtered through filter paper and the residue was washed with water and ethanol several times. The washed $Co_3V_2O_8$ powder was dried overnight at 60 °C. Following the drying procedure, the $Co_3V_2O_8$ powder was annealed at 400 °C for 4 h. After drying, the $Co_3V_2O_8$ powder was given the designation CVO_U_4h. The reaction mechanism for the formation of $Co_3V_2O_8$ nanowires is represented by reactions (1)–(4) [47]. A similar approach was used for the synthesis of $Co_3V_2O_8$ powder at 7 h, 10 h, and 16 h of hydrothermal reaction time, and the products were called CVO_U_7h, CVO_U_10h, and CVO_U_16h, respectively. $Co_3V_2O_8$ nanowires prepared for different hydrothermal reaction times were utilized for further characterization and study with similar nomenclature.

$$NH_2CONH_2 + H_2O \rightarrow 2NH_3 + CO_2 \quad (1)$$

$$NH_3 + H_2O \leftrightarrow NH_4^+ + OH^- \quad (2)$$

$$3Co^{2+} + 2VO_3^- + 4OH^- \rightarrow Co_3V_2O_8 \cdot H_2O + 2H_2O \quad (3)$$

$$Co_3V_2O_8 \cdot H_2O \rightarrow Co_3V_2O_8 + H_2O \quad (4)$$

2.3. Characterizations

X-ray diffraction (XRD; PANalytical, Almelo, The Netherlands) with CuK radiation was used to analyze the crystal structure and phase evolution of the microparticles. The surface morphology and elemental mapping of the nanoparticles were studied using field-emission scanning electron microscopy (FE-SEM; S-4800 HITACHI, Ltd., Tokyo, Japan). X-ray photoelectron spectroscopy (XPS; K-alpha, Thermo Scientific, Altrincham, UK) was used to analyze the chemical composition of the surface. To investigate element mapping and surface morphology further, high-resolution transmission electron microscopy (HRTEM; Tecnai F21, FEI Company, Hillsboro, OR, USA) was used. On a ZIVE SP5 (WonAtech, Seoul, South Korea) electrochemical workstation, electrochemical tests such as cyclic voltammetry (CV), electrochemical impedance spectroscopy (EIS), galvanostatic charge–discharge (GCD), linear sweep voltammetry (LSV), and cyclic stability measurements were performed. These measurements were performed using a three-electrode configuration consisting of a platinum counter electrode, the active material as a working electrode, and an Ag/AgCl reference electrode.

2.4. Electrode Preparation

The working electrode was made with extreme accuracy and attention to allow for a complete examination of the electrochemical performance of the active substance. Using an NMP solution, a slurry of the active material ($Co_3V_2O_8$), PVDF, and carbon black (CB) in an 80:10:10 ratio was made and employed to form the electrode. To achieve a clean electrode surface, the 1×2 cm^2 Ni-foam region was extensively cleaned with hydrochloric acid to remove any contaminants. The dried Ni-foam substrate was then utilized to prepare the final electrode. Prior to electrochemical investigation, the active material slurry was drop-cast onto the Ni substrate's 1×1 cm^2 active region and dried overnight at 60 °C. An identical procedure was used for preparing the working electrode and activated carbon electrode. Only activated carbon (AC) was used instead of $Co_3V_2O_8$ nanoparticles in the fabrication of the activated carbon (AC) electrode.

2.5. Electrochemical Measurement

The active electrode ($Co_3V_2O_8$) served as the working electrode in the three-electrode arrangement, platinum served as the counter electrode, and an Ag/AgCl electrode served as the reference electrode. The ZIVE-SP5 WonAtech workstation was used to conduct electrochemical tests with a 2M KOH electrolyte. A two-electrode design was prepared for the practical application. The $Co_3V_2O_8$ nanoparticles were synthesized after a 10-h hydrothermal process that was utilized to prepare two electrode configurations. For 20 min, both the CVO_U_10h electrode and the AC electrode were soaked in 2M KOH. Following that, the two electrodes were securely wrapped in paraffin paper, with wetted filter paper put between them. Using a workstation, the device was used to investigate cyclic voltammetry (CV), galvanostatic charge–discharge (GCD), electrochemical impedance spectroscopy (EIS), and electrode stability.

3. Result and Discussion

3.1. Physicochemical Properties of the $Co_3V_2O_8$ Nanoparticles

To investigate the phase evolution and crystalline structure of CVO nanowires, powder X-ray diffraction was used. Figure 1 depicts the XRD diffraction pattern of CVO nanowires generated by different hydrothermal reaction time frames ranging from 4 h to 16 h. Reflections were seen in the XRD spectra of all nanoparticles at 2θ angles of 21.15°, 27.42°, 31.37°, 32.99°, 35.82°, 36.81°, 48.35°, 48.84°, 57.41°, and 63.84°. These are clearly matched with the JCPDS database of 01-074-1487, which has an orthorhombic phase with lattice parameters of a = 8.30 Å, b = 11.50 Å, and c = 6.03 Å and 2θ angles of 22.47°, 23.98°, 39.74°, and 50.44° match with the JCPDS database of 00-022-0599, confirming the nanoparticles' mixed $Co_3V_2O_8$ phase growth. The peak intensity of the 211 plane grows as the hydrothermal reaction time increases, and the XRD findings agree with those of Avi et al. [10], Mushtaq et al. [48], and Samuel et al. [49].

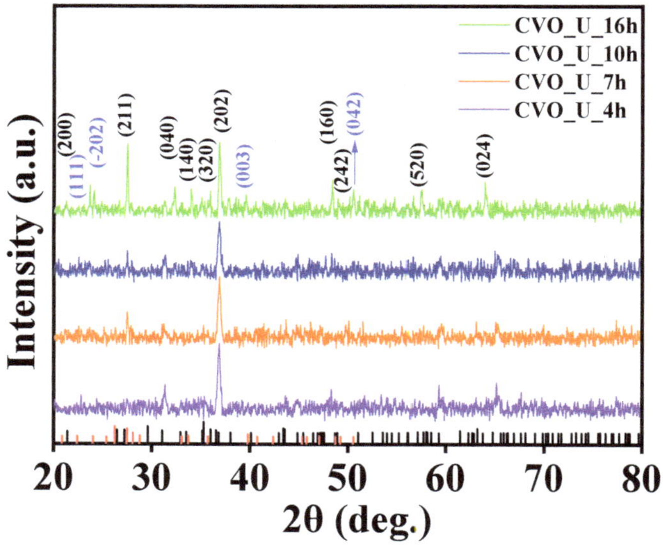

Figure 1. XRD spectra of $Co_3V_2O_8$ nanoparticles at 4 h, 7 h, 10 h, and 16 h of hydrothermal reaction time.

The elemental state, purity, and chemical composition of the prepared CVO_U_10h electrode were investigated through X-ray photoelectron spectroscopy (XPS). Figure 2a shows the survey scan spectra of the CVO_U_10h electrode over the range of 0–1200 eV, showing the presence of sharp peaks at 516.4 eV, 780.19 eV, and 529.93 eV and confirm the existence of vanadium (V2p), cobalt (Co2p), and oxygen (O1s) in the CVO_U_10h electrode. The elemental compositions of the XPS results were well matched with the EDX results, which confirm the formation of $Co_3V_2O_8$ stoichiometrically. Figure 2b shows the spin orbit narrowing two core-level spectra of Co $2p_{3/2}$ and Co $2p_{1/2}$. Co $2p_{3/2}$ and Co $2p_{1/2}$ were observed at binding energies of 780.01 eV and 795.81 eV, with a separation of 15.8 eV. The Co $2p_{3/2}$ and Co $2p_{1/2}$ spectra were split into two peaks, showing the presence of Co^{2+} and Co^{3+}. The peaks appearing at 779.81 eV and 794.81 eV show the presence of Co^{3+}, and the peaks observed at 781.23 eV and 796.60 eV show the presence of Co^{2+} in the CVO_U_10h electrode. Additionally, two shakeup satellite peaks were observed at 786.92 eV and 803.52 eV, which is in good agreement with the results [43]. The XPS study reveals the presence of Co ions with a Co^{2+} oxidation state [50,51].

The O 1s spectra of the CVO_U_10h electrode are shown in Figure 2c. Three peaks at binding energies of 529.87 eV (O_I), 530.18 eV (O_{II}), and 531.87 eV (O_{III}) are attributed to the O 1s spectra. The peak at 531.87 eV (O_{III}) can be ascribed to oxygen defects. The peak at 530.18 eV (O_{II}) is attributed to chemo-absorbed and physio-absorbed water molecules, and further peaks corresponding to the binding energy of 529.87 eV (O_I) are attributed to the metal–oxygen bond (Co-O and V-O) [43]. The deconvolution spectra of the V are shown in Figure 2d, attributed to the two peaks of V $2p_{3/2}$ and V $2p_{1/2}$ at binding energies of 516.41 eV and 524.11 eV. The peaks corresponding to 516.34 and 523.54 confirm the existence of the V^{+4} oxidation state, and the peaks attributed to 517.18 and 524.49 correspond to the V^{+5} oxidation state [50]. Overall, the XPS study confirmed the existence, chemical composition, and purity of Co, V, and O in the CVO_U_10h sample. All CVO_U_10h nanowire surface microstructures were examined using field emission scanning electron microscopy

(FESEM) and transmission electron microscopy (TEM). At various magnifications, Figure S1 depicts an FESEM micrograph of the CVO nanowires prepared over various hydrothermal reaction periods ranging from 4 h to 16 h. Figure S1(a1–a4) shows FESEM micrographs of CVO_U_4h nanowires obtained after 4 h of hydrothermal reaction time. Because of the formation of nanowires, these micrographs have spherical forms. The nanowires grow from the core of the sphere, with one end linked to the center and the other end free, like flower petals. Each nanowire is independent at the free end, which provides a better environment for electrolyte ion insertion and extraction during the electrochemical supercapacitor charging and discharging process. As the hydrothermal reaction period extends from 4 h to 7 h, some of the nanowires begin to cluster together and join together, as seen in Figure S1(b1–b4). Furthermore, additional CVO nanowires begin to form, influencing the electrode's cyclic voltammetry (CV) and galvanostatic charge–discharge (GCD) profiles. The joining of the nanowires forms a network, which improves electrical conductivity and facilitates ion movement throughout the charging and discharging processes. A distinct, uniform spherical shape with a swarm of linked nanowires is found after the 10 h hydrothermal reaction period. Clusters are formed by small bundles of interconnected nanowires (Figure 3a–d), with numerous bundles connected to one another while leaving vacant spaces between them, as shown in Figure S1(c1–c4). These void regions are advantageous for simple electrolyte ion entrance with few impediments, and their impacts may be seen in the electrochemical performance of the CVO_U_10h electrode. It has a wider area under the CV profile curve and takes longer to charge and discharge. The 10 h hydrothermal reaction time appears to be an important optimized parameter for achieving the desired microstructure of the CVO nanowires. Figure S1(d1–d4) shows an FESEM micrograph for 16 h of hydrothermal reaction, in which the surface microstructure is completely clumped together due to the increased thickness of the nanowires and reduced vacant space between the bundles. As shown in the CV and GCD profiles, the reduction in empty area results in a reduction in energy storage capacity. The schematic shown in Figure 4 depicts the growth mechanism of $Co_3V_2O_8$ nanoparticles during a 10 h hydrothermal reaction based on structural and surface microstructural investigations. Cobalt vanadium nanowires are prepared at the appropriate hydrothermal reaction time of 10 h, and their continued development leads to the formation of nanowire frameworks. Energy-dispersive X-ray spectroscopy (EDS) measurements were used to determine the elemental composition of all CVO nanowires. The EDS spectra of CVO_U_10h nanoparticles prepared at 10 h of hydrothermal reaction time are depicted in Figure 3e, and comparative EDS spectra for all nanoparticles are shown in Figure S2a–d. Sharp EDS peaks indicate the presence of cobalt, vanadium, and oxygen. The X-ray diffraction (XRD) and EDS data show that the intended phase and stoichiometric cobalt vanadium oxide were formed. Figure 3f–h displays the elemental mapping of the CVO_U_10h sample, which demonstrated that the Co, V, and O distributions in the nanoparticles were uniform. The surface microstructure of CVO_U_10h nanowires was investigated using transmission electron microscopy (TEM) to analyze their morphology, structural arrangement, and potential effects on electrochemical performance. Figure 5 displays TEM micrographs of CVO_U_10h nanoparticles at different magnifications, highlighting their distinct nanostructural characteristics.

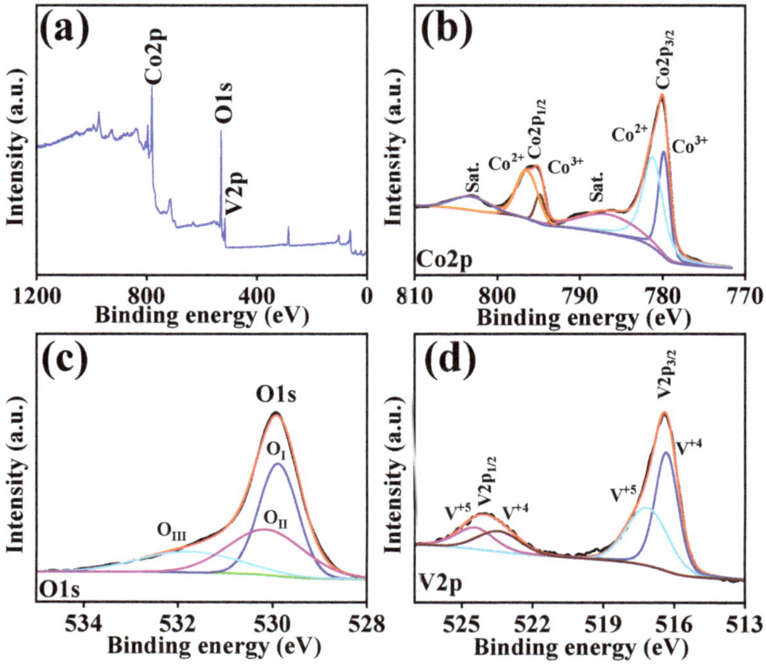

Figure 2. XPS spectra of CVO_U_10h nanoparticles: (**a**) survey, (**b**) cobalt, (**c**) oxygen, and (**d**) vanadium.

Figure 5a illustrates the formation of nanowires, which are observed to interconnect and assemble into bundles. This bundled configuration significantly enhances the network connectivity of the electrode material, thereby improving charge transport and electron mobility in electrochemical processes. The interconnected nanowire structure offers a continuous conductive pathway for electron flow, significantly enhancing the performance of the supercapacitor, especially regarding power density and cycle stability. The bundling effect preserves the mechanical integrity of the electrode and mitigates excessive degradation across multiple charge–discharge cycles. At higher magnifications (Figure 5b–d), the nanowires demonstrate minor splitting at their free ends, resulting in the development of voids or pores among the individual nanowires. The inter-nanowire gaps significantly influence the electrochemical behavior of the material by enhancing the available surface area for electrolyte interaction and facilitating rapid ion transport. The existence of these disparity areas facilitates the efficient diffusion of a substantial number of electrolyte ions into the electrode material, thereby significantly improving its charge storage capacity. Figure 5e,f displays high-resolution TEM images that illustrate the crystal plane spacing (d-spacing) of the CVO_U_10h electrode material. Figure 5e clearly displays well-defined crystal planes, indicating a high degree of crystallinity in the synthesized material. Fast Fourier transform (FFT) analysis (Figure 5h) revealed a d-spacing value of 0.315 nm for CVO_U_10h, aligning with the characteristic lattice spacing of $Co_3V_2O_8$. The crystallographic data offers direct evidence of the ordered atomic structure in the nanowires, essential for ensuring stable electrochemical performance during extended cycling. Figure 5g presents the selected area electron diffraction (SAED) pattern of CVO_U_10h nanoparticles. The distinct ring patterns observed in the SAED image confirm the material's polycrystalline nature. The observed diffraction rings align with various crystallographic planes of $Co_3V_2O_8$, demonstrating strong concordance with the X-ray diffraction (XRD) analysis. The correlation between the SAED and XRD results confirms that the synthesized nanoparticles possess a highly crystalline and polycrystalline structure, conducive to enhanced electrochemical activity. Surface examination of the CVO_U_10h nanowires was investigated using N_2 adsorption/desorption BET studies. Figure S3 depicts the N_2 adsorption/desorption isotherm with a typical Type I-V adsorption and desorption curve. CVO_U_10h nanowires have a surface area of 41.26 m^2/g. The inclusion of tiny nanoballs with empty spaces may give an enormous surface area, allowing for improved interaction between the electrode and the electrolyte. Tiny nanoparticles provide a large surface area to enhance the energy storage capacity of the CVO_U_10h electrode when compared to other electrodes.

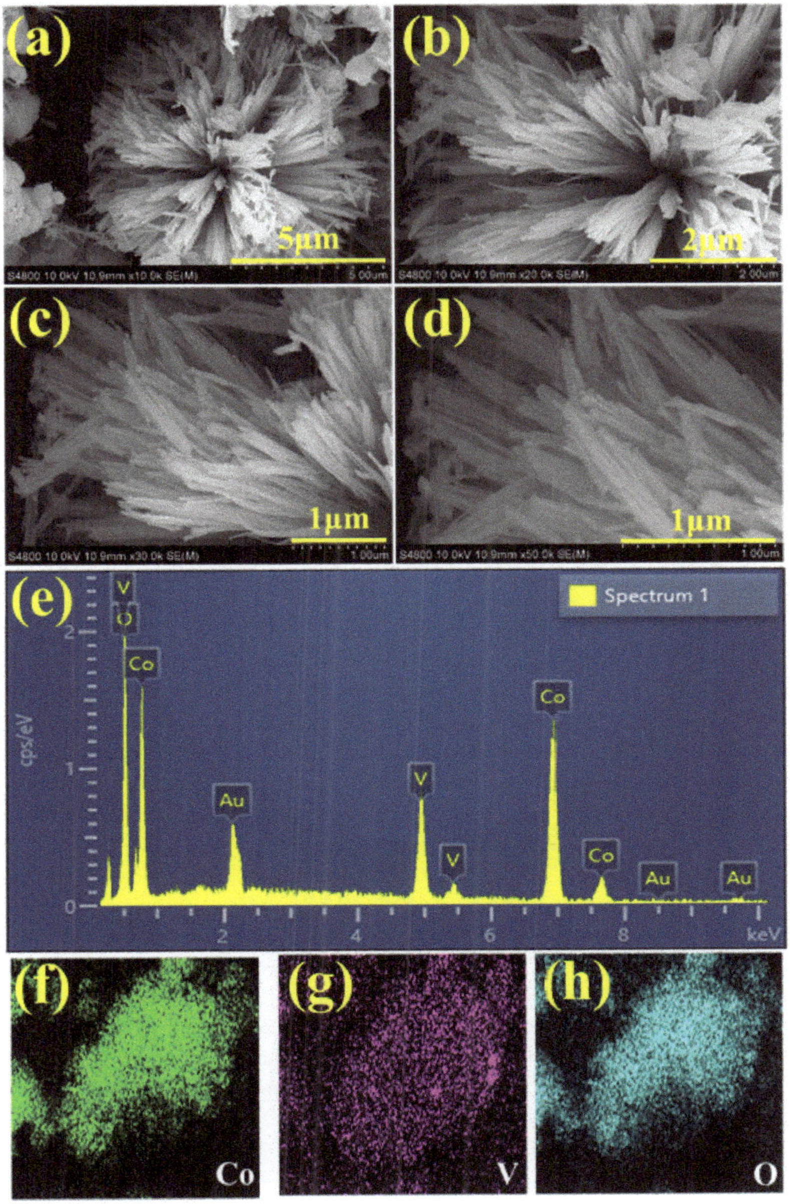

Figure 3. (**a–d**) FESEM micrographs at different magnifications, (**e**) EDS spectra, and (**f–h**) elemental mapping of CVO_U_10h nanowires.

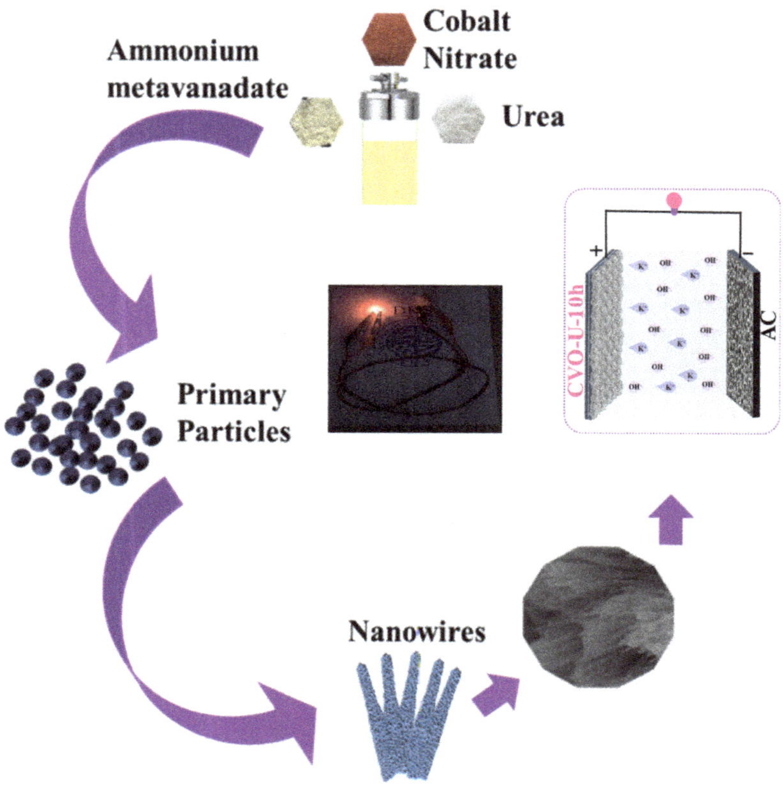

Figure 4. Schematic growth mechanism of $Co_3V_2O_8$ nanoparticles prepared using a hydrothermal approach.

3.2. Electrochemical Performance and Kinetics Analyses of the $Co_3V_2O_8$ Nanoparticles

The electrochemical analysis of all the CVO_U electrodes was carried out in a 2M KOH electrolyte using cyclic voltammetry and galvanostatic charge–discharge measurements. Figure S4 depicts the CV profiles of all CVO_U electrodes. The area under the curve grows with increasing hydrothermal reaction time, as shown in Figure 6a. The electrode's energy storage capacity rises with increasing area. All electrodes show distinct redox peaks, demonstrating charge storage via pseudocapacitive behavior. The redox peaks shift as the scan rate increases, indicating fast and reversible redox kinetics, a characteristic of pseudocapacitive behavior [52]. As shown by the CV profile, the hydrothermal reaction period influences the surface microstructure of CVO_U nanowires. Nanowires grow after 4 h of hydrothermal processing, giving additional active sites for redox reactions. With a 7-h reaction, nanowires link sideways, increasing interaction, as can be seen in the CV profile of the CVO_U_10h electrode, which has an expanded area in the CV profile curve, as shown in Figure S4c. Nanowire bundles with a large active surface area provide a grown conjunction and a large active surface area for a 10-h hydrothermal reaction. Figure 6a depicts the area of the CV profile impacted by the hydrothermal reaction time. When compared to the other electrodes, the area under the curve for the 10-h hydrothermal reaction was larger, indicating superior energy storage capabilities. In a 16-h reaction, however, the nanowire size increases, resulting in a loss in active surface area and storage capacity. The electrode prepared in a 10-h hydrothermal reaction has a void space with connected nanowire CVO nanoparticles which provide a high energy storage capacity, making it an attractive electrode material for energy storage applications. The CV profiles of all electrodes were employed to examine the kinematics of their electrochemical characteristics in more depth. The charge distribution was calculated using the 'b' value, which reflects two different charge storage processes, namely, capacitive and diffusive processes. In this case, 'b' = 1.0 for the capacitive process and 'b' = 0.5 for the diffusive process. Charge storage via diffusion and capacitive processes is indicated by values of 0.5 < b < 1.0. The following equations were used to calculate the b values from the CV profiles:

$$i = a \times v^b \tag{5}$$

$$\log(i) = b \cdot \log(v) + \log(a) \tag{6}$$

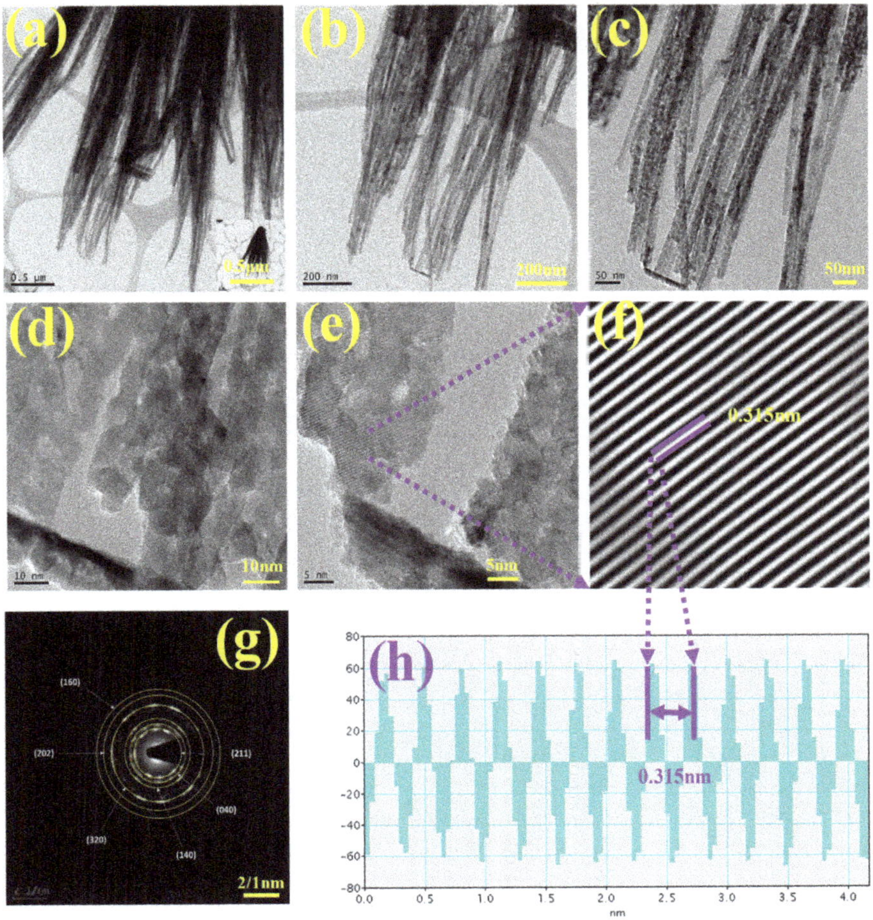

Figure 5. (**a–d**) TEM at different magnifications, (**e**,**f**) d-spacing image, (**g**) SEAD pattern, (**h**) FFT pattern for d-spacing CVO_U_10h nanoparticle.

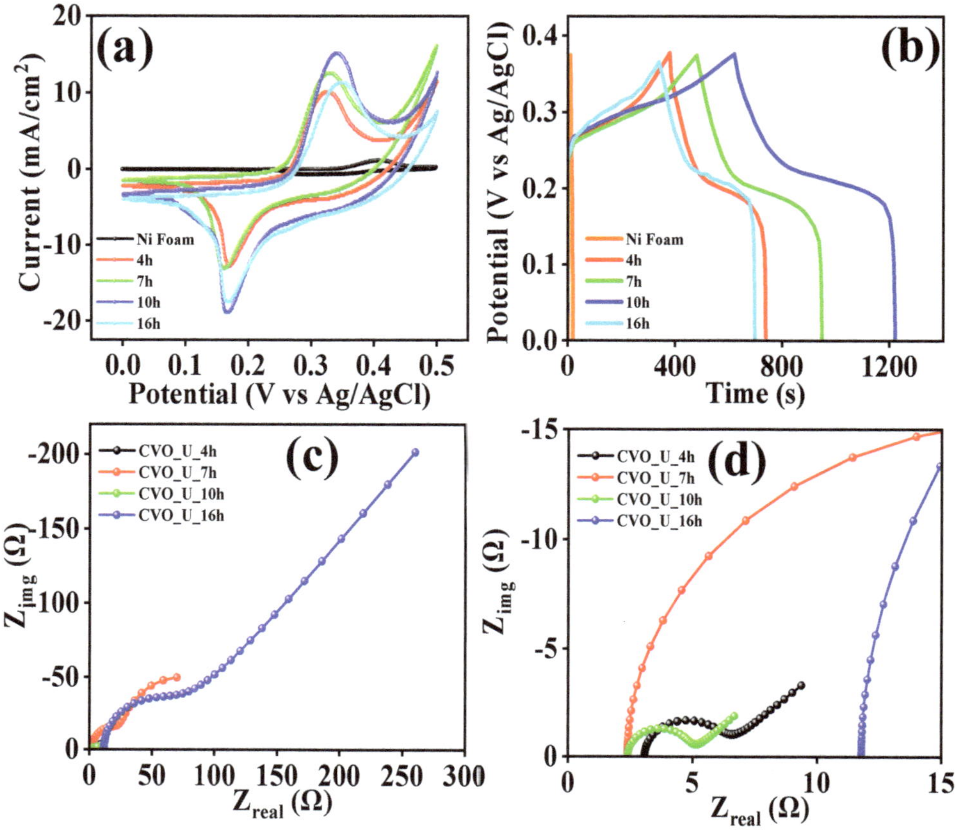

Figure 6. (**a**) Comparative CV profile, (**b**) comparative GCD profile, and (**c**,**d**) EIS spectra of $Co_3V_2O_8$ nanowires prepared at different hydrothermal reaction times.

The values were calculated using a log(i) vs. log(v) plot, as shown in Figure 7a,b, where 'v' is the scan rate and 'i' is the peak current corresponding to the scan rate. The 'b' values for the CVO_U_10h electrode are 0.44 and 0.39 for the cathode and anode, suggesting that the diffusion process dominates in the charge process. This might be ascribed to the bundle-like microstructures containing void spaces, which allow electrolyte ions to easily diffuse during the charge and discharge processes [53,54]. The computation of the b value reveals the combined capacitive and diffusion processes. At constant potential, the total current in a CV profile is caused by pseudo capacitance from diffusion-controlled contributions and double-layer capacitance from surface-controlled contributions. These contributions are calculated using the formulae below.

$$i(V) = k_c v + k_d v^{0.5};\ \frac{i(V)}{v^{0.5}} = k_c / v^{0.5} + k_d \quad (7)$$

Here, $k_c/v^{0.5}$ denotes capacitive current contributions and k_d denotes diffusion-controlled contributions. The values of k_c and k_d were calculated using Equation (7) with different scan rates. Figure 7c depicts the capacitive and diffusion-controlled current contributions of the CVO_U_10h electrode at a scan rate of 5 mV/s. The 'b' value of the CVO_U_10h electrode shows a larger diffusion contribution at 5 mV/s, implying a larger diffusion contribution, presumably owing to the bundle nanowire with empty space providing a higher diffusion contribution. Figure 7d shows the capacitive and diffusion-controlled contributions at various scan rates, with the diffusion-controlled contribution decreasing as the scan rate increases [53]. The 'b' values were calculated to demonstrate that the diffusion process outperforms the capacitive process, and the current contribution at constant potential also shows a current owing to the diffusion-controlled contribution. This indicates that the diffusion coefficient for the related electrode may be higher. The diffusion coefficients were computed using the cyclic voltammetry (CV) profile spectra provided in Figure 8a,b and Equation (8). The diffusion coefficient of the CVO_U_10h electrode is higher than that of the other electrode, measuring 4.2×10^{-7} cm^2/S for the anodic curve and 4.5×10^{-7} cm^2/S for the cathodic curve, respectively. The existence of vacuum spaces between the nanowire bundles may account for the high diffusion coefficient, allowing for effortless insertion and extraction of ions during the electrochemical reaction process. This demonstrates that the diffusion contribution for the CVO_U_10h electrode is higher than the capacitive contribution, and these results are compatible with the 'b' value and current contribution. The diffusion coefficient values are listed in Table 1.

$$i_p = 0.4463 \times A \times F \times C \times \sqrt{\frac{nFDv}{RT}} \quad (8)$$

$$i_p = 0.227AFCnk^0\left[-\left(\frac{nF\alpha}{RT}\right)\left(E_p - E^0\right)\right] \quad (9)$$

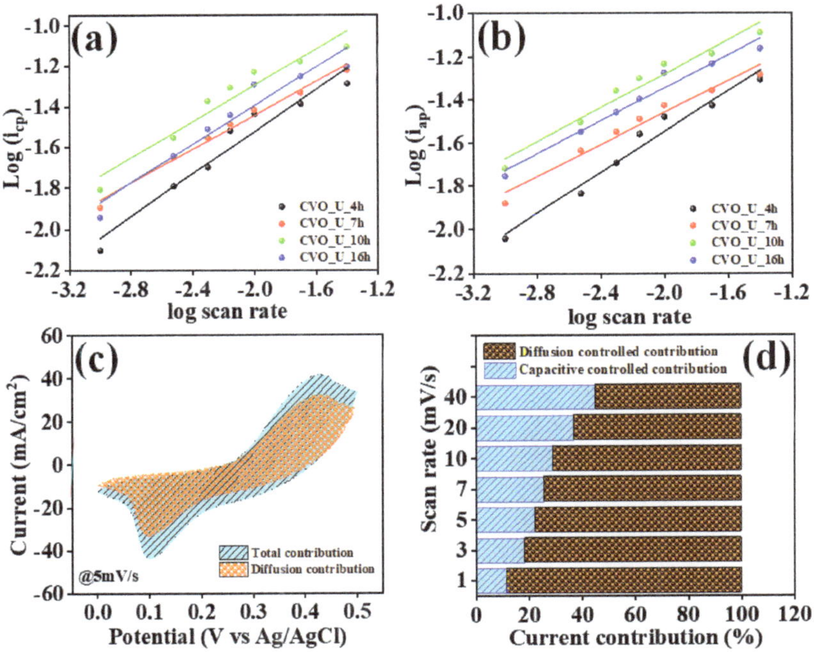

Figure 7. (**a**,**b**) Log(i) vs. log (v) for b values, (**c**) diffusive and capacitive current contributions at a 5 mV/s scan rate, and (**d**) diffusive and capacitive current contributions at different scan rates for CVO_U_10h nanoparticles.

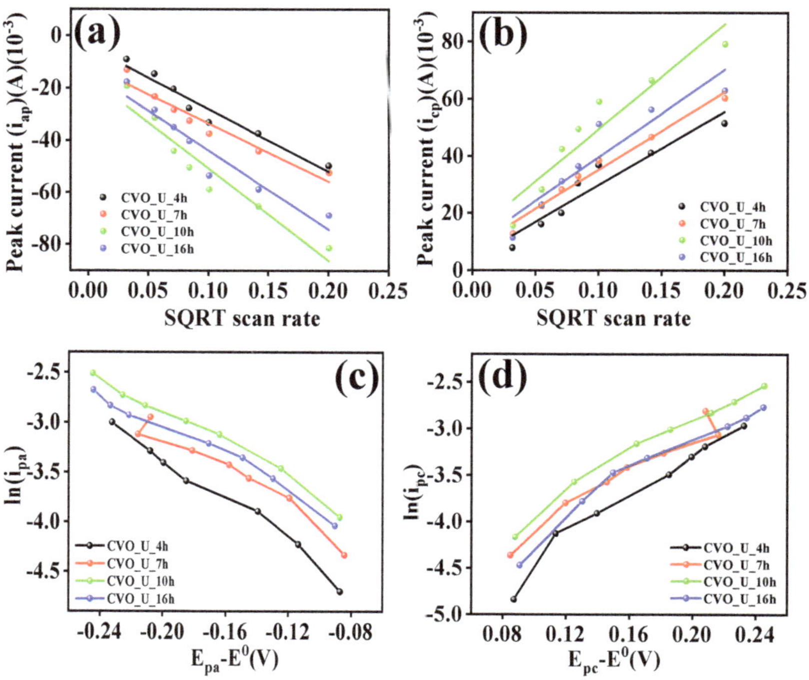

Figure 8. (**a**,**b**) Peak current (i_p) vs. SQRT of v for the diffusion coefficient, (**c**,**d**) ln (i) vs. E_p-E^0 for the transfer coefficient and standard rate constant.

Table 1. Transfer coefficients, diffusion coefficients, and standard rate constants of all $Co_3V_2O_8$ electrodes.

Sample Code	Transfer Coefficient (α)		Diffusion Coefficient (cm^2/S) $\times 10^{-7}$		Standard Rate Constant (k^0) (cm/S) $\times 10^{-5}$	
	Reduction	Oxidation	Reduction	Oxidation	Reduction	Oxidation
CVO_U_4h	0.30	0.27	1.98	2.30	7.49	6.58
CVO_U_7h	0.27	0.24	1.67	2.54	12.9	11.2
CVO_U_10h	0.25	0.22	4.26	4.51	18.6	14.2
CVO_U_16h	0.26	0.22	3.14	3.38	18.6	10.8

Here, A represents the electrode area, C represents the electrolyte concentration, v represents the scan rate, R represents the universal gas constant, T represents the absolute temperature, F represents Faraday's constant, D represents the diffusion constant, α represents the transfer coefficient, n represents the number of moles participating in the chemical reaction, E_p represents the peak potential, and E^0 represents the formal potential. The transfer rate constant (k^0) was explored to analyze electrochemical kinetics and establish the nature of the reaction process, whether reversible, irreversible, or quasi-reversible. As illustrated in Figure 8c,d, the k^0 values were determined using the cyclic voltammetry (CV) profile using the peak current and corresponding potential. Depending on the value of k^0, it may be feasible to identify whether the reaction is irreversible, reversible, or quasi-reversible. $k^0 > 10^{-1}$ for reversible chemical processes and $k^0 < 10^{-5}$ for irreversible chemical reactions. Chemical reactions with $10^{-1} < k^0 < 10^{-5}$ are categorized as quasi-reversible. All electrodes have k^0 values ranging between 10^{-1} and 10^{-5}, implying that the charging and discharging processes are quasi-reversible. The transfer coefficient α is computed and listed in Table 1, and outcomes vary between 0.22 and 0.30, implying that the reaction is quasi-reversible. The areal capacitance, areal energy density, and areal power density of all CVO_U electrodes were calculated from the GCD profile and Equations (10)–(12). Figure S5 shows the GCD profiles of all the electrodes. The charging and discharging times of all electrodes were altered in accordance with the surface microstructural changes induced by the hydrothermal reaction time. The surface microstructure of the nanowires exhibited a shorter discharging time when compared to the other electrode, as shown in Figure 6b. The nanowire microstructure demonstrated a progressive expansion as the hydrothermal reaction time reached 7 hours, resulting in extended charging and discharging durations, potentially attributed to the development of a wire network through lateral joining. When the hydrothermal reaction time is increased to 10 h, the nanowires begin to form bundles, and empty spaces arise between the bundles. The void spaces between the bundles are beneficial for charging and discharging the ions, and their effects were observed in the GCD profile shown in Figure 6a. The CVO_U_10h electrode showing the longer charging and discharging times demonstrated excellent areal energy density and power density. For the hydrothermal reaction time of 16 h, the thickness of the nanowires increased and the void spacing between the bundles decreased, decreasing the charging and discharging times. Overall, the electrode prepared with a 10 h hydrothermal reaction shows excellent charging and discharging times, possibly due to the uniform nanowire microstructure with bundle arrangement and void spaces between the bundles. The areal capacitance, areal energy density, and power density of the electrode were calculated using Equations (10)–(12) [55,56]. The CVO_U_10h electrode shows an areal capacitance of 4.67 F/cm^2 at a current density of 3 mA/cm^2, with an energy density of 94 μWh/cm^2 and a power density of 573 μW/cm^2. The rest of the electrodes areal capacitances, energy densities, and power densities are listed in Table 2. In an electrochemical cell, electrochemical impedance spectroscopy was employed for assessing solution resistance (R_s) and charge transfer resistance (R_{ct}). R_s is the solution resistance originating primarily from ion transfer resistance through an interface between the electrolyte and electrode material, estimated using a Nyquist plot at the low-frequency intercept on the x-axis. R_{ct} is the charge transfer resistance measured from the semicircle diameter in the EIS spectra. Figure 6c,d illustrates the EIS spectra of all CVO-U electrodes in the frequency range of 10^5 to 10^{-1} Hz. In the EIS investigation, the impact of hydrothermal reaction time on surface morphological tailoring was observed. At first, nanowires were observed in FESEM, and the electrode exhibited an R_s of 3.01 Ω and an R_{ct} of 3.91 Ω, which might be attributable to the nanowire microstructure of the CVO_U_4h electrode. Once the reaction period extended to 7 h, the surface microstructure was tailored to establishing nanowires that were connected laterally to each other, and some nanowires formed looping structures, resulting in a network-like microstructure. The series resistance of the electrode fell to 2.09 Ω owing to lateral linking; however, R_{ct} rose, probably due to the connection of the nanowire tips in the CVO_U_7h electrode. This impact was found in the GCD analysis, particularly the capacitance study. Following a 10-h hydrothermal reaction duration, nanowires continued to form bundles with the highest lateral linking and void spacing between bundles, and all wire tips were free. The series resistance (R_s) and charge transfer resistance (R_{ct}) of the CVO_U_10h electrode decreased to 2.01 Ω and 2.24 Ω, respectively, due to substantial lateral linking and free ends of the nanowires, and this impact was noticed in the GCD profile due to an increase in charging and discharging times. The tiny values of R_s and R_{ct} may improve the electrode's charging and discharging time. As the hydrothermal reaction time rose to 16 h, dense bundles with increasing nanowire thickness were identified, affecting ion diffusion, as well as R_s and R_{ct} values, and originating in a drop in specific capacitance in the GCD profile.

The CVO_U_10h electrode possessed the lowest R_s and R_{ct} values compared to the other electrodes, which might be due to laterally connected nanowires producing nano bundles with void spacing and demonstrating good supercapacitor performance. The values of R_s and R_{ct} are listed in Table 2.

$$\text{Areal Capacitance } (C_a) = \frac{T_d \times I_d}{dV \times A} \tag{10}$$

$$\text{Areal Energy Density } (ED_a) = \frac{C_a \times dV^2}{7200} \tag{11}$$

$$\text{Areal Power Density } (PD_a) = \frac{ED_a \times 3600}{T_d} \tag{12}$$

Table 2. Areal capacitance, energy density, power density, series resistance, and charge transfer resistance of $Co_3V_2O_8$ electrodes.

Sample Code	Areal Capacitance (C_a) F/cm^2	Areal Energy Density (ED_a) $\mu Wh/cm^2$	Areal Power Density (PD_a) $\mu W/cm^2$	R_s (Ω)	R_{ct} (Ω)
CVO_U_4h	2.86	56	567	3.01	3.91
CVO_U_7h	3.82	77	570	2.07	8.98
CVO_U_10h	4.67	94	573	2.02	2.21
CVO_U_16h	2.76	56	570	11.8	53

Here, C_a is the areal capacitance, T_d is the discharging time, I_d is the current density, dV is the potential window of the GCD profile, A is the area of the electrode, ED_a is the area energy density, and PD_a is the areal power density. The EIS spectra employed frequency-dependent real and imaginary impedance [22]. The following algorithms were employed to determine real and imaginary capacitance values:

$$C' = \frac{Z''(\omega)}{\omega|Z(\omega)|^2} \tag{13}$$

$$C'' = \frac{Z'(\omega)}{\omega|Z(\omega)|^2} \tag{14}$$

Here, $Z(\omega)$ denotes complex impedance, $Z'(\omega)$ is the real part of complex impedance, and $Z''(\omega)$ is the imaginary part of complex impedance and represents the 'f' frequency. Figure S6 illustrates C' and C'', which represent real and imaginary capacitance. The impact of the surface microstructure on C' and C'' was observed. The electrode CVO_U_10h exhibited the highest values for both C and C'' compared to other electrodes. This might be attributed to laterally linked nanowires with void spacing between nanowire bundles.

3.3. Electrochemical Properties of the CVO_U_10h//AC ASC Device

To examine the actual application of the designed CVO_U_10h electrode, an asymmetric supercapacitor with two electrodes, one coated with CVO_U_10h and a second with activated carbon, was fabricated. The electrode method of preparation was exactly the same for both electrodes, with the exception of the CVO_U_10h electrode, which used $Co_3V_2O_8$ nanoparticles as the active material, while the other electrode used activated carbon. The electrode arrangement was encased in paraffin paper, and the electrolyte was 2M KOH. The two electrodes were coated in paraffin paper and separated by filter paper. Figure 9g depicts a schematic of the two electrodes. Cyclic voltammetry (CV), the galvanostatic charge–discharge (GCD), and electrochemical impedance spectroscopy (EIS) were employed to look into the chemical characteristics of asymmetric supercapacitor (ASC) electrodes. The potential window of the ASC was maintained at 1.6V using the CV profile of CVO_U_10 and AC from Figure S7a. The plots for the CV, GCD, and EIS of AC are presented in Figure S7b, S7c, and S7d, respectively. Figure 9a depicts the CV profile of the ASC at various potentials ranging from 0.8 to 1.6 V. The potential window was raised to 1.6 V without significant polarization, and this potential was retained for future use. Figure 9b also shows the CV profile of the CVO_U_10h//AC at various scan rates (20–100 mV/s). GCD profiles of the CVO_U_10h//AC ASC at various potentials are displayed in Figure 9d,c, including 8 mA/cm^2 current density and various current densities with a constant potential of 1.6V. The areal capacitance (C_a), areal energy density (ED_a), and power density (PD_a) of the CVO_U_10h//AC ASC were calculated using GCD profiles at various current densities. Using Equations (10)–(12), the volumetric capacitance (C_v) was calculated to be 1.43 F/cm^3, and the volumetric energy density (ED_v) was found to be 511 μWh/cm^3 at a 40 mW/cm^3 volumetric power density (PD_v) and an areal capacitance of 230 mF/cm^2, with an areal energy density and power density 81 μWh/cm^3 at 6.4 mW/cm^3 of at an 8 mA/cm^2 current density. To calculate the volumetric capacitance, the volume of the CVO-U-10h//AC ASC was used instead of the area of the electrode. The obtained results are compared with the reported study in Table 3. Figure 9e depicts the electrochemical impedance spectroscopy of the CVO_U_10h//AC ASC before and after 15 K GCD cycles of stability testing. The CVO_U_10h//AC ASC's R_s value was 7.09 Ω before stability and grew to 11.03 Ω after stability. Figure 9f shows cyclic stability over 15 K cycles. The capacity retention of the CVO_U_10h//AC ASC electrode was reported to be 83.40% following stabilization. Overall, the CVO_U_10h//AC ASC electrode, constructed from $Co_3V_2O_8$ nanoparticles in a 10-h hydrothermal process, demonstrates good electrochemical performance in a three-electrode configuration. The ASC proves to be an efficient electrode material for supercapacitor applications. The practical performance of the CVO_U_10h electrode with two electrode configurations with multiple LEDs or a single LED glowing, respectively, is shown in Figure 9h,i.

Table 3. Comparison of CVO-U-10h//AC ACS with previous studies.

Sr. No.	Electrode Material	Areal Capacitance	Areal Energy Density	Areal Power Density	Reference
1.	CVO-U-10h//AC	230 mF/cm^2	81 μWh/cm^2	6.4 μW/cm^2	Present work
2.	NiO@N-C/CC//Fe_2O_3@N-C/CC	127.1 mF/cm^2	32 μWh/cm^2	0.8 μW/cm^2	[57]
3.	MoS_2-graphene bilayer/textile	63.73 mF/cm^2	35.40 μWh/cm^2	8.4 μW/cm^2	[58]
4.	$BiMn_xO_y$ ‖ PVA-KOH ‖ Bi_2MoO_6	52 mF/cm^2	18.5 μWh/cm^2	978.7 μW/cm^2	[59]
5.	V_2O_5//C	34.68 mF/cm^2	1.73 μWh/cm^2	0.072 μW/cm^2	[60]
6.	NP/MnO2700//AC	32.8 mF/cm^2	18.2 μWh/cm^2	65.6 μW/cm^2	[61]
7.	V_2O_5@PEDOT/graphene	22.4 mF/cm^2	0.18 μWh/cm^2	11.0 μW/cm^2	[62]

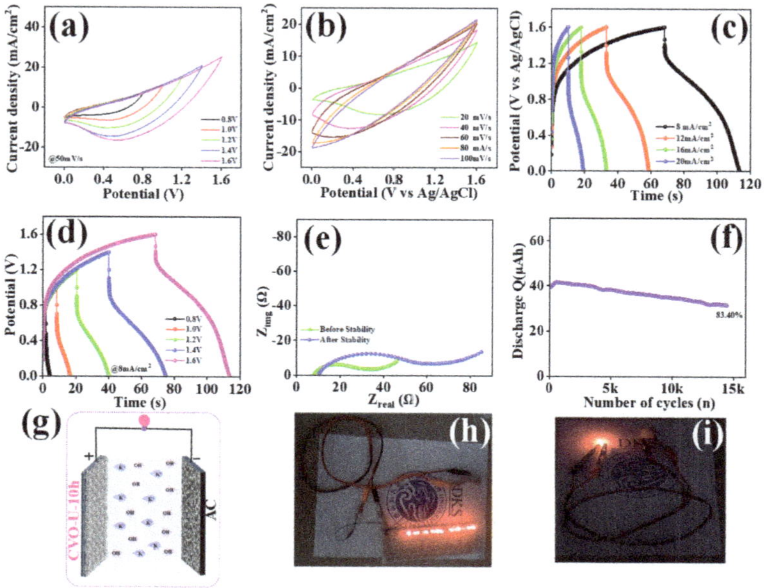

Figure 9. (**a**) CV profile at different potential windows with a constant scan rate, (**b**) CV profile at different scan rates with a constant potential window, (**c**) GCD profile at different current densities with a fixed potential window, (**d**) GCD profile at a fixed current density with different potential windows, (**e**) EIS spectra before and after stability over 15 K cycles, (**f**) cyclic stability for 15 K GCD cycles, (**g**) schematic representation of the ASC using CVO_U_10h and AC, (**h**,**i**) red LED glowing using the ASC of CVO_U_10h//AC.

3.4. Electrocatalysis Study of the CVO_U_10h Electrode

Figure 10a,b presents the linear sweep voltammetry (LSV) plot of the HER study as well as the LSV plot of the OER study, in addition to investigating the electrocatalytic properties of the CVO_U_10h electrode used to investigate the hydrogen evolution reaction (HER) and the oxygen evolution reaction (OER) in a 1M KOH electrolyte.The CVO_U_10h electrode exhibits an overpotential of 259 mV at a current density of 10 mA/cm^2 for HER and 578 mV at 30 mA/cm^2 for OER. The nanowire morphology with vacuum spaces, allowing for a larger number of active sites, might be responsible for the smaller overpotential. Tafel plots were generated employing the LSV profile in order to investigate the reaction kinetics of the CVO_U_10h electrode. Figure 10c,d shows the Tafel plots for HER and OER, respectively. The Tafel slope of the CVO_U_10h electrode for HER is 176 mV/dec, and it is 389 mV/dec for OER. The comparative study of the electrocatalytic performance of $Co_3V_2O_8$ nanoparticles is summarized in Table 4. Figure 10e,f shows the electrochemical impedance spectroscopy of the CVO_U_10h electrode for HER and OER. For HER and OER, the series resistance of the CVO_U_10h electrode is 3.74 Ω and 3.56 Ω, respectively, while the charge transfer resistance is 12.2 Ω and 1.54 Ω, respectively. The nanowire microstructure and lateral connections, which form a network with low resistance, are credited with the lower series resistance. As a result of its nanowire structure, the CVO_U_10h electrode has an outstanding overpotential and an acceptable Tafel slope. Figure S8a depicts total water splitting utilizing the CVO_U_10h electrode as both a cathode and anode for hydrogen and oxygen evolution. According to the LSV plot, the total potential required for an overall current density of 10 mA/cm^2 is 1.79 V, rising to 1.82 V after 8 h of stability. The long-term durability of the CVO_U_10h electrode was investigated using chronopotentiometry for 8 h at a current density of 10 mA/cm^2, as shown in Figure S8b. Overall, findings illustrate that CVO nanoparticles developed in a 10-h hydrothermal reaction function well as supercapacitors in both three-electrode and two-electrode configurations. Furthermore, the CVO_U_10h electrode exhibits excellent electrocatalytic activity, confirming CVO nanoparticles produced via a 10-h hydrothermal process are a suitable material for supercapacitor and electrocatalytic applications.

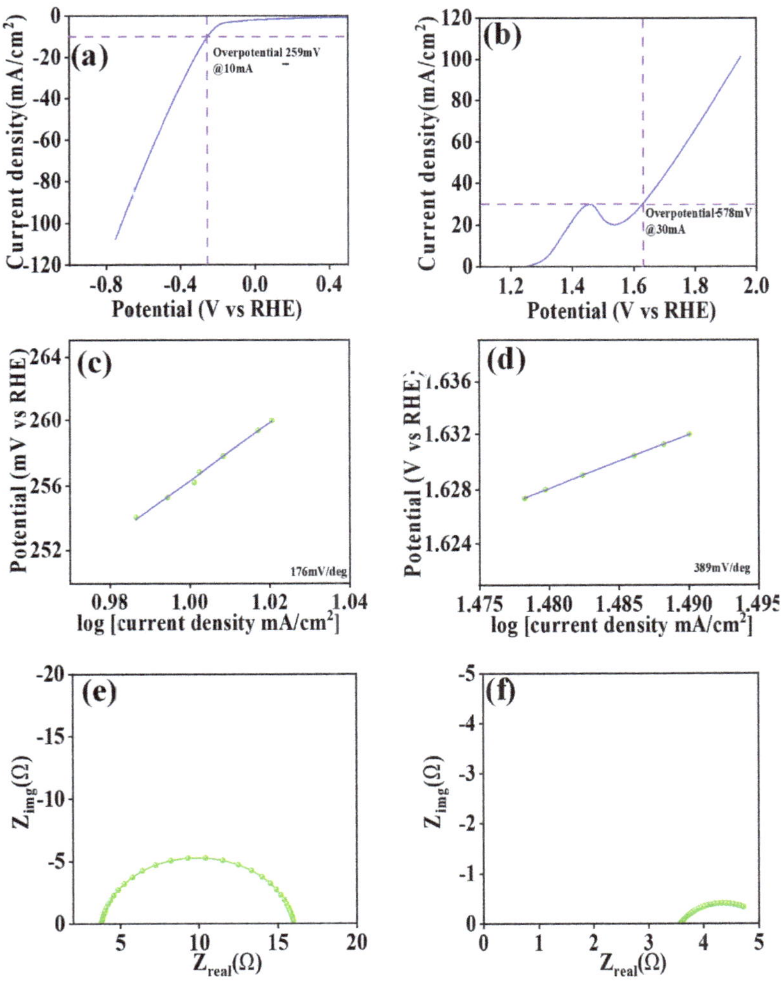

Figure 10. (**a**,**b**) LSV plot, (**c**,**d**) Tafel plot, (**e**,**f**) EIS spectra of OER and HER for the CVO_U_10h electrode.

Table 4. HER and OER Comparative study of CVO-U-10h electrode.

Electrode	Overpotential mV (HER)	Tafel mV/dec (HER)	Overpotential mV (OER)	Tafel mV/dec (OER)	Reference
$Co_3V_2O_8$	-	-	190	85	[63]
$CoO-V_2O_5$	297	159	429	175	[64]
$V_{8.3}$-Co-MOF	293	156	435	78	[65]
$Co_3V_2O_3$	278	180	318	130	[64]
$Co_3V_2O_8$	-	-	359	65	[66]
$V_{16.7}$-Co-MOF	271	146	413	77	[65]
CVO_U_10h	259	176	578 (30 mA/cm^2)	389	Present work

4. Conclusions

In conclusion, using the hydrothermal approach, nanowires of $Co_3V_2O_8$ nanoparticles were prepared. At 10-h hydrothermal reaction time, the laterally linked nanowires form clusters of nanowire bundles with void spaces between them. The prepared nanowire bundles with void spacing have a considerable areal capacitance of 4.67 F/cm^2, with energy density and power density of 94 µWh/cm^2 and 573 µW/cm^2, respectively, at 3 mA/cm^2 current density. Activated carbon (AC) and $Co_3V_2O_8$ nanoparticles prepared with a 10-h hydrothermal reaction time were employed to make an asymmetric supercapacitor (ASC). The device has a volumetric capacitance of 1.43 F/cm^3 and an areal capacitance of 781 mF/cm^2 in its two-electrode assembly. The ASC has an areal energy density of 81 µWh/cm^2 and an areal power density of 6.4 mW/cm^2, as well as a volumetric energy density of 511 µWh/cm^3 and a volumetric power density of 40 mW/cm^3 at 8 mA/cm^2 current density. Furthermore, electrocatalysis examinations show that electrocatalysis performs effectively in the hydrogen evolution process (HER), the oxygen evolution reaction (OER), and total water splitting. HER electrocatalysis reveals a 259 mV overpotential at a current density of 10 mA/cm^2, with a Tafel slope of 176 mV/dec. For OER, there is a 578 mV overpotential with a Tafel slope of 389 mV/dec at a current density of 30 mA/cm^2. The $Co_3V_2O_8$ electrode prepared with a 10-h hydrothermal reaction time acquires 1.78 V for overall water splitting, which rises marginally to 1.8 V after an 8-h stability period. This work demonstrates that nanowire bundles with void spaces prepared with a 10-h reaction time are an acceptable electrode material for energy storage in supercapacitors, as well as HER and OER electrocatalysis applications.

Supplementary Materials: The following supporting information can be downloaded at: https://www.mdpi.com/article/10.3390/batteries11040118/s1, Figure S1: FESEM micrographs of (a1–a4) CVO_U_4h, (b1–b4) CVO_U_7h, (c1–c4) CVO_U_10h, and (d1–d4) CVO_U_16h nanowires at different magnifications; Figure S2: EDS spectra of (a) CVO_U_4h, (b) CVO_U_7h, (c) CVO_U_10h, and (d) CVO_U_16h nanowires; Figure S3: BET nitrogen adsorption isotherm plot of CVO_U_10h nanowires; Figure S4: CV profiles of (a) CVO_U_4h, (b) CVO_U_7h, (c) CVO_U_10h, and (d) CVO_U_16h nanowires; Figure S5: (a) GCD profiles of (a) CVO_U_4h, (b) CVO_U_7h, (c) CVO_U_10h, and (d) CVO_U_16h nanowires; Figure S6: Real and imaginary capacitance of (a) CVO_U_4h, (b) CVO_U_7h, (c) CVO_U_10h, and (d) CVO_U_16h electrodes; Figure S7: (a) CV profile of the CVO-U-10h and activated carbon, (b) CV profile at different scan rates, (c) GCD profile at current density, and (d) EIS spectra of activated carbon; Figure S8: (a) LSV plot before and after stability, (b) stability test at 10 mA/cm^2 current density.

Author Contributions: Conceptualization, M.A.Y.; Supervision, D.K.S. All authors have read and agreed to the published version of the manuscript.

Funding: This work was supported by the Technology Innovation Program (#20010170), funded by the Ministry of Trade, Industry & Energy (MOTIE, Korea).

Data Availability Statement: The raw data supporting the conclusions of this article will be made available by the authors on request.

Conflicts of Interest: The authors declare no conflict of interest.

References

1. Lu, Z.; Hu, Z.; Xiao, L.; Xie, Y.; Li, N.; Xi, L.; Chen, W.; Xiao, J ; Zhu, Y. Battery-type Ni-Co-Se hollow microspheres cathode materials enabled by bifunctional N-doped carbon quantum dots with ultrafast electrochemical kinetics for hybrid supercapacitors. *Chem. Eng. J.* **2022**, *450*, 138347. [CrossRef]
2. Wan, L.; Wang, Y.; Jiang, D.; Zhang, Y.; Xie, M.; Chen, J.; Du, C. Constructing nickel sulfide @ nickel boride hybrid nanosheet arrays with crystalline/amorphous interfaces for supercapacitors. *J. Colloid Interface Sci.* **2023**, *649*, 815–825. [CrossRef] [PubMed]

3. Hu, L.; Li, M.; Wei, X.; Wang, H.; Wu, Y.; Wen, J.; Gu, W.; Zhu, C. Modulating interfacial electronic structure of CoNi LDH nanosheets with $Ti_3C_2T_x$ MXene for enhancing water oxidation catalysis. *Chem. Eng. J.* **2020**, *398*, 125605. [CrossRef]
4. Wu, H.; Zhang, X.; Xue, J.; Zhang, H.; Yang, L.; Li, S. Engineering active sites on hierarchical ZnNi layered double hydroxide architectures with rich Zn vacancies boosting battery-type supercapacitor performances. *Electrochim. Acta* **2021**, *374*, 137932. [CrossRef]
5. Lv, X.W.; Tian, W.W.; Yuan, Z.Y. Recent Advances in High-Efficiency Electrocatalytic Water Splitting Systems. *Electrochem. Energy Rev.* **2023**, *6*, 23. [CrossRef]
6. Liu, S.; Yang, J.; Chen, P.; Wang, M.; He, S.; Wang, L.; Qiu, J. Flexible Electrodes for Aqueous Hybrid Supercapacitors: Recent Advances and Future Prospects. *Electrochem. Energy Rev.* **2024**, *7*, 25. [CrossRef]
7. Wang, G.; Zhang, L.; Zhang, J. A review of electrode materials for electrochemical supercapacitors. *Chem. Soc. Rev.* **2012**, *41*, 797–828. [CrossRef]
8. Brousse, T.; Bélanger, D.; Long, J. To Be or Not To Be Pseudocapacitive? *J. Electrochem. Soc.* **2015**, *162*, A5185–A5189. [CrossRef]
9. Lan, Y.; Zhao, H.; Zong, Y.; Li, X.; Sun, Y.; Feng, J.; Wang, Y.; Zheng, X.; Du, Y. Phosphorization boosts the capacitance of mixed metal nanosheet arrays for high performance supercapacitor electrodes. *Nanoscale* **2018**, *10*, 11775–11781. [CrossRef]
10. Teli, A.M.; Bhat, T.S.; Beknalkar, S.A.; Mane, S.M.; Chaudhary, L.S.; Patil, D.S.; Pawar, S.A.; Efstathiadis, H.; Cheol Shin, J. Bismuth manganese oxide based electrodes for asymmetric coin cell supercapacitor. *Chem. Eng. J.* **2022**, *430*, 133138. [CrossRef]
11. Thirumal, V.; Yuvakkumar, R.; Senthil Kumar, P.; Ravi, G.; Velauthapillai, D. Direct growth of multilayered graphene nanofibers by chemical vapour deposition and their binder-free electrodes for symmetric supercapacitor devices. *Prog. Org. Coat.* **2021**, *161*, 106511. [CrossRef]
12. Rawat, P.; Kumar Sharma, P.; Malik, V.; Umapathi, R.; Kaushik, N.; Rhyee, J.S. Emergence of high-performing and ultra-fast 2D-graphene nano-biosensing system. *Mater. Lett.* **2022**, *308*, 131241. [CrossRef]
13. Tan, Y.B.; Lee, J.M. Graphene for supercapacitor applications. *J. Mater. Chem. A* **2013**, *1*, 14814–14843. [CrossRef]
14. Mao, B.S.; Wen, Z.; Bo, Z.; Chang, J.; Huang, X.; Chen, J. Hierarchical nanohybrids with porous CNT-networks decorated crumpled graphene balls for supercapacitors. *ACS Appl. Mater. Interfaces* **2014**, *6*, 9881–9889. [CrossRef]
15. Dubal, D.P.; Chodankar, N.R.; Kim, D.H.; Gomez-Romero, P. Towards flexible solid-state supercapacitors for smart and wearable electronics. *Chem. Soc. Rev.* **2018**, *47*, 2065–2129. [CrossRef]
16. Lin, J.; Xu, Y.; Wang, J.; Zhang, B.; Li, L.; Wang, X.; He, S.; Zhu, J. A pH-Tailored Anodic Deposition of Hydrous RuO_2 for Supercapacitors. *ChemistrySelect* **2019**, *4*, 8122–8128. [CrossRef]
17. Snook, G.A.; Kao, P.; Best, A.S. Conducting-polymer-based supercapacitor devices and electrodes. *J. Power Sources* **2011**, *196*, 1–12. [CrossRef]
18. Sethi, M.; Bhat, D.K. Facile solvothermal synthesis and high supercapacitor performance of $NiCo_2O_4$ nanorods. *J. Alloys Compd.* **2019**, *781*, 1013–1020. [CrossRef]
19. Yewale, M.A.; Kadam, R.A.; Kaushik, N.K.; Vattikuti, S.V.P.; Lingamdinne, L.P.; Koduru, J.R.; Shin, D.K. Hydrothermally synthesized microrods and microballs of NiCo2O4 for supercapacitor application. *Ceram. Int.* **2022**, *48*, 22037–22046. [CrossRef]
20. Deokate, R.J.; Kalubarme, R.S.; Park, C.J.; Lokhande, C.D. Simple Synthesis of $NiCo_2O_4$ thin films using Spray Pyrolysis for electrochemical supercapacitor application: A Novel approach. *Electrochim. Acta* **2017**, *224*, 378–385. [CrossRef]
21. Rajasekhara Reddy, G.; Ramesh Reddy, N.; Dillip, G.R.; Joo, S.W. In Situ Construction of Binder-Free Stable Battery-Type Copper Cobaltite and Copper Oxide Composite Electrodes for All-Solid-State Asymmetric Supercapacitors: Cation Concentration and Morphology-Dependent Electrochemical Performance. *Energy Fuels* **2022**, *36*, 5965–5978. [CrossRef]
22. Yewale, M.A.; Kadam, R.A.; Nakate, U.T.; Teli, A.M.; Kumar, V.; Beknalkar, S.A.; Jadhavar, A.A.; Kadam, S.L.; Shelke, N.T.; Shin, D.K. Sphere-shaped $CuCo_2O_4$ nanostructures battery type electrode for supercapacitor via hydrothermal synthesis approach. *Colloids Surf. A Physicochem. Eng. Asp.* **2023**, *679*, 132541. [CrossRef]
23. Alqahtani, D.M.; Zequine, C.; Ranaweera, C.K.; Siam, K.; Kahol, P.K.; Poudel, T.P.; Mishra, S.R.; Gupta, R.K. Effect of metal ion substitution on electrochemical properties of cobalt oxide. *J. Alloys Compd.* **2019**, *771*, 951–959. [CrossRef]
24. Yewale, M.A.; Jadhvar, A.A.; Kharade, R.B.; Kadam, R.A.; Kumar, V.; Nakate, U.T.; Shelke, P.B.; Bobade, D.H.; Teli, A.M.; Dhas, S.D.; et al. Hydrothermally synthesized $Ni_3V_2O_8$ nanoparticles with horny surfaces for HER and supercapacitor application. *Mater. Lett.* **2023**, *338*, 134033. [CrossRef]
25. Low, W.H.; Khiew, P.S.; Lim, S.S.; Siong, C.W.; Ezeigwe, E.R. Facile solvothermal designing of graphene/$Ni_3V_2O_8$ nanocomposite as electrode for high performance symmetric supercapacitor. *J. Alloys Compd.* **2018**, *768*, 995–1005. [CrossRef]
26. Liu, M.C.; Kong, L.B.; Kang, L.; Li, X.; Walsh, F.C.; Xing, M.; Lu, C.; Ma, X.J.; Luo, Y.C. Synthesis and characterization of $M_3V_2O_8$ (M = Ni or Co) based nanostructures: A new family of high performance pseudocapacitive materials. *J. Mater. Chem. A* **2014**, *2*, 4919–4926. [CrossRef]
27. Kadam, R.A.; Yewale, M.A.; Teli, A.M.; Annu; Nakate, U.T.; Kumar, V.; Kadam, S.L.; Shin, D.K. Bimetallic $Co_3V_2O_8$ microstructure: A versatile bifunctional electrode for supercapacitor and electrocatalysis applications. *Surf. Interfaces* **2023**, *41*, 103267. [CrossRef]

28. Luo, Z.; Liu, L.; Yang, X.; Luo, X.; Bi, P.; Fu, Z.; Pang, A.; Li, W.; Yi, Y. Revealing the Charge Storage Mechanism of Nickel Oxide Electrochromic Supercapacitors. *ACS Appl. Mater. Interfaces* **2020**, *12*, 39098–39107. [CrossRef]
29. Jadhav, A.L.; Jadhav, S.L.; Mandlekar, B.K.; Kadam, A.V. Effects of various molarities of nickel oxide on the aggregate 1D–3D structure and its electrochemical activity. *J. Alloys Compd.* **2022**, *925*, 166716. [CrossRef]
30. Dam, D.T.; Lee, J.M. Three-dimensional cobalt oxide microstructures with brush-like morphology via surfactant-dependent assembly. *ACS Appl. Mater. Interfaces* **2014**, *6*, 20729–20737. [CrossRef]
31. Ji, Y.; Fang, D.; Wang, C.; Zhou, Z.; Luo, Z.; Huang, J.; Yi, J. Cobalt-doped V_2O_5 nanowire arrays on Ti foil for enhanced lithium-ion storage. *J. Alloys Compd.* **2018**, *742*, 567–576. [CrossRef]
32. Cao, Y.; Yan, L.; Gang, H.; Wu, B.; Wei, D.; Wang, H. Large gap cobalt-vanadium oxide structure encapsulated in porous carbon for high performance capacitive deionization. *Sep. Purif. Technol.* **2023**, *306*, 122709. [CrossRef]
33. Saasa, V.; Malwela, T.; Lemmer, Y.; Beukes, M.; Mwakikunga, B. The hierarchical nanostructured Co-doped WO_3/carbon and their improved acetone sensing perfomance. *Mater. Sci. Semicond. Process.* **2020**, *117*, 105157. [CrossRef]
34. Liao, M.; Liu, Y.; Hu, Z.; Yu, Q. Novel morphologic Co_3O_4 of flower-like hierarchical microspheres as electrode material for electrochemical capacitors. *J. Alloys Compd.* **2013**, *562*, 106–110. [CrossRef]
35. Chen, J.; Wu, X.; Selloni, A. Electronic structure and bonding properties of cobalt oxide in the spinel structure. *Phys. Rev. B—Condens. Matter Mater. Phys.* **2011**, *83*, 245204. [CrossRef]
36. Gao, Z.; Yun, S.; Yang, C.; Zhang, Y.; Dang, J.; Yang, G.; Yang, T.; Qiao, D.; Wang, K. Niobium- and cobalt-modified dual-source-derived porous carbon with a honeycomb-like stable structure for supercapacitor and hydrogen evolution reaction. *J. Colloid Interface Sci.* **2023**, *639*, 33–48. [CrossRef]
37. Laursen, A.B.; Kegnæs, S.; Dahl, S.; Chorkendorff, I. Molybdenum sulfides—Efficient and viable materials for electro—And photoelectrocatalytic hydrogen evolution. *Energy Environ. Sci.* **2012**, *5*, 5577–5591. [CrossRef]
38. Simon, P.; Gogotsi, Y.; Dunn, B. Where do batteries end and supercapacitors begin? *Science* **2014**, *343*, 1210–1211. [CrossRef]
39. Liu, C.; Li, F.; Lai-Peng, M.; Cheng, H.M. Advanced materials for energy storage. *Adv. Mater.* **2010**, *22*, 28–62. [CrossRef]
40. Vrubel, H.; Moehl, T.; Grätzel, M.; Hu, X. Revealing and accelerating slow electron transport in amorphous molybdenum sulphide particles for hydrogen evolution reaction. *Chem. Commun.* **2013**, *49*, 8985–8987. [CrossRef]
41. Wang, H.; Zhang, H.; Zhang, D.; Chen, J.; Zhang, S.; Zhang, S.; Yu, J.; Wu, Q.; Li, Q. Toward Enhanced Electrochemical Performance by Investigation of the Electrochemical Reconstruction Mechanism in $Co_2V_2O_7$ Hexagonal Nanosheets for Hybrid Supercapacitors. *ACS Appl. Mater. Interfaces* **2022**, *14*, 8106–8114. [CrossRef] [PubMed]
42. Sun, H.; Chen, X.; Chai, H.; Wang, Y.; Jia, D.; Cao, Y.; Liu, A. 3D porous hydrated cobalt pyrovanadate microflowers with excellent cycling stability as cathode materials for asymmetric supercapacitor. *Appl. Surf. Sci.* **2019**, *469*, 118–124. [CrossRef]
43. Sharma, G.P.; Gupta, P.K.; Sharma, S.K.; Pala, R.G.S.; Sivakumar, S. Chalcogenide Dopant-Induced Lattice Expansion in Cobalt Vanadium Oxide Nanosheets for Enhanced Supercapacitor Performance. *ACS Appl. Energy Mater.* **2021**, *4*, 4758–4771. [CrossRef]
44. Thorat, G.M.; Jadhav, H.S.; Roy, A.; Chung, W.J.; Seo, J.G. Dual Role of Deep Eutectic Solvent as a Solvent and Template for the Synthesis of Octahedral Cobalt Vanadate for an Oxygen Evolution Reaction. *ACS Sustain. Chem. Eng.* **2018**, *6*, 16255–16266. [CrossRef]
45. Singh, D.K.; Yadav, M.; Ganesan, V.; Bhobe, P.A. Cox(VO)yOzNanocrystal-Integrated Covalent Organic Polymers as a Highly Active and Durable Catalyst for Electrochemical Water Oxidation: An Untold Role of the VO^{2+}/VO^{2+} Redox Couple. *ACS Appl. Energy Mater.* **2022**, *5*, 2805–2816. [CrossRef]
46. Han, J.; Liu, X.; Wan, H.; Wu, D.; Chen, G.; Li, J.; Cao, Y.; Ma, R. Composition Tuning of Ultrafine Cobalt-Based Spinel Nanoparticles for Efficient Oxygen Evolution. *ACS Sustain. Chem. Eng.* **2020**, *8*, 5534–5543. [CrossRef]
47. Fahimi, Z.; Moradlou, O. High-performance solid-state asymmetric supercapacitor based on $Co_3V_2O_8$/carbon nanotube nanocomposite and gel polymer electrolyte. *J. Energy Storage* **2022**, *50*, 104697. [CrossRef]
48. Mushtaq, M.W.; Imran, M.; Bashir, S.; Kanwal, F.; Mitu, L. Synthesis, structural and biological studies of cobalt ferrite nanoparticles. *Bulg. Chem. Commun.* **2016**, *48*, 565–570.
49. Samuel, V.; Navale, S.C.; Jadhav, A.D.; Gaikwad, A.B.; Ravi, V. Synthesis of ultrafine $BiMnO_3$ particles at 100 °C. *Mater. Lett.* **2007**, *61*, 1050–1051. [CrossRef]
50. Zhang, Q.; Pei, J.; Chen, G.; Bie, C.; Chen, D.; Jiao, Y.; Rao, J. $Co_3V_2O_8$ Hexagonal Pyramid with Tunable Inner Structure as High Performance Anode Materials for Lithium Ion Battery. *Electrochim. Acta* **2017**, *238*, 227–236. [CrossRef]
51. Teng, Y.; Li, Y.; Yu, D.; Meng, Y.; Wu, Y.; Zhao, X.; Liu, X. The Microwave-Assisted Hydrothermal Synthesis of CoV_2O_6 and $Co_3V_2O_8$ with Morphology Tuning by pH Adjustments for Supercapacitor Applications. *ChemistrySelect* **2019**, *4*, 956–962. [CrossRef]
52. Zheng, L.; Xiao, Z.; Wang, X.; Xi, Y.; Zhai, S.; Fu, X.; Hao, J.; An, Q.; Yang, D. Biomass-Assisted Construction of Carbon-Supported Fe–Co–Cu Trimetallic Oxides/Sulfides for Supercapacitors with Excellent Performance. *ACS Appl. Energy Mater.* **2023**, *6*, 10862–10873. [CrossRef]

53. Chu, W.; Shi, Z.; Hou, Y.; Ma, D.; Bai, X.; Gao, Y.; Yang, N. Trifunctional of Phosphorus-Doped $NiCo_2O_4$ Nanowire Materials for Asymmetric Supercapacitor, Oxygen Evolution Reaction, and Hydrogen Evolution Reaction. *ACS Appl. Mater. Interfaces* **2020**, *12*, 2763–2772. [CrossRef] [PubMed]
54. Kulurumotlakatla, D.K.; Raghavendra, K.V.G.; Vinodh, R.; Saeed, G.; Muralee Gopi, C.V.V. Facile hydrothermal synthesis of copper oxide microspheres decorated with nickel sulfide nanoparticles battery-type electrode materials for high-energy-density hybrid supercapacitors. *J. Energy Storage* **2025**, *110*, 115327. [CrossRef]
55. Patil, A.M.; Moon, S.; Jadhav, A.A.; Hong, J.; Kang, K.; Jun, S.C. Modifying Electronic Structure of Cation-Exchanged Bimetallic Sulfide/Metal Oxide Heterostructure through In Situ Inclusion of Silver (Ag) Nanoparticles for Extrinsic Pseudocapacitor. *Adv. Funct. Mater.* **2023**, *33*, 2305264. [CrossRef]
56. Patil, A.M.; Wang, J.; Li, S.; Hao, X.; Du, X.; Wang, Z.; Hao, X.; Abudula, A.; Guan, G. Bilateral growth of monoclinic WO_3 and 2D $Ti_3C_2T_x$ on 3D free-standing hollow graphene foam for all-solid-state supercapacitor. *Chem. Eng. J.* **2021**, *421*, 127883. [CrossRef]
57. Cai, D.; Du, J.; Zhu, C.; Cao, Q.; Huang, L.; Wu, J.; Zhou, D.; Xia, Q.; Chen, T.; Guan, C.; et al. Iron Oxide Nanoneedles Anchored on N-Doped Carbon Nanoarrays as an Electrode for High-Performance Hybrid Supercapacitor. *ACS Appl. Energy Mater.* **2020**, *3*, 12162–12171. [CrossRef]
58. Islam, M.R.; Afroj, S.; Karim, N. Scalable Production of 2D Material Heterostructure Textiles for High-Performance Wearable Supercapacitors. *ACS Nano* **2023**, *17*, 18481–18493. [CrossRef]
59. Balamurugan, K.; Velmurugan, R.; Subramanian, B. Effect of in situ annealing on pulsed laser ablated mixed metal oxide (BixMyOz; M=Mn, Mo) thin film electrodes for flexible hybrid supercapacitor devices. *Mater. Today Chem.* **2024**, *35*, 101908. [CrossRef]
60. Zhou, H.; Zheng, S.; Guo, X.; Gao, Y.; Li, H.; Pang, H. Ordered porous and uniform electric-field-strength micro-supercapacitors by 3D printing based on liquid-crystal V_2O_5 nanowires compositing carbon nanomaterials. *J. Colloid Interface Sci.* **2022**, *628*, 24–32. [CrossRef]
61. Mohanty, R.I.; Mukherjee, A.; Bhanja, P.; Jena, B.K. Novel microporous manganese phosphonate-derived metal oxides as prospective cathode materials for superior flexible asymmetric micro-supercapacitor device. *J. Energy Storage* **2023**, *72*, 108730. [CrossRef]
62. Wang, L.; Shu, T.; Guo, S.; Lu, Y.; Li, M.; Nzabahimana, J.; Hu, X. Fabricating strongly coupled V_2O_5@PEDOT nanobelts/graphene hybrid films with high areal capacitance and facile transferability for transparent solid-state supercapacitors. *Energy Storage Mater.* **2020**, *27*, 150–158. [CrossRef]
63. Gyanprakash, D.M.; Gupta, P.K.; Sharma, G.P.; Pala, R.G.S. Surface-enhanced OER activity in $Co_3V_2O_8$ using cyclic charge-discharge to balance electrocatalytic active site generation and degradation. *Electrochim. Acta* **2021**, *367*, 137538. [CrossRef]
64. Yang, Z.; Chen, H.; Xiang, M.; Yu, C.; Hui, J.; Dong, S. Coral reef structured cobalt-doped vanadate oxometalate nanoparticle for a high-performance electrocatalyst in water splitting. *Int. J. Hydrogen Energy* **2022**, *47*, 31566–31574. [CrossRef]
65. Fang, W.; Dang, J.; Hu, Y.; Wu, Y.; Xin, S.; Chen, B.; Zhao, H.; Li, Z. Electronic distribution tuning of vanadium-cobalt bimetallic MOFs for highly efficient hydrazine-assisted energy-saving hydrogen production. *Electrochim. Acta* **2023**, *439*, 141682. [CrossRef]
66. Xing, M.; Kong, L.B.; Liu, M.C.; Liu, L.Y.; Kang, L.; Luo, Y.C. Cobalt vanadate as highly active, stable, noble metal-free oxygen evolution electrocatalyst. *J. Mater. Chem. A* **2014**, *2*, 18435–18443. [CrossRef]

batteries

MDPI

Review

From Thermosetting Resins to Energy Devices: A Review on Polybenzoxazine-Derived Materials for Supercapacitors

Shakila Parveen Asrafali †, Thirukumaran Periyasamy † and Jaewoong Lee *

Department of Fiber System Engineering, Yeungnam University, Gyeongsan 38541, Republic of Korea; shakilaparveen@yu.ac.kr (S.P.A.); thirukumaran@ynu.ac.kr (T.P.)
* Correspondence: jaewlee@yu.ac.kr
† These authors contributed equally to this work.

Abstract

Polybenzoxazines (PBZs) have garnered significant attention as a versatile class of precursors for the development of advanced carbon-based materials, particularly in the field of electrochemical energy storage. This review comprehensively examines recent progress in the synthesis, structural design, and application of polybenzoxazine-derived materials for supercapacitor electrodes. Owing to their intrinsic nitrogen content, tunable functionality, and excellent thermal and mechanical stability, polybenzoxazines serve as ideal precursors for producing nitrogen-doped porous carbons with high surface areas and desirable electrochemical properties. This review discusses the influence of molecular design, polymerization conditions, and carbonization parameters on the resulting microstructure and performance of the materials. Furthermore, the electrochemical behavior of these materials in both electric double-layer capacitors (EDLCs) and pseudocapacitors is analyzed in detail. Challenges such as optimizing pore architecture, improving conductivity, and achieving scalable synthesis are also addressed. This article highlights emerging trends and offers perspectives on the future development of polybenzoxazine-derived materials for next-generation high-performance supercapacitors.

Keywords: polybenzoxazine; porous carbon; composites; enhanced supercapacitor performance

1. Introduction

The heavy dependence on traditional fossil fuels has caused significant environmental issues. Consequently, there is an urgent demand for clean energy conversion and storage solutions to enhance energy security and build sustainable, carbon-free power systems, which are essential steps toward achieving global carbon neutrality. Despite advancements, fossil fuels remain a dominant energy source, contributing to both energy shortages and environmental concerns. Researchers are increasingly investing in advanced materials to address the growing need for sustainable and renewable energy solutions. The scope covers advancements in flexible energy storage devices like capacitors, supercapacitors, and batteries, as well as energy conversion systems such as solar panels, fuel cells, thermoelectric modules, and wind energy technologies [1–7]. Notably, fuel cells and supercapacitors are gaining attention as prominent candidates for future energy storage and conversion solutions. Their high efficiency, near-zero greenhouse gas emissions, and ability to support clean, smokeless transportation make them particularly attractive. As dependence on fossil fuels persists, the demand for eco-friendly energy solutions—such as electrochemical supercapacitors and fuel cells—continues to rise. Additionally, the increasing global reliance on electronic and electrical devices has intensified scientific and technological interest in energy storage and renewable energy innovations [8–11].

Supercapacitors (SCs), or electrochemical capacitors, have emerged as promising energy storage devices due to their rapid charging/discharging, high power density, long cycle life, and excellent safety profile. Commonly used in electric vehicles, renewable energy systems, and aerospace, SCs complement or replace traditional batteries in various applications by offering lightweight design, fast response times, and extended operational life (>10^5 cycles) [12–15]. SCs are broadly classified into electrical double-layer capacitors (EDLCs) and pseudocapacitors (PCs). EDLCs store charge electrostatically using carbon-based materials like carbon nanotubes, graphene, and activated carbon. PCs, in contrast, utilize Faradaic redox reactions with materials such as conductive polymers or transition metal compounds to achieve higher capacitance. While SCs offer much higher power density (>10 $kW \cdot kg^{-1}$) than batteries, their energy density (~10 $W \cdot h \cdot kg^{-1}$) remains lower. To improve this, research focuses on optimizing electrode materials, electrolytes, and operating voltages [16–18], since specific energy (E) is given by Equation (1):

$$E = \frac{1}{2} \cdot C \cdot V^2 \quad (1)$$

A typical SC consists of two carbon-based electrodes, a separator, and an aqueous or organic electrolyte. Due to the cost of conventional carbon materials, heteroatom-doped porous carbons—especially those rich in nitrogen and oxygen—have gained attention as low-cost alternatives. These materials enhance electrostatic attraction and charge storage through surface doping, increasing capacitance and enabling high-performance systems [19–21]. To bridge the gap between energy and power densities, supercapacitors combine EDLC and pseudocapacitive mechanisms. This integration maximizes energy storage while retaining fast response and stability, making SCs a promising next step in supercapacitor technology [22–24].

Carbon materials, particularly activated carbon, are widely used as electrode components in electrical double-layer capacitors due to their high surface area and conductivity. Carbon particles, ranging in size from a few nanometers to micrometers, have gained widespread attention across various fields, including energy storage and conversion, biomedicine, catalysis, and adsorption [25–27]. Their controllable size, high surface-area-to-volume ratio, engineered morphology, and tunable composition make them highly versatile. These physicochemical properties can be precisely tailored through the careful selection of precursors and synthetic strategies. As a result, in the past decade, significant research efforts have been devoted to designing novel molecular precursors to develop functional carbon materials with enhanced properties [28–30]. Polymer precursors have emerged as promising candidates for deriving carbon frameworks due to their well-defined molecular architecture, versatile chemical composition, and ability to retain morphological integrity during carbonization. Their adaptability allows for the fabrication of carbon materials with tailored properties, making them attractive for various applications. Over the years, several polymer precursors—such as polyimide, poly(ether imide), polystyrene, polyaniline, and polyacrylonitrile—have been widely explored for their potential in carbon material synthesis [31,32]. Phenolic resins, such as resorcinol formaldehyde (RF) and phenol formaldehyde (PF) are among the most widely used precursors for synthesizing porous carbon spheres, particularly due to their high carbon yield, controllable cross-linking chemistry, excellent thermal stability and easy compatibility with templating or surfactant-assisted synthesis routes [33,34]. More recently, nitrogen-rich or heteroatom-doped carbon materials have been successfully obtained using an in situ process with polymers such as polyamines, polyamides, and polyacrylonitrile [35–37]. These nitrogen-doped carbon materials exhibit enhanced physicochemical properties, further expanding their application potential.

However, many of these polymers fail to achieve high carbon yields due to molecular fragmentation during carbonization [38,39].

As the core component of supercapacitors, electrode materials play a crucial role in determining performance. The most commonly used electrode materials include porous carbon materials, metal oxides, and conductive polymers. Among these, metal oxides exhibit the highest specific capacitance but suffer from high costs and poor cycle stability. Similarly, conductive polymers offer excellent capacitance and conductivity but are limited by poor thermal stability and mechanical deformability, restricting their practical applications. In contrast, porous carbon materials have gained significant attention due to their exceptional electrical and thermal conductivity, high-temperature resistance, and strong corrosion resistance, making them highly suitable for energy storage applications [40–43]. Polybenzoxazine (PBZ) has recently emerged as an advanced class of phenolic thermosetting polymers, synthesized via the ring-opening polymerization of the oxazine moiety in benzoxazine monomers (Figure 1). This process occurs through thermal curing without requiring a catalyst or curing agent, allowing PBZs to form inter- and intramolecular hydrogen bonds. Due to their unique structural and chemical properties, PBZs are widely utilized in coatings, low-dielectric materials, and aerospace applications. The use of amine precursors in benzoxazine synthesis enriches the resulting polymer network with nitrogen heteroatoms, making PBZs highly suitable for producing heteroatom-doped carbon materials. PBZs also offer several advantageous characteristics, including molecular design flexibility, high char yield (35–80 wt.%), high cross-linking density, and excellent dimensional stability with minimal volume shrinkage. These materials exhibit remarkable thermal resistance, high glass transition temperatures (T_g), and flame-retardant properties, making them one of the most promising polymeric precursors for carbon material synthesis [44–47]. To obtain heteroatom-doped carbon from PBZ precursors, various advanced synthetic approaches have been employed, including soft templating via organic–organic self-assembly, hard templating, and modified sol–gel methods. Additionally, carbon materials derived from PBZs—such as foams, nanofibers, nanospheres, aerogels, and xerogels—have been synthesized using carbonization and graphitization techniques [48–51]. These materials possess controlled pore distribution, pseudocapacitive properties, and high charge stability, making them highly relevant for energy storage applications, particularly in supercapacitors and batteries [52,53].

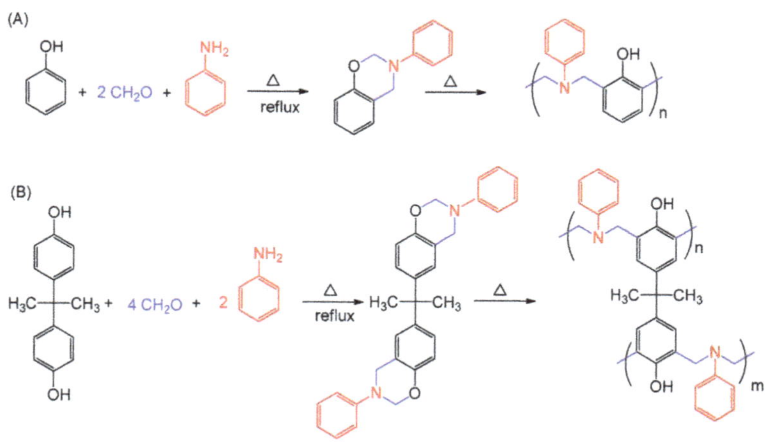

Figure 1. Synthesis and heat-triggered ring-opening polymerization of BZ monomers: (**A**) type P-a and (**B**) type B-a.

This review details the use of PBZs in the development of advanced carbon materials for supercapacitors, highlighting the role of structured polymer frameworks in energy storage applications. Additionally, it explores the combination of PBZ-based materials with metal oxides and conductive polymers to create hybrid structures that bridge the gap between capacitors and batteries, offering high power density along with improved energy density. This review discusses the results of the literature reports in several sections. The first section deals about the supercapacitor performance of novel benzoxazines and their copolymers synthesized using several raw materials such as eugenol, apigenin, vanillin, melamine, furfurylamine, tetraethylene pentaamine, and boric acid. This section highlights the importance of several dopants into the carbon framework, like nitrogen, oxygen, boron and phosphorous in enhancing the supercapacitor performance. In the next section, the role of metal oxides along with PBZ carbon is discussed in detail. The formation of PBZ carbon with metal oxides—Ni, Mn, Co and bimetaloxides—NiMn and NiCo and the effect of carbonization temperature and redox active electrolytes in enhancing the electrochemical performance have been discussed in detailed. The third sections deal with PBZ composites with CNTs, graphitic carbon nitride, silica and graphene oxide. Here, the structure stability of the composites and the enhanced conductivity of the composites in enhancing the supercapacitor performance have been discussed. Finally, it concludes with a discussion on challenges, future perspectives, and potential avenues for improving performance through hybridization, scalable synthesis, and structural design innovations.

2. Supercapacitor Performance of Novel Benzoxazines and Their Copolymers

Recent research highlights the promise of novel benzoxazine-based materials and their copolymers in enhancing the performance of supercapacitors. By tailoring monomer structures or blending them into copolymers, enhanced electrochemical properties, such as higher capacitance, better cycle life, and increased energy density could be achieved. Thirukumaran et al. [38] introduced nitrogen-enriched mesoporous carbon ropes (NMCRs), produced from polybenzoxazine—an advanced thermosetting polymer—as active electrode materials for supercapacitors. The benzoxazine monomer was synthesized using eugenol and tetraethylenepentamine. The NMCRs were generated via straightforward carbonization followed by activation using aqueous KOH. Their research highlighted the significant influence of the nitrogen-rich polybenzoxazine precursor's composition on boosting electrochemical performance. The final NMCRs contained notable nitrogen (5.26 wt.%) and oxygen (7.31 wt.%) levels [38], attributed due to the intrinsic heteroatoms present in the polybenzoxazine [54,55].

The Brunauer–Emmett–Teller (BET) surface analysis revealed that the NMCRs possessed a surface area of approximately 300 m^2/g, with an average pore size of 3.0 nm and a pore volume of 0.003 cm^3/g. High-resolution imaging revealed a unique rope-like porous morphology, which contributes to the enhanced surface area and offers additional electrochemically active sites [56,57]. Elemental mapping confirmed the distribution of carbon, nitrogen, and oxygen elements within the NMCR structure (Figure 2). Enhanced electrochemical behavior was achieved through modifications to both the surface properties and electrical conductivity of the carbon material [58,59]. Performance testing indicated a specific capacitance of 60 F/g in a 2 M KOH solution under a current density of 1 A/g. This improved result is attributed to the material's distinct porous framework and excellent conductivity.

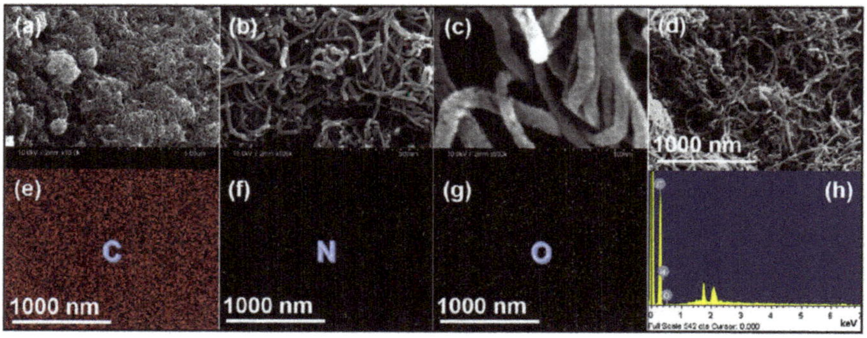

Figure 2. (**a–d**) FESEM images of N-MCRs at varying magnifications [(**a**): 5 µm, (**b**): 500 nm, (**c**): 100 nm and (**d**): 1000 nm]; (**e–g**) elemental distribution maps highlighting C, N, and O presence; (**h**) EDX spectrum analysis of the synthesized N-MCRs. Reproduced from [38].

In a related study by Thirukumaran et al. [60], a cost-efficient and straightforward method (involving Mannich condensation, thermal polymerization, carbonization and activation) was introduced to synthesize nitrogen-doped mesoporous carbon using a novel polybenzoxazine precursor. Two carbon structures were fabricated from the same precursor, made from apigenin, furfuryl amine, and formaldehyde. The product obtained through thermal treatment was named APFC-N (apigenin-furfurylamine-based carbon obtained through normal carbonization), while the one prepared via an aerogel route was termed APFC-G (apigenin-furfurylamine-based carbon obtained through gelation). Nitrogen adsorption–desorption studies revealed that both materials exhibited characteristics of type I and IV isotherms, indicating a considerable presence of mesopores—particularly pronounced in APFC-G [61]. BET analysis showed surface areas of 248 m^2/g for APFC-N and 635 m^2/g for APFC-G. APFC-G displayed a narrow pore size distribution between 20–50 nm, as well as notable macropores with voids larger than 0.5 µm, making it well suited for electrochemical use (Figure 3a,b). Electrochemical evaluation of APFC-G, which was activated in KOH at 900 °C, revealed a peak specific capacitance of 120 F/g in a 1 M H_2SO_4 electrolyte at a current density of 0.5 A/g (Figure 4). The high performance is credited to its oxygen- and nitrogen-rich surface and large surface area [62,63]. Impressively, it retained excellent stability, with no meaningful loss in capacitance observed over 25,000 charge–discharge cycles, highlighting its long-term durability. Moreover, the APFC-G electrode displays a low solution resistance of 2.2 Ω, indicating good electronic conductivity.

Thubsuang et al. [64] synthesized carbon microspheres based on polybenzoxazine (PBZ) using a simple preparation method (Mannich polycondensation) that utilized a mixture of formaldehyde and dimethylformamide (DMF) as solvents. Successful formation of PBZ microspheres was achieved at F/DMF weight ratios of 0.4 and 0.6. Upon carbonization, the materials displayed a high nitrogen content, and structural analysis confirmed an amorphous nature with regions of partial graphitization. Notably, the carbon derived from a 0.4 F/DMF ratio showed a superior specific capacitance of 275.1 F g^{-1}, outperforming the reference carbon (PBZ carbon synthesized using pure DMF) (198.9 F g^{-1}) when tested at 0.05 A g^{-1}. This performance boost was linked to a synergistic effect of electric double-layer capacitance and pseudocapacitance, the latter being enhanced by nitrogen and oxygen surface groups [65,66]. After CO_2 activation, the specific surface area of the 0.4 ratio sample increased markedly from 349 m^2 g^{-1} to 859 m^2 g^{-1}, resulting in a further rise in capacitance to 424.7 F g^{-1}—more than double that of the reference. These results indicate that an F/DMF ratio of 0.4 is optimal for producing high-performance carbon microspheres, with improvements mainly driven by surface chemistry and increased porosity [67,68].

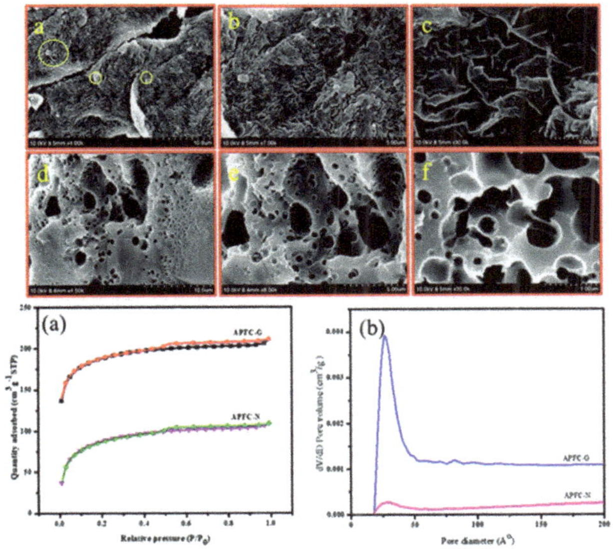

Figure 3. (**Top**) FESEM micrographs of the synthesized samples: (**a**–**c**) APFC-N and (**d**–**f**) APFC-G, captured at varying magnifications [(**a**,**d**): 10 μm; (**b**,**e**): 5 μm; and (**c**,**f**): 1 μm]. (**Bottom**) (**a**) N_2 adsorption isotherms and (**b**) pore size distributions of APFC-N and APFC-G samples. Reproduced from [60].

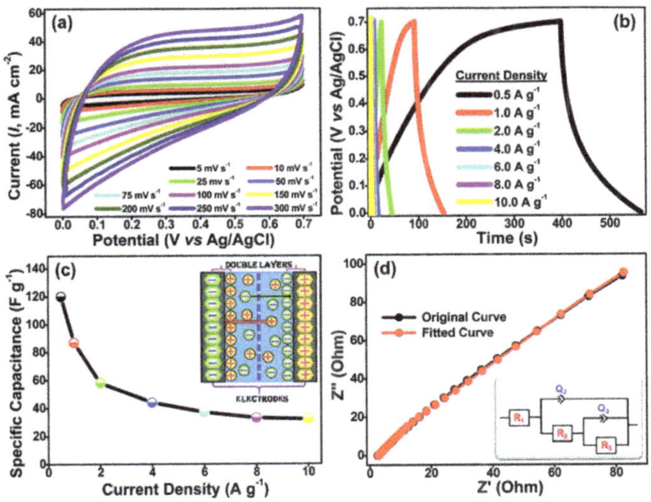

Figure 4. (**a**) Cyclic voltammetry curves recorded at multiple scan rates, (**b**) galvanostatic charge–discharge profiles under varying current densities, (**c**) plot of specific capacitance versus current density [inset: schematic illustration of an electric double-layer capacitor (EDLC)], and (**d**) Nyquist plot from electrochemical impedance spectroscopy with curve fitting [inset: corresponding equivalent circuit model, R_1 = solution resistance, R_2 = charge-transfer resistance, R_3 = Warburg resistance] for the synthesized APFC-G material. Reproduced from [60].

Meanwhile, Mohammed et al. [44] developed a new type of porous organic polymer (POP) by coupling Cr-TPA-4BZ-Br_4 with tetraethynylpyrene (Py-T) (Figure 5). The precursor benzoxazine monomers—Cr-TPA-4BZ (Dibenzo-crown ether-triphenylamine-based benzoxazine) and its brominated version Cr-TPA-4BZ-Br_4—were synthesized through a conventional three-step pathway, involving CH=N formation, sodium borohydride reduction, and a Mannich-type condensation. These monomers were then linked via a Sonogashira cross-coupling reaction to create Cr-TPA-4BZ-Py-POP. Upon further carbonization, this yielded poly(Cr-TPA-4BZ-Py-POP)-800, a carbonaceous material with graphite-like domains. This material showed a high CO_2 adsorption capacity of 4.4 mmol g^{-1} at 273 K. Electrochemical evaluations revealed a significant specific capacitance of 397.2 F g^{-1} at a current density of 0.5 A g^{-1}, along with excellent stability, retaining 94% of its capacitance over extended cycling. The enhanced performance was largely attributed to an optimized ratio (C = 73.5%, N = 8.9% and O = 17.6%) of heteroatom doping [69,70], which contributed to improved structural features and electrochemical behavior.

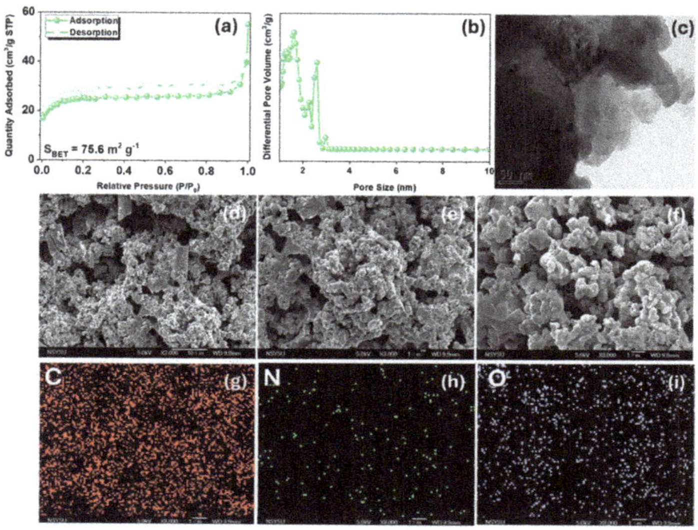

Figure 5. (**a**,**b**) BET isotherm and pore distribution analysis; (**c**) TEM image; (**d**–**f**) SEM micrographs at different magnifications [(**d**): 10 μm, (**e**): 1 μm and (**f**): 1 μm]; (**g**–**i**) elemental mapping (EDS-SEM) of poly(Cr-TPA-4BZ-Py-POP)-800. Reproduced from [44].

Wan et al. [71] developed nitrogen-doped porous carbon materials with high surface areas and well-defined micro-/mesoporous architectures by employing a novel nitrile-functionalized polybenzoxazine. The process combined a soft-templating strategy (Surfactant F127 was used as soft templating agent) with KOH chemical activation. Both the use of the soft template and the activation temperature were found to significantly affect the resulting pore structure and surface characteristics [72]. Compared to the primarily microporous NPC-0, the hierarchical pore systems in NPC-1 and NPC-2 demonstrated superior capacitive behavior. This enhancement was linked to a large surface area (up to 2036 m^2/g), balanced pore distribution, robust electrical conductivity, and the inclusion of nitrogen (2.33–5.32 wt.%) and oxygen (10.26–14.22 wt.%) functionalities within the carbon framework [73,74]. Notably, NPC-2, activated at 700 °C, delivered the highest specific capacitance of 362.4 F/g at a current density of 1 A/g in a KOH aqueous solution (Figure 6).

The material also exhibited excellent rate performance and long-term durability, preserving 94.7% of its capacitance after 5000 charge–discharge cycles. Moreover, the solid adsorbent NPC-1, thermally activated at 600 °C, achieved the highest CO_2 adsorption capacity under 1 bar—recording 6.20 mmol g^{-1} at 0 °C and 3.95 mmol g^{-1} at 25 °C. It also demonstrated high CO_2/N_2 selectivity and could be reused effectively without performance loss. These exceptional properties, both as supercapacitor electrodes and CO_2 adsorbents, were primarily ascribed to their extensive surface area, tailored pore architecture, superior electrical conductivity, and the incorporation of nitrogen functionalities within the carbon framework [75,76]. In a related study, Liu et al. [77] developed nitrogen and oxygen co-doped porous carbon materials (NOPC-x and NOPC-bis-CN-x) [NOPC and NOPC-bis-CN are carbon materials derived from different benzoxazine monomers, Boz-Va (benzoxazine synthesized from vanillin, aniline and paraformaldehde) and Boz-bis-VaCN (benzoxazine synthesized from diaminodiphenyl ether, 2-(4-hydroxy-3-methoxybenzylidene) malanonitrile and paraformaldehyde)] derived from bio-based polybenzoxazines through a soft-template strategy. They found that increasing the cyano content in the monomers improved surface area, pore structure, and graphitization level. Notably, the NOPC-bis-CN-3 variant achieved a remarkable surface area of 2347 $m^2\ g^{-1}$ and abundant mesopores (20–40 nm), along with a higher density of electrochemically active nitrogen [N-1 (pyridinic N), N-2 (pyridonic N), and N-3 (quarternary N)] and oxygen (O-2, O-3) species [78,79]. Electrochemical testing confirmed the superior performance of the NOPC-bis-CN-x series over the NOPC-x group. Specifically, NOPC-bis-CN-3 delivered the highest specific capacitance of 167.3 F g^{-1} at 1 A g^{-1}, retained more than 80% of its capacitance at 10 A g^{-1}, and demonstrated strong pseudocapacitive behavior and excellent cycling durability. The symmetric device assembled with this material showed minimal decline in energy density—from 10.8 to 9.0 Wh kg^{-1}—as power density increased from 45 to 4500 W kg^{-1} (see Figure 7). The convergence of outstanding electrochemical features, eco-friendly origins, and straightforward synthesis renders NOPC-bis-CN-3 a top contender for next-generation supercapacitor applications.

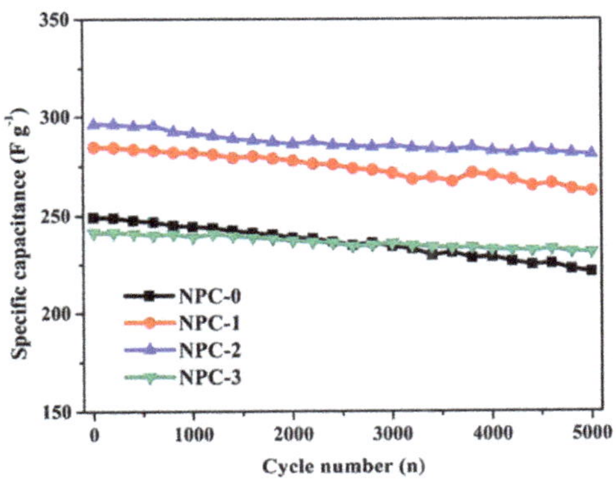

Figure 6. Change in specific gravimetric capacitance over cycling for the four NPC electrodes at 10 A g^{-1} current density. Reproduced from [71].

M.M. Samy and colleagues [46] developed an innovative bio-based benzoxazine monomer, termed VFBZ-CN, synthesized through the condensation reaction of vanillin, formaldehyde, and furfurylamine—compounds derived from renewable natural sources. Characterization via differential scanning calorimetry (DSC) and thermogravimetric analysis (TGA) indicated that VFBZ-CN underwent thermal curing at a relatively low onset temperature of 196 °C, a property attributed to cyano functionalities enhancing oxazine ring-opening. Upon curing at 250 °C, the resulting polymer displayed impressive thermal resilience, with a 10% weight loss temperature (Td_{10}) of 379 °C, surpassing traditional benzoxazine analogs. Electrochemical assessments revealed that poly(VFBZ-CN)-800 achieved a gravimetric capacitance of 506 F g^{-1}, significantly exceeding that of poly(VFBZ-CN)-700, which reached only 171 F g^{-1} at 0.5 A g^{-1} in an alkaline KOH medium. Moreover, the poly(VFBZ-CN)-800 electrode retained 99.43% of its capacitance after 2000 cycles at a high current density of 10 A g^{-1}, underscoring its cycling durability. This enhanced electrochemical efficiency was primarily ascribed to its porous carbon architecture and elevated nitrogen/oxygen content [80,81]. Apart from energy storage, these N- and O-enriched microporous carbons derived from VFBZ-CN also exhibited strong CO_2 adsorption capacity, reinforcing their multifunctionality and alignment with sustainable technologies.

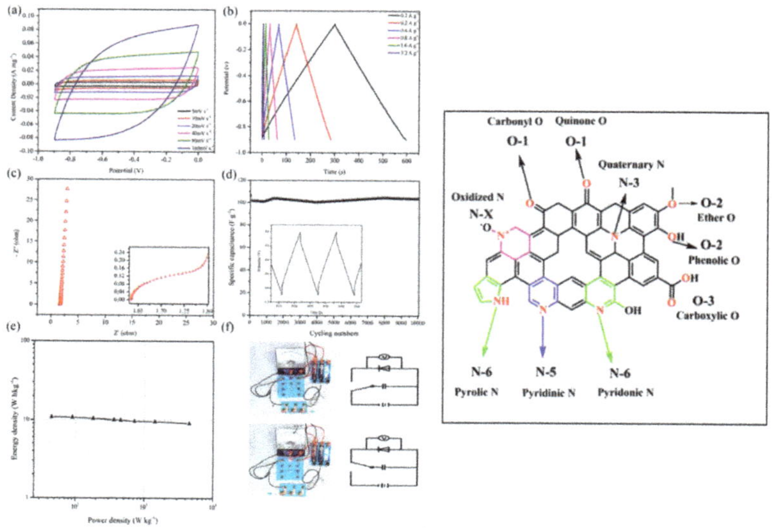

Figure 7. (**Left**) Electrochemical characterization of the assembled symmetric two-electrode supercapacitor: (**a**) cyclic voltammetry (CV) curves measured at various scan rates; (**b**) galvanostatic charge–discharge (GCD) profiles at different current densities; (**c**) electrochemical impedance spectroscopy (EIS) represented by Nyquist plots from 100 kHz to 10 mHz; (**d**) cycling stability at 2 A g^{-1} (inset shows selected GCD segments); (**e**) Ragone plot of the NOPC-bis-CN-3-based device; (**f**) charging of two series-connected symmetric supercapacitors over 30 s, and illumination of a red LED (1.5 V) powered by the series-connected devices. (**Right**) Illustration of potential nitrogen and oxygen functional groups on the carbon surface. Reproduced from [77].

Zhang and colleagues [82] developed nitrogen and phosphorus co-doped carbon materials (C/P-Cs) derived from nonporous polybenzoxazine, using polybenzoxazine as the carbon base and melamine polyphosphate as a dual-source dopant for nitrogen and phosphorus. These carbon structures exhibited notably high atomic concentrations of nitrogen (5.5%) and phosphorus (5.1%), which played a key role in enhancing their electrochemical properties—specifically in terms of capacitance, charge–discharge efficiency, and operational stability as supercapacitor electrodes. Among them, the optimized variant (C/P-20-1) delivered an impressive specific capacitance of 203.0 F g^{-1} at a current density of 0.5 A g^{-1}, and still maintained 173.2 F g^{-1} under a much higher load of 20 A g^{-1}. The electrode's durability was also remarkable, preserving 90.1% of its capacitance over 5000 cycles at 5 A g^{-1}. When configured into a symmetric supercapacitor, this material demonstrated a peak energy density of 11.45 Wh kg^{-1} at a power density of 50 W kg^{-1}, and a maximum power output of 25 kW kg^{-1} at an energy density of 5.55 Wh kg^{-1}. Even after undergoing 10,000 cycles of charging and discharging at 5 A g^{-1}, the device retained 79.9% of its original capacitance. The superior electrochemical behavior of the C/P-20-1 electrode and the symmetric device was attributed to a combination of factors: (i) the presence of functional nitrogen and phosphorus species (such as pyridinic and pyrrolic N), which boosted pseudocapacitive effects and overall charge storage [41]; (ii) nitrogen doping, which enhanced electrical conductivity and charge transfer rate [83]; and (iii) phosphorus doping, which notably contributed to the material's long-term cycling resilience [84].

Wang et al. [40] developed a range of oxygen-rich porous carbon materials by copolymerizing a diacetal-type benzoxazine (ACE-a) with melamine in varying proportions, utilizing a straightforward, template-free synthesis approach. The inclusion of melamine introduced a substantial amount of reactive C–N bonds into the network during polymer formation, which contributed to the creation of abundant micropores during the carbonization and activation phases [85], thanks to melamine's self-sacrificing behavior (see Figure 8). This decomposition process not only promoted micropore generation but also increased the oxygen content in the resulting carbon, improved its surface wettability, and lowered internal resistance. Among the materials synthesized, CA3MK showed outstanding characteristics, including a high BET surface area of 1383.9 m^2 g^{-1}, a large pore volume of 0.748 cm^3 g^{-1}, and a notable surface oxygen content of 66.2%. Electrochemical evaluations revealed impressive performance, with a specific capacitance of 430 F g^{-1} in 0.5 M H_2SO_4 and 194 F g^{-1} in 6 M KOH at 0.5 A g^{-1} (Figure 9). Moreover, the material demonstrated excellent cycling stability and strong performance at high current densities. These results underscore the promise of oxygen-functionalized porous carbons in supercapacitor technologies, driven by their high surface area, well-developed porosity, and abundant oxygen-containing groups [86,87].

Bai et al. [88] developed boron and nitrogen co-doped porous carbons (BNPC-X) by carbonizing boron-containing polybenzoxazines followed by chemical activation. In this synthesis, the benzoxazine resin functioned as both the carbon and nitrogen precursor, while boric acid provided the boron dopant. Among the series, BNPC-0.15 showed notable elemental incorporation, containing 2.97 wt.% boron and 2.43 wt.% nitrogen, with a homogeneous spatial distribution (Figure 10). This material delivered a high specific capacitance of 286 F g^{-1} at 0.05 A g^{-1} and retained a commendable 174 F g^{-1} at 1 A g^{-1}, maintaining 92% of its initial capacitance after 1000 cycles in 6 M KOH. The synergistic effects of boron and nitrogen functionalities enhanced electrolyte interaction and introduced additional redox activity, leading to improved electrochemical performance [89,90]. These features highlight BNPC-X as a strong candidate for advanced supercapacitor electrode applications.

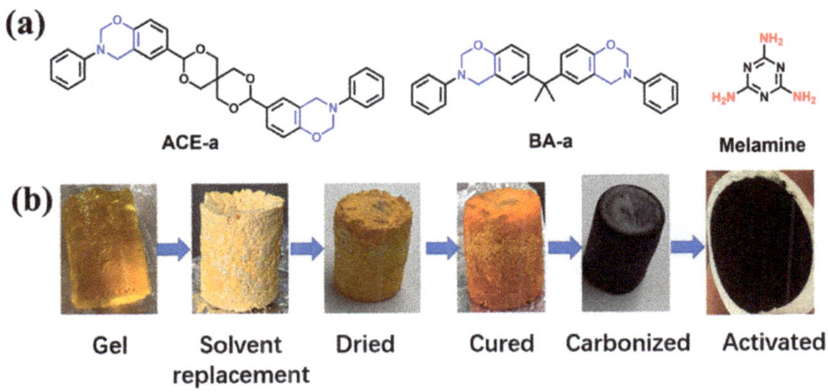

Figure 8. (**a**) Chemical structures of ACE-a, BA-a and melamin; (**b**) preparation process of porous carbon materials. Reproduced from [40].

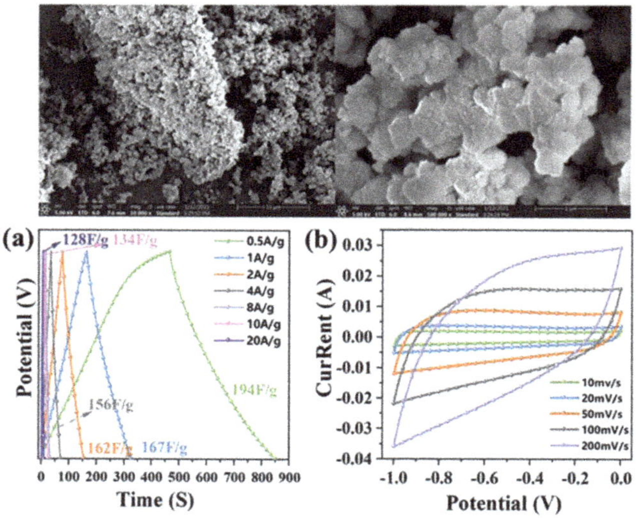

Figure 9. (**Top**) SEM micrographs of CA3MK material captured at various magnifications (10 µm and 1 µm). (**Bottom**) (**a**) Galvanostatic charge–discharge profiles of the CA3MK electrode at multiple current densities; (**b**) cyclic voltammetry curves of the same electrode recorded at different scan rates. Reproduced from [40].

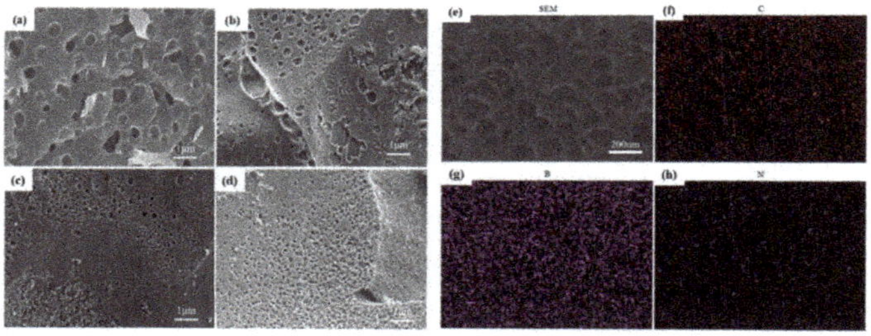

Figure 10. SEM micrographs of (**a**) NPC, (**b**) BNPC-0.05, (**c**) BNPC-0.10, and (**d**) BNPC-0.15. Panels (**e**–**h**) show the elemental mapping of BNPC-0.15 obtained via energy-dispersive X-ray spectroscopy (EDS). Reproduced from [88].

3. Supercapacitor Performance of Polybenzoxazine/Bimetal Oxides

The integration of polybenzoxazine-based carbons with bimetallic oxides is one of the emerging combinations that has gained notable attention. This synergy merges the high surface area and conductivity of carbon materials with the redox-active sites provided by bimetallic oxides, leading to significant performance enhancements in electrochemical energy storage. Thirukumaran et al. [91] developed a hierarchical framework in which electroactive materials are securely integrated onto a carbon scaffold, promoting optimal exposure of active sites and facilitating efficient ion and electron transport. The carbon base is derived from polybenzoxazine (Pbz), synthesized from a benzoxazine monomer. The carbonized Pbz features a rigid network enriched with nitrogen and oxygen heteroatoms, which contributes to enhanced supercapacitor (SC) performance. Furthermore, incorporation of pseudocapacitive metal hydroxides like $Ni(OH)_2$ and $Mn(OH)_2$ supports reversible Faradaic reactions. The final products—nitrogen-rich porous carbon (NRPC), along with its composites NRPC/Mn, NRPC/Ni, and NRPC/NiMn—demonstrate pore volumes between 0.18 and 0.42 $cm^3\ g^{-1}$. The NRPC displayed a tangled porous network containing a variety of pore sizes, including prominent macropores. SEM analysis revealed that the NRPC/Mn composite developed flake- or petal-like formations while NRPC/Ni exhibited a deformed petal structure with sharp, spike-like tips. These morphological elements were embedded within the carbon framework. Interestingly, co-incorporation of Mn and Ni gave rise to a vertically oriented, three-dimensional flower-like architecture formed by interlinked petal structures. These petals were anchored by a hollow central core, contributing to mechanical robustness and offering abundant electroactive sites conducive to efficient ion storage in supercapacitor applications [92,93] (Figure 11).

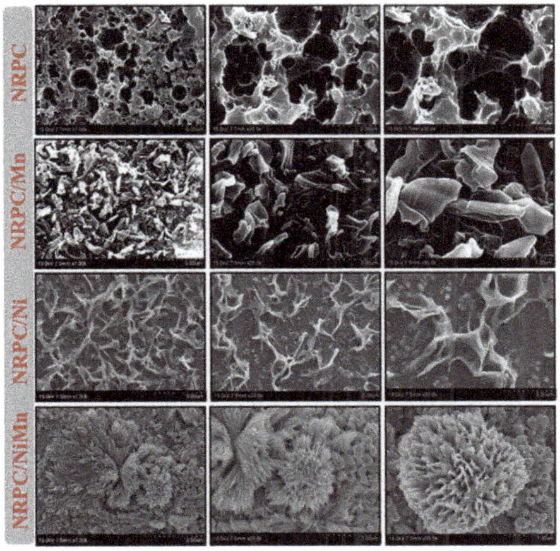

Figure 11. SEM images of NRPC, NRPC/Mn, NRPC/Ni, and NRPC/NiMn at different magnifications (5 μm, 2 μm and 1 μm). Reproduced from [91].

In contrast to NRPC containing only monometallic components, the NRPC/NiMn composite exhibits a notably higher specific surface area of 365 m^2 g^{-1} and a greater pore volume of 0.42 cm^3 g^{-1}, attributed to its distinctive flower-like architecture. Electrochemical measurements reveal a significant specific capacitance of 1825 F g^{-1} for NRPC/NiMn, retaining 78% of its initial capacity after 2500 charge–discharge cycles. This performance is driven by the synergistic effects of bimetallic active sites and heteroatom doping, which collectively promote strong pseudocapacitive behavior. Additionally, the porous carbon framework acts as a highly conductive matrix, enhancing electron mobility and supporting electrochemical double-layer capacitance (EDLC) through efficient ion transport and surface adsorption. Asrafali et al. [15] designed a novel supercapacitor configuration integrating heteroatom-enriched carbon, bimetallic oxide nanostructures, and redox-active electrolytes to significantly boost electrochemical efficiency. The system leverages dual pseudocapacitive mechanisms: one originating from the electrode material—specifically, nitrogen and oxygen-doped carbon derived from polybenzoxazine blended with NiCo bimetallic oxides—and the other from the electrolyte, utilizing iodide-containing redox-active species such as KI and RbI. The polymer precursor, synthesized via Mannich condensation of eugenol and ethylenediamine, yields a heteroatom-doped carbon (HC) upon calcination at 800 °C. The resulting HC/NiCo@800 °C structure features a hierarchical assembly of interlinked 3D and ultrathin 2D morphologies [94], as confirmed by SEM and TEM imaging. The flower-like architecture, formed without observable thermal degradation, provides a high surface area, interconnected pore networks, and abundant electroactive sites, all of which contribute to improved electrolyte accessibility and charge storage kinetics [95,96]. Remarkably, devices tested with redox-active RbI and KI electrolytes achieved elevated specific capacitances of 2334 F g^{-1} and 2076 F g^{-1} at 1 A g^{-1}, respectively. The asymmetric device HC/NiCo@800 °C//HC operating in RbI demonstrated a peak specific capacitance of 232 F g^{-1} with 89.04% retention after 5000 cycles, along with a low solution resistance (Rs) of 0.75 Ω and charge transfer resistance (Rct) of 0.77 Ω (Figure 12). The superior electrochemical response in RbI is credited to the rapid transport dynamics of Rb^+ and I^- ions, which enhance ion–solvent and solvent–solvent interactions. This configuration achieved an impressive energy density of 96.57 Wh kg^{-1} while maintaining a power density of 850 W kg^{-1}. The integration of functionalized porous carbon, dual-metal oxides, and redox-active electrolytes demonstrates a promising strategy for developing high-performance supercapacitors with excellent durability, high capacitance, and energy output, paving the way for broader practical applications.

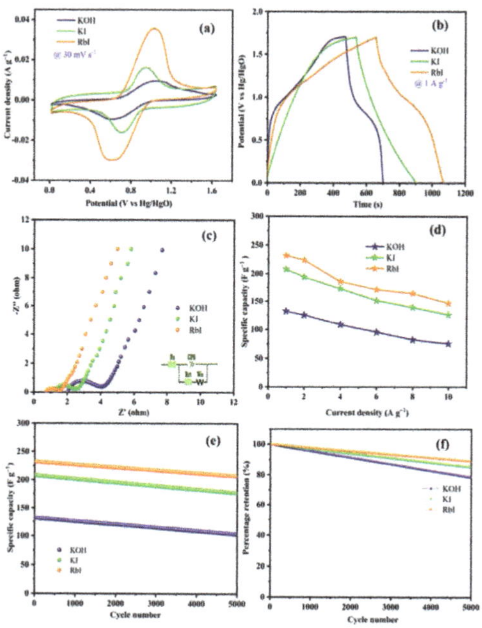

Figure 12. Electrochemical characterization of the asymmetric HC/NiCo@800//HC device, including (**a**) CV, (**b**) GCD, (**c**) EIS spectra, (**d**) specific capacitance versus current density, (**e**) cycle count, and (**f**) cyclic stability. Reproduced from [15].

4. Supercapacitor Performance of Polybenzoxazine Composites

Polybenzoxazine composites, particularly when combined with other functional materials like carbon nanotubes (CNTs), graphitic carbon nitride (g-C_3N_4), graphene oxide (GO) and silica (SiO_2), enable the development of high-capacitance, stable, and conductive electrode architectures. Ge et al. [97] introduced a straightforward yet efficient approach for producing highly porous carbon nanofiber (CNF) membranes with superior mechanical flexibility and multifunctional performance. Their method involved multicomponent electrospinning followed by in situ polymerization using polybenzoxazine (PBZ) as a novel carbon precursor. The incorporation of $SnCl_2$ enhanced both the spinnability and thermal stability of the precursor nanofibers (Figures 13 and 14). During the carbonization process, SnO_2 nanoclusters—ranging from 20 to 40 nm—were uniformly embedded throughout the carbon network and anchored onto the nanofiber surfaces via in situ synthesis. This led to the formation of a heterogeneous nanostructure that imparted a plasticizing effect, enabling the resulting SnO_2/CNF membrane to withstand large deformations while preserving its structural integrity. The membrane exhibited a high specific surface area of 1415 m^2/g and a pore volume of 0.82 cm^3/g. It also achieved an energy density of 16.25 Wh/kg and a power density of 1.03 kW/kg, significantly surpassing traditional porous CNF electrodes. Moreover, the membrane maintained excellent electrochemical performance under bending, highlighting its potential for use as flexible electrodes in wearable energy storage systems. The enhanced performance was primarily due to (i) the strong interfacial interaction between the SnO_2 clusters and the carbon matrix, which minimized resistance, and (ii) the well-developed porosity that facilitated rapid ion transport [98,99].

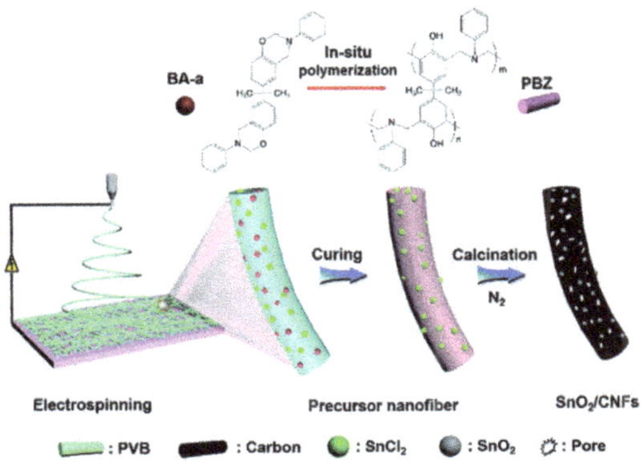

Figure 13. Diagram depicting the synthesis pathway of porous SnO2/CNF membranes obtained from PBZ. Reproduced from [97].

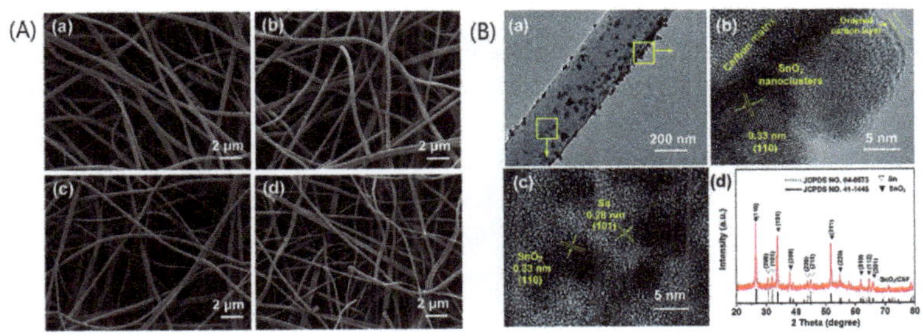

Figure 14. (**A**) SEM images of SnO2/CNFs at various carbonization temperatures: (**a**) 650 °C, (**b**) 750 °C, (**c**) 850 °C, and (**d**) 950 °C. (**B**) Corresponding (**a**) TEM image, (**b**) and (**c**) HR-TEM images, and (**d**) XRD patterns of SnO2/CNFs. Reproduced from [97].

Wu et al. [34] developed uniform porous yolk-shell carbon nanospheres (PYCNs) via a two-step coating strategy, using resorcinol–formaldehyde (RF) resin spheres as the initial core material. RF resin was selected for its affordability, high carbon yield, and structural stability. The process involved coating the RF spheres with two layers: an inner dense silica layer and an outer composite shell made of polybenzoxazine and silica (PB/SiO_2). Tetraethyl orthosilicate (TEOS) served as the silica source, while a mixture of resorcinol, formaldehyde, and ethylenediamine (EDA) was used for the PB layer. Transmission electron microscopy revealed that the final carbon shell measured about 20 nm in thickness, enclosing a central void of approximately 40 nm, with a carbon core approximately 600 nm in diameter—matching the original size of the solid carbon spheres (SCS). Surface area analysis showed that SCS possessed a higher specific surface area of 606 $m^2\ g^{-1}$ and a micropore volume of 0.25 cm^3 g^{-1}, in contrast to PYCNs, which exhibited values of 486 m^2 g^{-1} and 0.15 cm^3 g^{-1}, respectively. Despite the slightly lower surface area, the yolk-shell design, mesoporosity, and nitrogen doping in PYCNs significantly enhanced ion transport and diffusion [100,101]. Consequently, PYCNs outperformed SCS in supercapacitor applications, achieving a specific capacitance of 236 F g^{-1} at a current density of 0.5 A g^{-1}, compared to 176 F g^{-1} for SCS.

Du et al. [102] introduced a co-assembly strategy for producing nitrogen-doped hollow mesoporous carbon spheres (N-HMCS) with tunable morphologies. This method employed cetyl-3-methyl ammonium bromide (CTAB) and tetraethyl orthosilicate (TEOS) in conjunction with polybenzoxazine (PBZ), synthesized from phenol, formaldehyde, and ethylenediamine, via electrostatic interactions. In the process, CTAB served as a structural template, TEOS functioned as a silica precursor, and the PF oligomers with ethylenediamine acted as carbon and nitrogen sources within a modified Stöber synthesis. Ethylenediamine also facilitated both TEOS hydrolysis and PF oligomer polymerization. The resulting core-shell particles displayed morphology-dependent surface features—silica@PB-0.1 had a smooth, spherical structure with a distinct core-shell interface, while silica@PB-0.4 featured a rougher surface with prominent protrusions [103,104]. Among the derived carbon spheres, N-HMCS-0.1 delivered the highest specific capacitance of 307 F g^{-1}, outperforming its counterparts (198, 206, and 192 F g^{-1} for N-HMCS-0.05, 0.2, and 0.4, respectively) (Figure 15). Moreover, N-HMCS-0.1 maintained 83% capacitance retention at elevated current densities. A symmetric supercapacitor assembled using N-HMCS-0.1 and a 6 M KOH electrolyte attained an energy density of 11.2 Wh kg^{-1} at 660.8 W kg^{-1}, and 10 Wh kg^{-1} at 9000.5 W kg^{-1}. The superior electrochemical performance of N-HMCS-0.1 is ascribed to its fine particle size, thin shell, and high surface area, which together enhance charge storage and ion accessibility [105,106]. Additionally, the mesoporous architecture ensures efficient ion transport, while the nitrogen functionalities promote pseudocapacitive behavior through redox activity [107,108].

Wan et al. [109] developed a method for producing graphene oxide (GO) and nitrogen-doped porous carbon (NC) nanocomposites aimed at enhancing supercapacitor electrode performance. The synthesis involved a polybenzoxazine (PBZ)-based ring-opening polymerization followed by KOH activation, leveraging both hydrogen bonding and covalent interactions between GO and benzoxazine. Nanocomposites with different GO loadings were fabricated, revealing that the inclusion of GO significantly altered the surface characteristics, porosity, and electrical conductivity of the materials, thereby improving their electrochemical behavior [110–113]. Among the samples, the electrode containing 1.29 wt.% GO exhibited the highest specific capacitance of 405.6 F g^{-1} at 1.0 A g^{-1} in a 6 M KOH electrolyte. This electrode also showed excellent rate performance (267.8 F g^{-1} at 40 A g^{-1}) and strong cycling durability, retaining 95.8% of its capacitance after 5000 cycles. Moreover, symmetric supercapacitor devices assembled with the GO/NC material in 1 M Na_2SO_4 operated within a broad 1.8 V voltage range, delivering an energy density of 38.6 Wh kg^{-1} at 180 W kg^{-1} and sustaining 19.9 Wh kg^{-1} at a power density of 32.4 kW kg^{-1}.

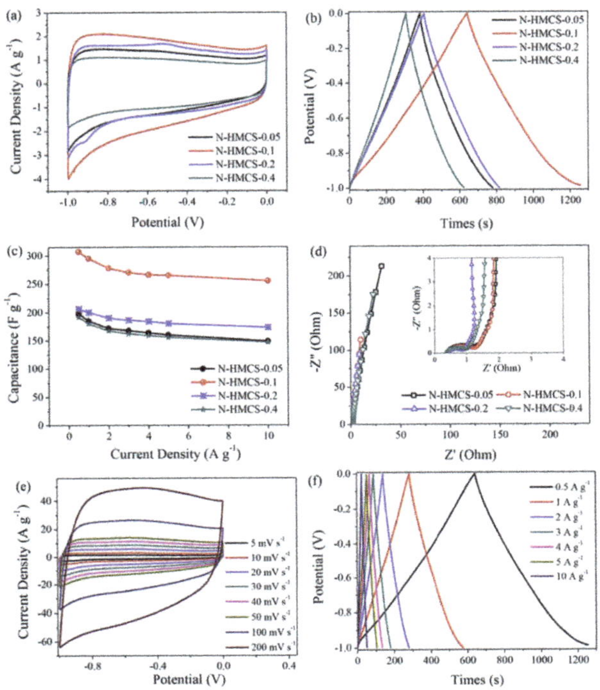

Figure 15. Electrochemical assessment of N-HMCS samples in a three-electrode setup: (**a**) CV profiles at a scan rate of 5 mV s^{-1}, (**b**) typical GCD profiles at a current density of 0.5 A g^{-1}, (**c**) specific capacitances at varying GCD current densities, (**d**) Nyquist plots with fitting curves and their corresponding high-frequency ranges (inset), (**e**) CV profiles at different scan rates, and (**f**) GCD profiles at various current densities for N-HMCS-0.1. Reproduced from [102].

Selvaraj et al. [10] reported the synthesis of a novel quinoline-based Mannich-type benzoxazine monomer (Q-xda), derived from the reaction of 8-hydroxyquinoline, xylylenediamine, and paraformaldehyde. This monomer was used to fabricate high-performance carbon-based materials for energy storage. Composites were formulated by incorporating graphitic carbon nitride (GCN) at varying concentrations (5, 10, and 15 wt.%) into the poly(Q-xda) matrix. The presence of GCN notably improved the thermal resistance and char yield of the composites [114,115]. Electrochemical tests confirmed pseudocapacitive behavior, with poly(Q-xda) + 15 wt% GCN showing the highest specific capacitance of 370 F g^{-1} at a current density of 6 A g^{-1}. The composites with 5 and 10 wt.% GCN exhibited capacitances of 294 and 310 F g^{-1}, respectively, while the pristine poly(Q-xda) showed a lower value of 216 F g^{-1}. The poly(Q-xda) + 15 wt% GCN sample also demonstrated superior charge transport characteristics, with a reduced charge transfer resistance (20.8 Ω) compared to that of the unmodified polymer (26.0 Ω), and maintained 96.2% of its capacitance after 2000 cycles. In a related study by the same group [9], a facile and scalable approach was introduced for producing nitrogen-rich porous carbon (NRPC), which was further integrated with graphitic carbon nitride and magnetite ($g\text{-}C_3N_4/Fe_3O_4$) to fabricate a functional nanocomposite. The difunctional benzoxazine monomer (NP-ha), synthesized from nonylphenol, hexane-1,6-diamine, and paraformaldehyde, served as the precursor. The process involved direct carbonization, KOH activation, and subsequent hydrothermal treatment to embed Fe_3O_4 and $g\text{-}C_3N_4$ into the NRPC matrix. Characterization using FE-SEM and HR-TEM confirmed uniform dispersion of the nanophases [116]. (Figure 16) The resulting composite exhibited a high specific surface area of 479.6 m^2 g^{-1} and demonstrated effective pseudocapacitive behavior, highlighting its potential for use in advanced supercapacitor applications.

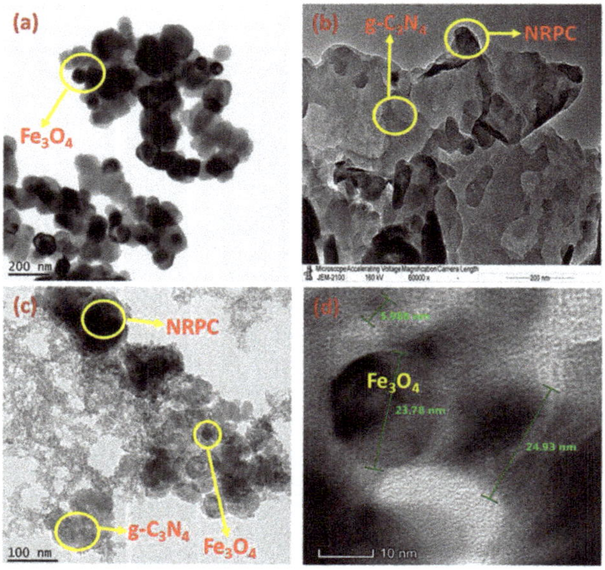

Figure 16. HR-TEM images of (**a**) Fe_3O_4, (**b**) NRPC/g-C_3N_4, and (**c**,**d**) NRPC/g-C_3N_4/Fe_3O_4-0.1. Reproduced from [9].

Compared to the Fe_3O_4 and g-C_3N_4/Fe_3O_4 electrodes, the NRPC/g-C_3N_4/Fe_3O_4 electrode exhibited lower charge transfer resistance and higher capacitance. Specifically, the NRPC/g-C_3N_4/Fe_3O_4 electrode achieved the highest specific capacitance of 385 F g^{-1} at 1 A g^{-1}, outperforming Fe_3O_4 (112 F g^{-1}) and g-C_3N_4/Fe_3O_4 (150 F g^{-1}). Additionally, the cycling efficiency of the NRPC/g-C_3N_4/Fe_3O_4 electrode remained at 94.3% after 2000 cycles. These results demonstrate that the incorporation of NRPC into g-C_3N_4/Fe_3O_4 significantly enhances its potential for use in high-performance supercapacitors [117,118].

5. Supercapacitor Performance of Polybenzoxazine with Other Polymers

Polybenzoxazine-based thermosets, known for their high nitrogen content and substantial char yield, are gaining attention as sustainable precursors for nitrogen-doped carbon materials. A key consideration in upcycling these materials is the reduction in both energy input and processing costs. The combination of PBZ with other functional polymers, including polyaniline (PANI), polypyrrole (PPy), or porous organic polymers (POPs), leads to multifunctional electrode materials that deliver a balanced combination of energy and power densities, making them suitable for both commercial and high-performance supercapacitor applications. In line with this goal, Sharma et al. [119] demonstrated a method for producing carbon materials under relatively mild carbonization conditions, eliminating the need for chemical activation. Electrochemical analysis using a three-electrode system revealed that the carbon derived from guaiacol-based polybenzoxazine (C-GP81), which features 6.4% nitrogen incorporation, delivered a notable specific capacitance of 700 F g^{-1} at 10 A g^{-1}. This indicates excellent charge storage and rate capability, making it a strong candidate for supercapacitor electrode applications.

The material also achieved an energy density of 48 Wh kg^{-1} at a power density of 8400 W kg^{-1} in the same three-electrode setup. Its performance in an acidic medium is credited to the synergistic effects of a well-developed surface area and a favorable composition of nitrogen (pyridinic, pyrrolic, and graphitic) and oxygen (quinone) functional groups [41,120]. When evaluated in a symmetric supercapacitor device, C-GP81 delivered a specific capacitance of 76 F g^{-1} at 0.5 A g^{-1}, which gradually declined to 40% after 10,000 cycles, reflecting moderate cycling stability. The device also reached a peak energy density of 10 Wh kg^{-1} at a power density of 2400 W kg^{-1} (Figure 17). Overall, these findings underscore the promise of eco-friendly carbonization strategies for transforming polybenzoxazine resins into high-performance, heteroatom-enriched carbon materials for energy storage applications [121,122].

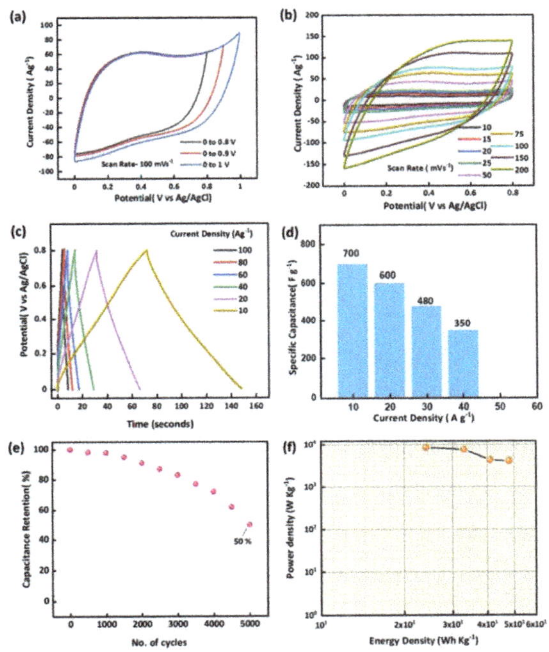

Figure 17. Electrochemical performance of the carbon material C-GP81 as an active material on carbon cloth, measured in a three-electrode system: (**a**) CV curves at a scan rate of 100 mV s^{-1} across different potential windows, (**b**) CV curves at a selected potential window with varying scan rates, (**c**) GCD profiles at different current densities, (**d**) variation in specific capacitance with changing current density, (**e**) long-term stability over 5000 cycles, and (**f**) Ragone plot showing energy and power densities. Reproduced from [119].

Tiwari et al. [25] leveraged the molecular design flexibility of polybenzoxazine to fabricate polybenzoxazine colloidal spheres using phloroglucinol, polyethylenimine, and formaldehyde as multifunctional precursors, employing a simple template-free extended sol–gel method (Figure 18). The kinetics of particle formation were controlled by adjusting synthesis parameters such as the stoichiometric ratio of the reactants, precursor concentration, and solvent ratio, which influenced the morphology, particle size, and heteroatom content in the polymeric particles [123]. The heteroatom-doped carbon spheres, containing 28% nitrogen and 20.5% oxygen (as determined by CHNS analysis), were obtained by subjecting the polybenzoxazine particles to moderate carbonization conditions, resulting in a partial graphitic structure.

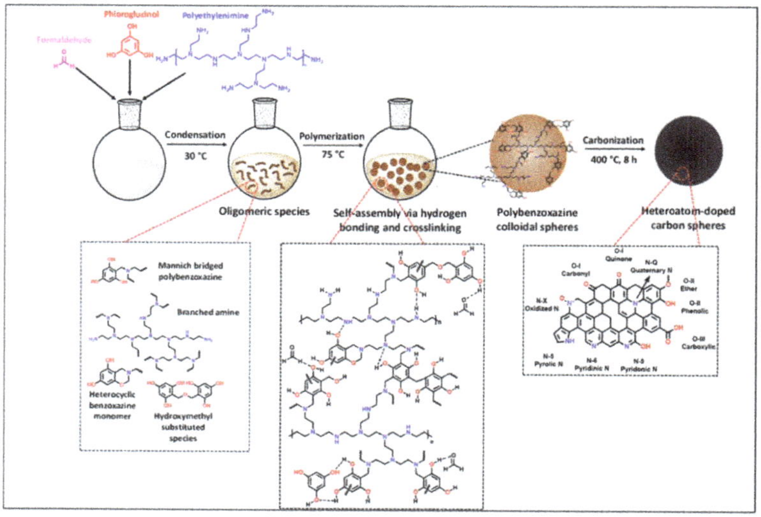

Figure 18. Schematic diagram showing the N, O-co-doped carbon particles obtained from polybenzoxazine colloidal spheres using a template-free extended sol–gel process. Reproduced from [25].

The uniform morphology, coupled with a significant BET surface area of 221 m^2 g^{-1} and high heteroatom content in the resulting carbon structure, highlights their potential as active materials for supercapacitor electrodes [124,125]. The N, O co-doped carbon structure exhibited an impressive specific capacitance of 728 F g^{-1} at a current density of 10 A g^{-1}, resulting in a maximum energy density of 56 W-h kg^{-1} and a maximum power density of 14,246 W kg^{-1}. Further electrochemical performance evaluation was conducted using a flexible all-solid-state symmetric two-electrode system to simulate real-time conditions. The active material displayed a notable specific capacitance of 50.3 F g^{-1} at 0.2 A g^{-1}. Additionally, the material maintained approximately 86% capacitance retention after 2500 cycles in the asymmetric two-electrode configuration, confirming the reversible nature of the device with negligible changes in resistance before and after cyclic tests. These results underscore the potential of fabricating high-performance supercapacitors by precisely modulating the surface characteristics and functionality of polybenzoxazine precursors through the careful selection of multifunctional precursors in the extended sol–gel method [126,127].

Murugan et al. [1] reported the development of a series of polybenzoxazine-co-copper metal-organic frameworks (PABz-co-Cu MOFs) and their covalently cross-linked membrane composites with poly(imidazole-diphosphoric acid) (PIDPA) in various weight ratios (80/20, 60/20, 50/50, 40/60, and 20/80%). These hybrid membranes were fabricated using a sequential thermal curing method at different temperatures, resulting in a networked polymer structure (Figure 19). Among the formulations, the 50/50 wt.% PABz-co-Cu MOFs-graft-PIDPA membrane displayed the best performance as a high-temperature proton exchange membrane fuel cell (HT-PEMFC) material, outperforming both pristine PA-PIDPA and unmodified PABz-co-Cu MOFs. The enhanced performance was attributed to the presence of large voids between the particles, which effectively retained phosphoric acid (PA), crucial for proton conduction [128,129]. At the optimal 50/50 ratio, the PA-doped membrane achieved a proton conductivity of 7.57×10^{-2} S cm^{-1}, an open-circuit voltage (OCV) of 0.91 V, and a peak power density of 0.729 W cm^{-2}—significantly higher than those of the pure PABz-co-Cu MOFs membrane (3.21×10^{-2} S cm^{-1} conductivity, 0.43 V OCV, and 0.357 W cm^{-2} power density).

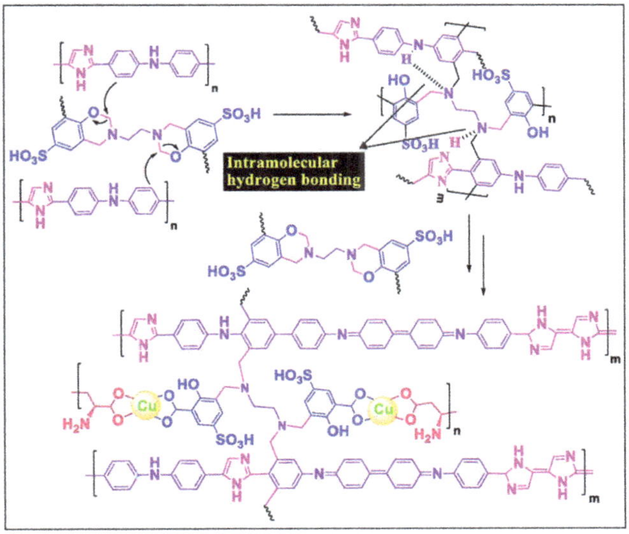

Figure 19. Proposed reaction pathway between ABz-co-Cu MOFs and PIDPA leading to the formation of covalently crosslinked PABz-co-Cu MOFs-graft-PIDPA. Reproduced from [1].

Furthermore, the 50/50 membrane composite showed a specific capacitance of 387 F g^{-1} at a current density of 1 A g^{-1}, exceeding the 187 F g^{-1} achieved by the unmodified MOFs. This research offers a cost-efficient strategy to design covalently bonded PABz-co-Cu MOFs-graft-PIDPA networks enriched with imidazole, diphosphoric acid, and 5-sulfo salicylic acid functional groups, showcasing strong potential for both HT-PEMFC and supercapacitor technologies. Table 1 shows the list of different carbon materials synthesized from PBZ and their properties.

PBZ naturally contain nitrogen and oxygen, making them ideal for heteroatom doping during pyrolysis. This enhances Faradaic activity, wettability, and electrical conductivity, boosting pseudocapacitance and electrolyte access. With rising interest in green chemistry, PBZ precursors are increasingly synthesized from biomass sources like natural phenols and amines. Advanced methods, such as foaming and templating, enable the creation of hierarchically porous structures (micro-, meso-, and macropores), which improve ion transport and boost power and energy density. PBZ-derived carbons are often composited with graphene, CNTs, or conductive polymers to enhance conductivity and mechanical strength, supporting the development of flexible, wearable supercapacitors. However, the field presents contradictions: Some studies credit micropores with high EDLC [59], while others highlight mesopores for better ion compatibility [130]. The effect of KOH activation is debated—beneficial in some cases [15], but linked to pore collapse in others [59]. There is also no agreement on the best electrolyte type, or whether co-doping (e.g., N and S) improves or destabilizes performance. Key research gaps include: understanding capacity fading (e.g., heteroatom leaching, pore blockage), limited data on cycling beyond 10,000 cycles, and a lack of studies on all-solid-state or stretchable devices. Additionally, there is minimal insight into the cost, scalability, and environmental impact of PBZ synthesis, calling for life cycle and techno-economic assessments.

Table 1. Comparative data showing the different synthesis process for carbon materials and their properties.

Carbon Materials	Synthesis Method	Properties
Nitrogen-enriched mesoporous carbon ropes, NCMR [38]	Carbonization and KOH activation.	SA = 300 $m^2\ g^{-1}$; pore size = 3 nm; pore volume = 0.003 $cm^3\ g^{-1}$; C_s = 60 F g^{-1} @ 1 A g^{-1} (2 M KOH)
Apigenin and furfurylamine-based Bzo, APFC-N and APFC-G [60]	Gelation, calcination and KOH activation.	SA = 248 $m^2\ g^{-1}$ (APFC-N) and 635 $m^2\ g^{-1}$ (APFC-G); pore size = 2–5 nm; C_s (APFC-G) = 120 F g^{-1} @ 0.5 A g^{-1} (1 M H_2SO_4)
Carbon microspheres [64]	Gelation, carbonization and CO_2 activation.	SA = 859 $m^2\ g^{-1}$; pore size = 1 nm; C_s = 424.7 F g^{-1} @ 0.5 A g^{-1} (1 M H_2SO_4)
Porous organic polymer, Cr-TPA-4Bz-PY-POP [44]	Sonogashira–Hagihara cross-coupling and carbonization.	Pore size = 4.17 nm; C_s = 397.2 F g^{-1} @ 0.5 A g^{-1} (1 M KOH)
Nitrogen-doped porous carbon, NPC-2 [71]	Soft templating method and KOH activation.	SA = 2036 $m^2\ g^{-1}$; C_s = 362.4 F g^{-1} @ 1 A g^{-1} (1 M KOH)
Nitrogen and oxygen-doped porous carbon, NOPC-bis-CN-3 [77]	Soft templating method and KOH activation.	SA = 2347 $m^2\ g^{-1}$; pore size = 20–40 nm; C_s = 167.3 F g^{-1} @ 1 A g^{-1} (6 M KOH)
Vanillin-malonitrile-based PBz, poly(VFBZ-CN) 800 [46]	Carbonization and KOH activation.	SA = 560 $m^2\ g^{-1}$; C_s = 506 F g^{-1} @ 0.5 A g^{-1} (1 M KOH)
Nitrogen and phosphorous co-doped carbon, C/P-20-1 [82]	Carbonization.	SA = 29 $m^2\ g^{-1}$; C_s = 203 F g^{-1} @ 0.5 A g^{-1} (1 M H_2SO_4)
Diacetyl-type Bzo carbon, CA3MK [40]	Gelation and curing.	SA = 1383.9 $m^2\ g^{-1}$; pore volume = 0.748 $cm^3\ g^{-1}$; C_s = 430 F g^{-1} (0.5 M H_2SO_4) and 194 F g^{-1} (6 M KOH) @ 0.5 A g^{-1}
Boron and nitrogen co-doped porous carbon, BNPC-0.15 [88]	Carbonization and KOH activation.	C_s = 286 F g^{-1} @ 0.5 A g^{-1} (6 M KOH)
Nitrogen rich porous carbon, NRPC/NiMn [91]	Carbonization, KOH activation and hydrothermal reaction.	SA = 365 $m^2\ g^{-1}$; pore volume = 0.42 $cm^3\ g^{-1}$; C_s = 1825 F g^{-1} @ 1 A g^{-1} (1 M KOH)
Hetero atom-doped carbon, HC/NiCo@800C [15]	Carbonization, KOH activation and hydrothermal reaction.	C_s = 2334 F g^{-1} @ 1 A g^{-1} (1 M RbI) and 2076 F g^{-1} @ 1 A g^{-1} (1 M KI)
Carbon nano fibers, SnO_2/CNF [97]	Template polymerization using PVB, electrospinning and carbonization.	SA = 1415 $m^2\ g^{-1}$; pore volume = 0.82 $cm^3\ g^{-1}$; C_s = 118 F g^{-1} @ 0.5 A g^{-1} (2 M HCl)
Porous yolk shell nanospheres, CPYCNs [34]	Layer-by-layer coating, ultra-sonication and carbonization.	SA = 486 $m^2\ g^{-1}$; pore volume = 0.15 $cm^3\ g^{-1}$; C_s = 236 F g^{-1} @ 0.5 A g^{-1} (6 M KOH)
Nitrogen-doped hollow mesoporous carbon spheres, N-HMCS (0.1) [102]	Stobber synthesis and polymerization.	SA = 636 $m^2\ g^{-1}$; pore volume = 1.60 $cm^3\ g^{-1}$; C_s = 307 F g^{-1} @ 0.5 A g^{-1} (6 M KOH)
Nitrogen-doped porous carbon/graphene oxide composites, GO/NC-2 [109]	Ring opening polymerization and KOH activation.	SA = 1345.8 $m^2\ g^{-1}$; pore volume = 0.53 $cm^3\ g^{-1}$; C_s = 405.6 F g^{-1} @ 1 A g^{-1} (6 M KOH)
Quinoline-based PBz/graphitic carbon nitride, poly(Q-xda) + 15 wt.% GCN [10]	Pyrolysis and ring-opening polymerization.	C_s = 370 F g^{-1} @ 6 A g^{-1} (1 M KOH)
Nitrogen rich porous carbon/graphitic carbon nitride/magnetite, NRPC/g-C_3N_4/Fe3O4 [9]	Carbonization, KOH activation, sonication and ageing.	SA = 497.6 $m^2\ g^{-1}$; C_s = 385 F g^{-1} @ 1 A g^{-1} (1 M KOH)
Guaiacol-based PBz carbon, C-GP81 [119]	Carbonization.	Pore size = 200–300 µm; C_s = 700 F g^{-1} @ 10 A g^{-1} (0.1 M H_2SO_4)
Hetero-doped carbon spheres, C-P-PEI [25]	Sol-gel method and carbonization.	SA = 221 $m^2\ g^{-1}$; pore size = 5.1 nm; pore volume = 0.28 $cm^3\ g^{-1}$; C_s = 728 F g^{-1} @ 10 A g^{-1} (0.1 M H_2SO_4)
PBz and poly(imidazole diphosphoric acid)-based carbon, PABz-co-Cu MOFs-graft-PIPDA (50/50) [1]	Thermal curing.	C_s = 387 F g^{-1} @ 1 A g^{-1} (3 M KOH)

SA = surface area; C_s = specific capacitance.

6. Conclusions and Future Directions

Polybenzoxazine-derived supercapacitors offer a promising alternative for energy storage applications due to their tunable properties, high stability, and potential for large-scale production. The research landscape on polybenzoxazine-derived supercapacitors is rapidly evolving, focusing on (i) *Sustainable materials:* The integration of bio-derived PBZs into supercapacitor electrodes is a major step toward sustainable energy storage solutions, aligning with the principles of green chemistry. These materials are gaining attention due to their eco-friendliness, cost-effectiveness, and excellent electrochemical properties, making them ideal candidates for next-generation supercapacitor electrodes. (ii) *Enhanced performance:* Heteroatom doping involves introducing elements like nitrogen (N), oxygen (O), sulfur (S), phosphorus (P), or boron (B) into the carbon framework derived from polybenzoxazines. This process modifies the electronic structure, improves charge distribution, and enhances redox activity, leading to higher specific capacitance and energy storage efficiency. Nitrogen and oxygen doping introduce functional groups, such as -NH_2, -OH, -C=O, and -COOH, that participate in Faradaic redox reactions, significantly boosting capacitance. Research results show that N-doped carbon materials show up to a 50% increase in capacitance compared to undoped carbons. Particularly, graphitic-N and pyridinic-N doping have been shown to improve conductivity and charge transfer rates. Oxygen and sulfur doping improve surface hydrophilicity, ensuring better electrolyte penetration and reducing charge transfer resistance. (iii) *Composite materials* combine PBZ-derived carbon with other high-performance materials like graphene and metal oxides to leverage synergistic effects that boost energy storage capabilities. Metal oxides (MnO_2, Fe_2O_3) and conducting polymers (polyaniline, polypyrrole) introduce Faradaic charge storage mechanisms, significantly increasing capacitance. Graphene reinforce the PBZ-derived carbon matrix, preventing electrode degradation over repeated charge–discharge cycles. Future work may explore scalable manufacturing techniques for PBZ-based supercapacitor materials and their integration into next-generation energy storage systems. These innovations will be crucial in meeting the increasing demand for high-performance, environmentally friendly, and cost-effective supercapacitors.

Future research should aim at low-cost, eco-friendly, and scalable polymerization and carbonization techniques, possibly incorporating bio-based benzoxazine monomers and solvent-free processes. It should emphasize controllable synthesis of multi-scale porous architectures (micro-, meso-, and macropores) to optimize ion transport and electrolyte accessibility while preserving high surface areas. Addressing degradation mechanisms and ensuring thermal and electrochemical stability over extended cycles remain crucial for commercial viability. Investigating the integration of polybenzoxazine-based materials into flexible or solid-state supercapacitors could expand their application in wearable and portable electronics. By focusing on these areas, researchers can unlock the full potential of polybenzoxazine-derived materials in next-generation supercapacitor technologies.

Author Contributions: Conceptualization—T.P. and S.P.A.; methodology—T.P. and S.P.A.; validation— J.L.; formal analysis—T.P. and S.P.A.; investigation—J.L.; resources—J.L.; data curation—T.P. and S.P.A.; writing—original draft preparation—T.P. and S.P.A.; writing—review and editing—S.P.A., T.P. and J.L.; visualization—J.L.; supervision—J.L.; project administration—J.L.; funding acquisition—J.L. All authors have read and agreed to the published version of the manuscript.

Funding: This research received no external funding.

Acknowledgments: This work was supported by the 2025 Yeungnam University Research Grant.

Conflicts of Interest: The authors declare no conflict of interest.

References

1. Murugan, E.; Munusamy, K.; Babu, A.V. Development of Aryl Ether-Free Cross-Linked Polymer Membranes for Sustainable Electrochemical Energy Conversion and Storage Applications. *Chem. Eng. J.* **2024**, *501*, 157473. [CrossRef]
2. Raja, M.; Sadhasivam, B.; Janraj Naik, R.; Dhamodharan, R.; Ramanujam, K. A Chitosan/Poly(Ethylene Glycol)-Ran-Poly(Propylene Glycol) Blend as an Eco-Benign Separator and Binder for Quasi-Solid-State Supercapacitor Applications. *Sustain. Energy Fuels* **2019**, *3*, 760–773. [CrossRef]
3. Murugan, E.; Govindaraju, S.; Santhoshkumar, S. Hydrothermal Synthesis, Characterization and Electrochemical Behavior of $NiMoO_4$ Nanoflower and $NiMoO_4$/RGO Nanocomposite for High-Performance Supercapacitors. *Electrochim. Acta* **2021**, *392*, 138973. [CrossRef]
4. Kesava, M.; Velautham, S.; Krishnan, S.; Kannaiyan, D. Graphene Nanosheets Dispersed Hydrophobic and Flexible Aliphatic Chain Containing Multifunctional Poly(Benzoxazines) Nanocomposites for Medium Temperature Proton Exchange Membrane Fuel Cell Applications. *Int. J. Energy Res.* **2022**, *46*, 18162–18178. [CrossRef]
5. Cao, X.; Cui, L.; Liu, B.; Liu, Y.; Jia, D.; Yang, W.; Razal, J.M.; Liu, J. Reverse Synthesis of Star Anise-like Cobalt Doped Cu-MOF/Cu_{2+1} O Hybrid Materials Based on a $Cu(OH)_2$ Precursor for High Performance Supercapacitors. *J. Mater. Chem. A* **2019**, *7*, 3815–3827. [CrossRef]
6. Kesava, M.; Dinakaran, K. SnO_2 Nanoparticle Assisted Enhanced Proton Exchange Membrane Fuel Cell Performance of Sulfuric Acid-Doped Porous Poly (Triphenylpyridine-Aliphatic Ethers). *J. Phys. Chem. C* **2021**, *125*, 130–142. [CrossRef]
7. Lade, H.; Kumar, V.; Arthanareeswaran, G.; Ismail, A.F. Sulfonated Poly(Arylene Ether Sulfone) Nanocomposite Electrolyte Membrane for Fuel Cell Applications: A Review. *Int. J. Hydrogen Energy* **2017**, *42*, 1063–1074. [CrossRef]
8. Le, Q.B.; Nguyen, T.H.; Fei, H.; Bubulinca, C.; Munster, L.; Bugarova, N.; Micusik, M.; Kiefer, R.; Dao, T.T.; Omastova, M.; et al. Electrochemical Performance of Composite Electrodes Based on RGO, Mn/Cu Metal–Organic Frameworks, and PANI. *Sci. Rep.* **2022**, *12*, 664. [CrossRef]
9. Selvaraj, K.; Yu, B.; Spontón, M.E.; Kumar, P.; Veerasamy, U.S.; Arulraj, A.; Mangalaraja, R.V.; Almarhoon, Z.M.; Sayed, S.R.M.; Kannaiyan, D. Nonylphenol Polybenzoxazines-Derived Nitrogen-Rich Porous Carbon (NRPC)-Supported g-C_3N_4/Fe_3O_4 Nanocomposite for Efficient High-Performance Supercapacitor Application. *Soft Matter* **2024**, *20*, 7957–7969. [CrossRef]
10. Selvaraj, K.; Spontón, M.E.; Estenoz, D.A.; Casarino, A.F.; Veerasamy, U.S.; Kumar, M.; Al-Mohaimeed, A.M.; Al-onazi, W.A.; Kannaiyan, D. Development of Quinoline-Based Heteroatom Polybenzoxazines Reinforced Graphitic Carbon Nitride (GCN) Carbonisation Composites for Emerging Supercapacitor Applications. *Soft Matter* **2024**, *20*, 1210–1223. [CrossRef]
11. Wang, S.; Ma, J.; Shi, X.; Zhu, Y.; Wu, Z.S. Recent Status and Future Perspectives of Ultracompact and Customizable Micro-Supercapacitors. *Nano Res. Energy* **2022**, *1*, e9120018. [CrossRef]
12. Amir, M.; Deshmukh, R.G.; Khalid, H.M.; Said, Z.; Raza, A.; Muyeen, S.M.; Nizami, A.S.; Elavarasan, R.M.; Saidur, R.; Sopian, K. Energy Storage Technologies: An Integrated Survey of Developments, Global Economical/Environmental Effects, Optimal Scheduling Model, and Sustainable Adaption Policies. *J. Energy Storage* **2023**, *72*, 108694. [CrossRef]
13. Wang, H.; Shao, Y.; Mei, S.; Lu, Y.; Zhang, M.; Sun, J.K.; Matyjaszewski, K.; Antonietti, M.; Yuan, J. Polymer-Derived Heteroatom-Doped Porous Carbon Materials. *Chem. Rev.* **2020**, *120*, 9363–9419. [CrossRef]
14. Rajendran, K.; Lolupiman, K.; Okhawilai, M.; Therese, H.A.; Kheawhom, S.; Tan, P.; Qin, J. Synthesis, Formation Mechanism and Supercapacitor Performance of MoS_2/Mo_2C/C Nanofibers. *J. Alloys Compd.* **2024**, *980*, 173549. [CrossRef]
15. Asrafali, S.P.; Periyasamy, T.; Kim, S.C. Enhanced Electrochemical Performance of HC/NiCo@800C//HC Using Redox-Active Electrolytes Showing Increased Energy Density. *J. Alloys Compd.* **2024**, *972*, 172753. [CrossRef]
16. Zhu, X.; Ji, C.; Meng, Q.; Mi, H.; Yang, Q.; Li, Z.; Yang, N.; Qiu, J. Freeze-Tolerant Hydrogel Electrolyte with High Strength for Stable Operation of Flexible Zinc-Ion Hybrid Supercapacitors. *Small* **2022**, *18*, 2200055. [CrossRef]
17. Silva, R.J.; Klobukoski, V.; de Paula, J.I.S.; Riegel-Vidotti, I.C.; Vidotti, M. Assembly of Symmetric Supercapacitor Based on Alginate Hydrogel Electrolyte and Polyaniline Modified Electrodes. *Electrochim. Acta* **2022**, *429*, 140914. [CrossRef]
18. Liu, J.; Ye, Z.; Hu, X.; Ahmed, S.; Song, S. High-Performance Na-Ion Conducting Polymer Gel Membrane for Supercapacitor Applications. *ACS Appl. Polym. Mater.* **2022**, *4*, 280–288. [CrossRef]
19. Jin, T.; Su, J.; Luo, Q.; Zhu, W.; Lai, H.; Huang, D.; Wang, C. Preparation of N,P Co-Doped Porous Carbon Derived from Daylily for Supercapacitor Applications. *ACS Omega* **2022**, *7*, 37564–37571. [CrossRef]
20. Zhang, P.; Mu, J.; Guo, Z.; Wong, S.I.; Sunarso, J.; Zhao, Y.; Xing, W.; Zhou, J.; Zhuo, S. Watermelon Peel-Derived Heteroatom-Doped Hierarchical Porous Carbon as a High-Performance Electrode Material for Supercapacitors. *ChemElectroChem* **2021**, *8*, 1196–1203. [CrossRef]
21. Ghosh, S.; Barg, S.; Jeong, S.M.; Ostrikov, K. Heteroatom-Doped and Oxygen-Functionalized Nanocarbons for High-Performance Supercapacitors. *Adv. Energy Mater.* **2020**, *10*, 2001239. [CrossRef]
22. Zuo, W.; Li, R.; Zhou, C.; Li, Y.; Xia, J.; Liu, J. Battery-Supercapacitor Hybrid Devices: Recent Progress and Future Prospects. *Adv. Sci.* **2017**, *4*, 1600539. [CrossRef]

23. Lukatskaya, M.; Dunn, B.; Gogotsi, Y. Multidimensional materials and device architectures for future hybrid energy storage. *Nat. Commun.* **2016**, *7*, 12647. [CrossRef]
24. Wang, D.-G.; Liang, Z.; Gao, S.; Qu, C.; Zou, R. Metal-organic framework-based materials for hybrid supercapacitor application. *Coord. Chem. Rev.* **2020**, *404*, 213093. [CrossRef]
25. Tiwari, I.; Tanwar, V.; Ingole, P.P.; Nebhani, L. Heteroatom-Enriched Carbon Particles Derived from Multifunctional Polybenzoxazine Particles for High-Performance Supercapacitors. *ACS Appl. Energy Mater.* **2024**, *7*, 7185–7204. [CrossRef]
26. Ye, X.; Fan, Q.; Shang, L.; Ye, F. Adsorptive Carbon-Based Materials for Biomedical Applications. *Eng. Regen.* **2022**, *3*, 352–364. [CrossRef]
27. Saida, T.; Sakakibara, K.; Igami, R.; Maruyama, T. Synthesis of a Pt/Carbon-Sphere Catalyst and Evaluation of Its Oxygen Reduction Reaction Activity in Acidic Environments. *Energy Fuels* **2022**, *36*, 1027–1033. [CrossRef]
28. Sobczuk, K.S.; Pełech, I.; Narkiewicz, U.; Staciwa, P.; Sibera, D.; Moszyński, D. The Influence of the Synthesis PH on the Morphology and Adsorption Properties of Carbon Spheres. *Appl. Surf. Sci.* **2023**, *608*, 155196. [CrossRef]
29. Kakani, V.; Ramesh, S.; Yadav, H.M.; Bathula, C.; Basivi, P.K.; Palem, R.R.; Kim, H.S.; Pasupuletti, V.R.; Lee, H.; Kim, H. Hydrothermal Synthesis of CuO@MnO_2 on Nitrogen-Doped Multiwalled Carbon Nanotube Composite Electrodes for Supercapacitor Applications. *Sci. Rep.* **2022**, *12*, 12951. [CrossRef]
30. Bîru, E.I.; Gârea, S.A.; Iovu, H. Developing Polybenzoxazine Composites Based on Various Carbon Structures. *Macromol. Chem. Phys.* **2019**, *220*, 1800322. [CrossRef]
31. Le, T.H.; Yoon, H. Strategies for Fabricating Versatile Carbon Nanomaterials from Polymer Precursors. *Carbon* **2019**, *152*, 796–817. [CrossRef]
32. Al Aiti, M.; Jehnichen, D.; Fischer, D.; Brünig, H.; Heinrich, G. On the Morphology and Structure Formation of Carbon Fibers from Polymer Precursor Systems. *Prog. Mater. Sci.* **2018**, *98*, 477–551. [CrossRef]
33. Liu, J.; Qiao, S.Z.; Liu, H.; Chen, J.; Orpe, A.; Zhao, D.; Lu, G.Q. Extension of the Stöber Method to the Preparation of Monodisperse Resorcinol-Formaldehyde Resin Polymer and Carbon Spheres. *Angew. Chem.—Int. Ed.* **2011**, *50*, 5947–5951. [CrossRef]
34. Wu, H.; Qin, Y.; Zong, S.; Hu, Y.; Xaba, M.S.; Liu, X.; Chen, A. Porous Yolk–Shell-Structured Carbon Nanospheres for Electrochemical Energy Storage. *J. Mater. Sci. Mater. Electron.* **2020**, *31*, 13321–13329. [CrossRef]
35. Konnola, R.; Anirudhan, T.S. Efficient Carbon Dioxide Capture by Nitrogen and Sulfur Dual-Doped Mesoporous Carbon Spheres from Polybenzoxazines Synthesized by a Simple Strategy. *J. Environ. Chem. Eng.* **2020**, *8*, 103614. [CrossRef]
36. Inagaki, M.; Toyoda, M.; Soneda, Y.; Morishita, T. Nitrogen-Doped Carbon Materials. *Carbon* **2018**, *132*, 104–140. [CrossRef]
37. Leng, J.; Wang, Z.; Wang, J.; Wu, H.H.; Yan, G.; Li, X.; Guo, H.; Liu, Y.; Zhang, Q.; Guo, Z. Advances in Nanostructures Fabricated: Via Spray Pyrolysis and Their Applications in Energy Storage and Conversion. *Chem. Soc. Rev.* **2019**, *48*, 3015–3072. [CrossRef]
38. Thirukumaran, P.; Atchudan, R.; Parveen, A.S.; Lee, Y.R.; Kim, S.C. Polybenzoxazine Originated N-Doped Mesoporous Carbon Ropes as an Electrode Material for High-Performance Supercapacitors. *J. Alloys Compd.* **2018**, *750*, 384–391. [CrossRef]
39. Yang, Y.; Du, H.; Sun, D.; Lu, C.; Lu, C.; Gao, J.; Xu, C.; Ma, X. Boosting Capacitive Performance of S-Doped Carbon Fibers via Substrate-Oriented Activation Methodology. *Ind. Eng. Chem. Res.* **2025**, *64*, 2745–2757. [CrossRef]
40. Wang, H.; Wang, P.; Li, J.; Ran, Q. Facile Preparation and Improved Electrochemical Performance of Oxygen-Enriched Porous Carbon Materials Based on Diacetal-Containing Polybenzoxazine. *Macromol. Mater. Eng.* **2023**, *308*, 2200508. [CrossRef]
41. Hulicova-Jurcakova, D.; Kodama, M.; Shiraishi, S.; Hatori, H.; Zhu, Z.H.; Lu, G.Q. Nitrogen-Enriched Nonporous Carbon Electrodes with Extraordinary Supercapacitance. *Adv. Funct. Mater.* **2009**, *19*, 1800–1809. [CrossRef]
42. Wang, L.; Sun, J.; Zhang, H.; Xu, L.; Liu, G. Preparation of Benzoxazine-Based N-Doped Mesoporous Carbon Material and Its Electrochemical Behaviour as Supercapacitor. *J. Electroanal. Chem.* **2020**, *868*, 114196. [CrossRef]
43. Cheng, Q.; Tang, J.; Ma, J.; Zhang, H.; Shinya, N.; Qin, L.C. Graphene and Carbon Nanotube Composite Electrodes for Supercapacitors with Ultra-High Energy Density. *Phys. Chem. Chem. Phys.* **2011**, *13*, 17615–17624. [CrossRef] [PubMed]
44. Mohamed, M.G.; Su, B.X.; Kuo, S.W. Robust Nitrogen-Doped Microporous Carbon via Crown Ether-Functionalized Benzoxazine-Linked Porous Organic Polymers for Enhanced CO_2 Adsorption and Supercapacitor Applications. *ACS Appl. Mater. Interfaces* **2024**, *16*, 40858–40872. [CrossRef]
45. Zhang, K.; Liu, Y.; Ishida, H. Polymerization of an AB-Type Benzoxazine Monomer toward Different Polybenzoxazine Networks: When Diels-Alder Reaction Meets Benzoxazine Chemistry in a Single-Component Resin. *Macromolecules* **2019**, *52*, 7386–7395. [CrossRef]
46. Samy, M.M.; Mohamed, M.G.; Kuo, S.W. Directly Synthesized Nitrogen-and-Oxygen–Doped Microporous Carbons Derived from a Bio-Derived Polybenzoxazine Exhibiting High-Performance Supercapacitance and CO_2 Uptake. *Eur. Polym. J.* **2020**, *138*, 109954. [CrossRef]
47. Mukherjee, S.; Amarnath, N.; Lochab, B. Oxazine Ring-Substituted 4th Generation Benzoxazine Monomers & Polymers: Stereoelectronic Effect of Phenyl Substituents on Thermal Properties. *Macromolecules* **2021**, *54*, 9510–9525. [CrossRef]

48. Pei, F.; An, T.H.; Zang, J.; Zhao, X.J.; Fang, X.L.; Zheng, M.S.; Dong, Q.F.; Zheng, N.F. From Hollow Carbon Spheres to N-Doped Hollow Porous Carbon Bowls: Rational Design of Hollow Carbon Host for Li-S Batteries. *Adv. Energy Mater.* **2016**, *6*, 1502539. [CrossRef]
49. Zhao, J.; Gilani, M.R.H.S.; Lai, J.; Nsabimana, A.; Liu, Z.; Luque, R.; Xu, G. Autocatalysis Synthesis of Poly(Benzoxazine-Co-Resol)-Based Polymer and Carbon Spheres. *Macromolecules* **2018**, *51*, 5494–5500. [CrossRef]
50. Majumdar, D.; Mandal, M.; Bhattacharya, S.K. V_2O_5 and Its Carbon-Based Nanocomposites for Supercapacitor Applications. *ChemElectroChem* **2019**, *6*, 1623–1648. [CrossRef]
51. Chen, N.; Younis, A.; Huang, S.; Chu, D.; Li, S. Advanced Three-Dimensional Hierarchical Pr_6O_{11}@Ni-Co Oxides-Based Core-Shell Electrodes for Supercapacitance Application. *J. Alloys Compd.* **2019**, *783*, 772–778. [CrossRef]
52. Das, T.K.; Ghosh, P.; Das, N.C. Preparation, Development, Outcomes, and Application Versatility of Carbon Fiber-Based Polymer Composites: A Review. *Adv. Compos. Hybrid Mater.* **2019**, *2*, 214–233. [CrossRef]
53. Zhang, Z.; Deng, S.; Wang, D.; Qing, Y.; Yan, G.; Li, L.; Wu, Y. Low-Tortuosity Carbon Electrode Derived from Wood@ZIF-67 for Supercapacitor Applications. *Chem. Eng. J.* **2023**, *454*, 140410. [CrossRef]
54. Wang, S.; Li, W.C.; Zhang, L.; Jin, Z.Y.; Lu, A.H. Polybenzoxazine-based monodisperse carbon spheres with low-thermal shrinkage and their CO_2 adsorption properties. *J. Mater. Chem. A* **2014**, *2*, 4406–4412. [CrossRef]
55. Wang, S.; Li, W.C.; Hao, G.P.; Hao, Y.; Sun, Q.; Zhang, X.Q.; Lu, A.H. Temperature-Programmed Precise Control over the Sizes of Carbon Nanospheres Based on Benzoxazine Chemistry. *J. Am. Chem. Soc.* **2011**, *133*, 15304–15307. [CrossRef]
56. Lu, N.; He, G.; Liu, J.; Liu, G.; Li, J. Combustion synthesis of graphene for water treatment. *Ceram. Int.* **2018**, *44*, 2463–2469. [CrossRef]
57. Nagaraju, P.; Alsalme, A.; Alswieleh, A.; Jayavel, R. Facile in-situ microwave irradiation synthesis of TiO_2/graphene nanocomposite for high-performance supercapacitor applications. *J. Electroanal. Chem.* **2018**, *808*, 90–100. [CrossRef]
58. Li, Y.-Y.; Li, Y.-L.; Liu, L.-N.; Xu, Z.-W.; Xie, G.; Wang, Y.; Zhao, F.-G.; Gao, T.; Li, W.-S. Microporous N- and O-Codoped Carbon Materials Derived from Benzoxazine for Supercapacitor Application. *Inorganics* **2023**, *11*, 269. [CrossRef]
59. Shaer, C.; Oppenheimer, L.; Lin, A.; Ishida, H. Advanced Carbon Materials Derived from Polybenzoxazines: A Review. *Polymers* **2021**, *13*, 3775. [CrossRef]
60. Thirukumaran, P.; Atchudan, R.; Parveen, A.S.; Lee, Y.R.; Kim, S.C. The Synthesis of Mechanically Stable Polybenzoxazine-Based Porous Carbon and Its Application as High-Performance Supercapacitor Electrodes. *New J. Chem.* **2021**, *45*, 8738–8746. [CrossRef]
61. Xu, X.; Zhou, J.; Nagaraju, D.H.; Jiang, L.; Marinov, V.R.; Lubineau, G.; Alshareef, H.N.; Oh, M. Flexible, Highly Graphitized Carbon Aerogels Based on Bacterial Cellulose/Lignin: Catalyst-Free Synthesis and its Application in Energy Storage Devices. *Adv. Funct. Mater.* **2015**, *25*, 3193–3202. [CrossRef]
62. Jin, Y.; Tian, K.; Wei, L.; Zhang, X.; Guo, X. Hierarchical porous microspheres of activated carbon with a high surface area from spores for electrochemical double-layer capacitors. *J. Mater. Chem. A* **2016**, *4*, 15968–15979. [CrossRef]
63. Sudhan, N.; Subramani, K.; Karnam, M.; Ilayaraja, N.; Sathish, M. Biomass-Derived Activated Porous Carbon from Rice Straw for a High-Energy Symmetric Supercapacitor in Aqueous and Non-aqueous Electrolytes. *Energy Fuels* **2017**, *31*, 977–985. [CrossRef]
64. Thubsuang, U.; Chotirut, S.; Thongnok, A.; Promraksa, A.; Nisoa, M.; Manmuanpom, N.; Wongkasemjit, S.; Chaisuwan, T. Facile Preparation of Polybenzoxazine-Based Carbon Microspheres with Nitrogen Functionalities: Effects of Mixed Solvents on Pore Structure and Supercapacitive Performance. *Front. Chem. Sci. Eng.* **2020**, *14*, 1072–1086. [CrossRef]
65. Lei, W.; Guo, J.; Wu, Z.; Xuan, C.; Xiao, W.; Wang, D. Highly nitrogen and sulfur dual-doped carbon microspheres for supercapacitors. *Sci. Bull.* **2017**, *62*, 1011–1017. [CrossRef] [PubMed]
66. Wang, Y.; Yan, X.; Tu, M.; Cheng, J.; Zhang, J. Resin-derived activated carbons with in-situ nitrogen doping and high specific surface area for high-performance supercapacitors. *Mater. Lett.* **2017**, *191*, 178–181. [CrossRef]
67. Zhu, D.; Jiang, J.; Sun, D.; Qian, X.; Wang, Y.; Li, L.; Wang, Z.; Chai, X.; Gan, L.; Liu, M. A general strategy to synthesize high-level N-doped porous carbons via Schiff-base chemistry for supercapacitors. *J. Mater. Chem. A* **2018**, *6*, 12334–12343. [CrossRef]
68. Song, Z.; Duan, H.; Zhu, D.; Lv, Y.; Xiong, W.; Cao, T.; Li, L.; Liu, M.; Gan, L. Ternary-doped carbon electrodes for advanced aqueous solid-state supercapacitors based on a "water-in-salt" gel electrolyte. *J. Mater. Chem. A* **2019**, *7*, 15801–15811. [CrossRef]
69. Li, Q.; Lu, T.; Wang, L.; Pang, R.; Shao, J.; Liu, L.; Hu, X. Biomass based N-doped porous carbons as efficient CO_2 adsorbents and high-performance supercapacitor electrodes. *Sep. Purif. Technol.* **2021**, *275*, 119204. [CrossRef]
70. Li, J.; Zou, Y.; Xiang, C.; Xu, F.; Sun, L.; Li, B.; Zhang, J. Osmanthus fragrans-derived N-doped porous carbon for supercapacitor applications. *J. Energy Storage* **2021**, *42*, 103017. [CrossRef]
71. Wan, L.; Wang, J.; Sun, Y.; Feng, C.; Li, K. Polybenzoxazine-Based Nitrogen-Containing Porous Carbons for High-Performance Supercapacitor Electrodes and Carbon Dioxide Capture. *RSC Adv.* **2015**, *5*, 5331–5342. [CrossRef]
72. Wang, J.C.; Liu, Q. An efficient one-step condensation and activation strategy to synthesize porous carbons with optimal micropore sizes for highly selective CO_2 adsorption. *Nanoscale* **2014**, *6*, 4148–4156. [CrossRef] [PubMed]
73. Hao, G.P.; Li, W.C.; Qian, D.; Lu, A.H. Rapid Synthesis of Nitrogen-Doped Porous Carbon Monolith for CO_2 Capture. *Adv. Mater.* **2010**, *22*, 853–857. [CrossRef] [PubMed]

74. Sun, L.; Tian, C.; Fu, Y.; Yang, Y.; Yin, J.; Wang, L.; Fu, H. Nitrogen-Doped Porous Graphitic Carbon as an Excellent Electrode Material for Advanced Supercapacitors. *Chem.–Eur. J.* **2014**, *20*, 564–574. [CrossRef]
75. Xu, B.; Zheng, D.; Jia, M.; Cao, G.; Yang, Y. Nitrogen-doped porous carbon simply prepared by pyrolyzing a nitrogen-containing organic salt for supercapacitors. *Electrochim. Acta* **2013**, *98*, 176–182. [CrossRef]
76. Chen, L.F.; Huang, Z.H.; Liang, H.W.; Yao, W.T.; Yu, Z.Y.; Yu, S.H. Flexible all-solid-state high-power supercapacitor fabricated with nitrogen-doped carbon nanofiber electrode material derived from bacterial cellulose. *Energy Environ. Sci.* **2013**, *6*, 3331–3338. [CrossRef]
77. Liu, Y.; Cao, L.; Luo, J.; Peng, Y.; Ji, Q.; Dai, J.; Zhu, J.; Liu, X. Biobased Nitrogen- and Oxygen-Codoped Carbon Materials for High-Performance Supercapacitor. *ACS Sustain. Chem. Eng.* **2019**, *7*, 2763–2773. [CrossRef]
78. Huang, C.H.; Zhang, Q.; Chou, T.C.; Chen, C.M.; Su, D.S.; Doong, R.A. Three dimensional hierarchically ordered porous carbons with partially graphitic nanostructures for electrochemical capacitive energy storage. *ChemSusChem* **2012**, *5*, 563–571. [CrossRef]
79. Chen, L.F.; Zhang, X.D.; Liang, H.W.; Kong, M.G.; Guan, Q.F.; Chen, P.; Wu, Z.Y.; Yu, S.H. Synthesis of nitrogen-doped porous carbon nanofibers as an efficient electrode material for supercapacitors. *ACS Nano* **2012**, *6*, 7092–7102. [CrossRef]
80. Li, J.G.; Lee, P.Y.; Ahmed, M.M.M.; Mohamed, M.G.; Kuo, S.W. Varying the hydrogen bonding strength in phenolic/PEO-b -PLA blends provides mesoporous carbons having large accessible pores suitable for energy storage. *Macromol. Chem. Phys.* **2020**, *221*, 2000040. [CrossRef]
81. Mohamed, M.G.; Hung, W.S.; EL-Mahdy, A.F.M.; Ahmed, M.M.M.; Dai, L.; Chen, T.; Kuo, S.W. High-molecular-weight PLA-b-PEO-b-PLA triblock copolymer templated large mesoporous carbons for supercapacitors and CO_2 capture. *Polymers* **2020**, *12*, 1193. [CrossRef]
82. Zhang, M.; Yang, C.; Wang, Y.; Gao, F.; Cheng, J.; Zhang, J. High-Performance Supercapacitor Based on Nitrogen and Phosphorus Co-Doped Nonporous Polybenzoxazine-Based Carbon Electrodes. *J. Electrochem. Soc.* **2018**, *165*, A3313–A3320. [CrossRef]
83. Lee, Y.-H.; Lee, Y.-F.; Chang, K.-H.; Hu, C.-C. Synthesis of N-doped carbon nanosheets from collagen for electrochemical energy storage/conversion systems. *Electrochem. Commun.* **2011**, *13*, 50. [CrossRef]
84. Wang, C.; Sun, L.; Zhou, Y.; Wan, P.; Zhang, X.; Qiu, J. P/N co-doped microporous carbons from H_3PO_4-doped polyaniline by in situ activation for supercapacitors. *Carbon* **2013**, *59*, 537. [CrossRef]
85. Bardestani, R.; Patience, G.S.; Kaliaguine, S. Experimental methods in chemical engineering: Specific surface area and pore size distribution measurements—BET, BJH, and DFT. *Can. J. Chem. Eng.* **2019**, *97*, 2781–2791. [CrossRef]
86. Liu, Y.; Yao, Z.; Zhou, J.; Jin, L.; Wei, B.O.; He, X. Facile synthesis of MOF-derived concave cube nanocomposite by self-templated toward lightweight and wideband microwave absorption. *Carbon* **2022**, *186*, 574–588. [CrossRef]
87. Chen, Y.; Zhang, Z.; Huang, Z.; Zhang, H. Effects of oxygen-containing functional groups on the supercapacitor performance of incompletely reduced graphene oxides. *Int. J. Hydrogen Energy* **2017**, *42*, 7186–7194. [CrossRef]
88. Bai, L.; Ge, Y.; Bai, L. Boron and Nitrogen Co-Doped Porous Carbons Synthesized from Polybenzoxazines for High-Performance Supercapacitors. *Coatings* **2019**, *9*, 657. [CrossRef]
89. Puthusseri, D.; Aravindan, V.; Madhavi, S.; Ogale, S. 3D micro-porous conducting carbon beehive by single step polymer carbonization for high performance supercapacitors: The magic of in situ porogen formation. *Energy Environ. Sci.* **2014**, *7*, 728–735. [CrossRef]
90. Wang, Y.G.; Song, Y.F.; Xia, Y.Y. Electrochemical capacitors: Mechanism, materials, systems, characterization and applications. *Chem. Soc. Rev.* **2016**, *45*, 5925–5950. [CrossRef]
91. Periyasamy, T.; Asrafali, S.P.; Kim, S.C. Nitrogen-Rich Porous Carbon/NiMn Hybrids as Electrode Materials for High-Performance Supercapacitors. *ACS Appl. Energy Mater.* **2022**, *5*, 15605–15614. [CrossRef]
92. Aasen, D.A.; Shen, Y.; Ivey, D.G. Zn-Based Oxides Anchored to Nitrogen-Doped Carbon Nanotubes as Efficient Bifunctional Catalysts for Zn-Air Batteries. *ChemElectroChem.* **2020**, *7*, 2283. [CrossRef]
93. Chen, X.; Chen, X.; Zhang, F.; Yang, Z.; Huang, S. One-pot hydrothermal synthesis of reduced graphene oxide/carbon nanotube/α-$Ni(OH)_2$ composites for high performance electrochemical supercapacitor. *J. Power Sources* **2013**, *243*, 555–561. [CrossRef]
94. Thirukumaran, P.; Atchudan, R.; Shakila Parveen, A.; Santhamoorthy, M.; Ramkumar, V.; Kim, S.-C. N-Doped Mesoporous Carbon Prepared from a Polybenzoxazine Precursor for High Performance Supercapacitors. *Polymers* **2021**, *13*, 2048. [CrossRef] [PubMed]
95. Zhu, G.; Wen, H.; Ma, M.; Wang, W.; Yang, L.; Wang, L.; Shi, X.; Cheng, X.; Sun, X.; Yao, Y. A self-supported hierarchical Co-MOF as a supercapacitor electrode with ultrahigh areal capacitance and excellent rate performance. *Chem. Commun.* **2018**, *54*, 10499–10502. [CrossRef] [PubMed]
96. Yang, Y.; Zhu, T.; Chi, C.; Liu, L.; Zheng, J.; Gong, X. All-Solid-State Asymmetric Supercapacitors with Novel Ionic Liquid Gel Electrolytes. *ACS Appl. Electron. Mater.* **2020**, *2*, 3906–3914. [CrossRef]
97. Ge, J.; Qu, Y.; Cao, L.; Wang, F.; Dou, L.; Yu, J.; Ding, B. Polybenzoxazine-Based Highly Porous Carbon Nanofibrous Membranes Hybridized by Tin Oxide Nanoclusters: Durable Mechanical Elasticity and Capacitive Performance. *J. Mater. Chem. A* **2016**, *4*, 7795–7804. [CrossRef]

98. Ma, C.; Li, Y.J.; Shi, J.L.; Song, Y.; Liu, L. High-performance supercapacitor electrodes based on porous flexible carbon nanofiber paper treated by surface chemical etching. *Chem. Eng. J.* **2014**, *249*, 216–225. [CrossRef]
99. Ike, I.S.; Sigalas, I.; Iyuke, S. Understanding performance limitation and suppression of leakage current or self-discharge in electrochemical capacitors: A review. *Phys. Chem. Chem. Phys.* **2016**, *18*, 661–680. [CrossRef]
100. Wang, G.; Wang, R.; Liu, L.; Zhang, H.; Du, J.; Zhang, Y.; Liu, M.; Liang, K.; Chen, A. Synthesis of hollow mesoporous carbon spheres via Friedel-Crafts reaction strategy for supercapacitor. *Mater. Lett.* **2017**, *197*, 71–74. [CrossRef]
101. Chen, A.; Xia, K.; Zhang, L.; Yu, Y.; Li, Y.; Sun, H.; Wang, Y.; Li, Y.; Li, S. Fabrication of Nitrogen-Doped Hollow Mesoporous Spherical Carbon Capsules for Supercapacitors. *Langmuir* **2016**, *32*, 8934–8941. [CrossRef]
102. Du, J.; Chen, A.; Liu, L.; Li, B.; Zhang, Y. N-Doped Hollow Mesoporous Carbon Spheres Prepared by Polybenzoxazines Precursor for Energy Storage. *Carbon* **2020**, *160*, 265–272. [CrossRef]
103. Ding, B.; Fan, Z.; Lin, Q.; Wang, J.; Chang, Z.; Li, T.; Henzie, J.; Kim, J.; Dou, H.; Zhang, X.; et al. Confined pyrolysis of ZIF-8 polyhedrons wrapped with graphene oxide nanosheets to prepare 3D porous carbon heterostructures. *Small Methods* **2019**, *3*, 1900277. [CrossRef]
104. Sobrinho, R.A.L.; Andrade, G.R.S.; Costa, L.P.; de Souza, M.J.B.; de Souza, A.; Gimenez, I.F. Ordered micro-mesoporous carbon from palm oil cooking waste via nanocasting in HZSM-5/SBA-15 composite: Preparation and adsorption studies. *J. Hazard. Mater.* **2019**, *362*, 53–61. [CrossRef] [PubMed]
105. Zhou, H.; Zhou, Y.; Li, L.; Li, Y.; Liu, X.; Zhao, P.; Gao, B. Amino acid protic ionic liquids: Multifunctional carbon precursor for N/S codoped hierarchically porous carbon materials toward supercapacitive energy storage. *ACS Sustain. Chem. Eng.* **2019**, *7*, 9281–9290. [CrossRef]
106. Shen, H.; Xia, X.; Ouyang, Y.; Jiao, X.; Mutahir, S.; Mandler, D.; Hao, Q. Preparation of high specific capacitance biomass based porous carbons for its application in supercapacitors. *ChemElectroChem* **2019**, *15*, 3599–3605. [CrossRef]
107. Liu, F.; Gao, Y.; Zhang, C.; Huang, H.; Yan, C.; Chu, X.; Xu, Z.; Wang, Z.; Zhang, H.; Xiao, X.; et al. Highly microporous carbon with nitrogen-doping derived from natural biowaste for highperformance flexible solid-state supercapacitor. *J. Colloid Interface Sci.* **2019**, *548*, 322–332. [CrossRef]
108. Lin, Y.; Chen, Z.; Yu, C.; Zhong, W. Heteroatom-doped sheet-like and hierarchical porous carbon based on natural biomass small molecule peach gum for high performance supercapacitors. *ACS Sustain. Chem. Eng.* **2019**, *7*, 3389–3403. [CrossRef]
109. Wan, L.; Du, C.; Yang, S. Synthesis of Graphene Oxide/Polybenzoxazine-Based Nitrogen-Containing Porous Carbon Nanocomposite for Enhanced Supercapacitor Properties. *Electrochim. Acta* **2017**, *251*, 12–24. [CrossRef]
110. Zhang, H.; Bhat, V.V.; Gallego, N.C.; Contescu, C.I. Thermal Treatment Effects on Charge Storage Performance of Graphene-Based Materials for Supercapacitors. *ACS Appl. Mater. Interfaces* **2012**, *4*, 3239–3246. [CrossRef]
111. Xu, J.; Gai, S.; He, F.; Niu, N.; Gao, P.; Chen, Y.; Yang, P. A Sandwich-Type Three-Dimensional Layered Double Hydroxide Nanosheet Array/Graphene Composite: Fabrication and High Supercapacitor Performance. *J. Mater. Chem. A* **2014**, *2*, 1022–1031. [CrossRef]
112. Shi, K.; Zhitomirsky, I. Polypyrrole Nanofiber-Carbon Nanotube Electrodes for Supercapacitors with High Mass Loading Obtained Using an Organic Dye as a Co-Dispersant. *J. Mater. Chem. A* **2013**, *1*, 11614–11622. [CrossRef]
113. Kotz, R.; Carlen, M. Principles and Applications of the Electrochemical Capacitors. *Electrochim. Acta* **2000**, *45*, 2483–2498. [CrossRef]
114. Mohamed, M.; Gamal, M.; Kuo, S. Pyrene-functionalized tetraphenylethylene polybenzoxazine for dispersing single-walled carbon nanotubes and energy storage. *Compos. Sci. Technol.* **2020**, *199*, 108360. [CrossRef]
115. Wen, Y.; Liu, X.; Wen, X.; Chen, X.; Szymanska, K.; Dobrzynska, R.; Mijowska, E. Na3PO4 assistant dispersion of nano-$CaCO_3$ template to enhance electrochemical interface: N/O/P co-doped porous carbon hybrids towards high-performance flexible supercapacitors. *Compos. Part B* **2020**, *199*, 108256. [CrossRef]
116. Yang, Y.; Jiang, K.; Guo, J.; Li, J.; Peng, X.; Hong, B.; Wang, X.; Ge, H. Facile fabrication of Au/Fe_3O_4 nanocomposites as excellent nanocatalyst for ultrafast recyclable reduction of 4-nitropheol. *Chem. Eng. J.* **2020**, *381*, 122596. [CrossRef]
117. Li, H.; Lun, N.; Bai, Y.J. N-doped carbon-coated TiN exhibiting excellent electrochemical performance for supercapacitors. *Electrochim. Acta* **2017**, *257*, 56–63.
118. Wang, S.; Wang, B.; He, S.; Wang, Y.; Cheng, J.; Li, Y. Enhancing the photovoltaic performance of planar heterojunction perovskite solar cells via introducing binary-mixed organic electron transport layers. *New J. Chem.* **2023**, *47*, 5048–5055. [CrossRef]
119. Sharma, P.; Tanwar, V.; Tiwari, I.; Ingole, P.P.; Nebhani, L. Sustainable Upcycling of Nitrogen-Enriched Polybenzoxazine Thermosets into Nitrogen-Doped Carbon Materials for Contriving High-Performance Supercapacitors. *Energy Fuels* **2023**, *37*, 7445–7467. [CrossRef]
120. Wu, J.; Zhang, D.; Wang, Y.; Hou, B. Electrocatalytic Activity of Nitrogen-Doped Graphene Synthesized via a One-Pot Hydrothermal Process towards Oxygen Reduction Reaction. *J. Power Sources* **2013**, *227*, 185–190. [CrossRef]
121. Bhattacharjya, D.; Kim, M.-S.; Bae, T.-S.; Yu, J.-S. High Performance Supercapacitor Prepared from Hollow Mesoporous Carbon Capsules with Hierarchical Nanoarchitecture. *J. Power Sources* **2013**, *244*, 799–805. [CrossRef]

122. Lau, S.C.; Lim, H.N.; Ravoof, T.B.S.A.; Yaacob, M.H.; Grant, D.M.; MacKenzie, R.C.I.; Harrison, I.; Huang, N.M. A Three-Electrode Integrated Photo-Supercapacitor Utilizing Graphene-Based Intermediate Bifunctional Electrode. *Electrochim. Acta* **2017**, *238*, 178–184. [CrossRef]
123. Zhao, J.; Niu, W.; Zhang, L.; Cai, H.; Han, M.; Yuan, Y.; Majeed, S.; Anjum, S.; Xu, G. A Template-Free and Surfactant-Free Method for High-Yield Synthesis of Highly Monodisperse 3-Aminophenol-Formaldehyde Resin and Carbon Nano/Microspheres. *Macromolecules* **2013**, *46*, 140–145. [CrossRef]
124. Silvestre-Albero, A.M.; Juárez-Galán, J.M.; Silvestre-Albero, J.; Rodríguez-Reinoso, F. Low-Pressure Hysteresis in Adsorption: An Artifact? *J. Phys. Chem. C* **2012**, *116*, 16652–16655. [CrossRef]
125. Wen, Y.; Chi, L.; Wenelska, K.; Wen, X.; Chen, X.; Mijowska, E. Eucalyptus Derived Heteroatom-Doped Hierarchical Porous Carbons as Electrode Materials in Supercapacitors. *Sci. Rep.* **2020**, *10*, 14631. [CrossRef] [PubMed]
126. Zhang, Y.; Li, G.Y.; Lv, Y.; Wang, L.Z.; Zhang, A.Q.; Song, Y.H.; Huang, B.L. Electrochemical Investigation of MnO_2 Electrode Material for Supercapacitors. *Int. J. Hydrogen Energy* **2011**, *36*, 11760–11766. [CrossRef]
127. Khan, M.S.; Shakya, P.; Bhardwaj, N.; Jhankal, D.; Sharma, A.K.; Banerjee, M.K.; Sachdev, K. Chemical Vapor Deposited Graphene-Based Quasi-Solid-State Ultrathin and Flexible Sodium-Ion Supercapacitor. *J. Electrochem. Sci. Eng.* **2022**, *12*, 799–813. [CrossRef]
128. Li, D.-J.; Lei, S.; Wang, Y.-Y.; Chen, S.; Kang, Y.; Gu, Z.-G.; Zhang, J. Helical carbon tubes derived from epitaxial Cu-MOF coating on textile for enhanced supercapacitor performance. *Dalton Trans.* **2018**, *47*, 5558–5563. [CrossRef] [PubMed]
129. Chameh, B.; Pooriraj, M.; Keyhan, M.; Moradi, M. Cu-MOF-derived CuO/NiO/$Ni_3(VO_4)_2$ composite materials with improved electrochemical performance for supercapacitor. *J. Mater. Sci Mater. Electron.* **2023**, *34*, 525. [CrossRef]
130. Deng, Y.; Xie, Y.; Zou, K.; Ji, X. Review on recent advances in nitrogen-doped carbons: Preparations and applications in supercapacitors. *J. Mater. Chem. A* **2016**, *4*, 1144–1173. [CrossRef]

batteries

MDPI

Review

Control Algorithms for Ultracapacitors Integrated in Hybrid Energy Storage Systems of Electric Vehicles' Powertrains: A Mini Review

Florin Mariasiu

Automotive Engineering and Transports Department, Technical University of Cluj-Napoca, Bdul, Muncii 103-105, RO-400641 Cluj-Napoca, Romania; florin.mariasiu@auto.utcluj.ro

Abstract

The integration of ultracapacitors into the propulsion systems and implicitly into the hybrid energy storage systems (HESSs) of electric vehicles offers significant prospects for increasing performance, improving efficiency and extending the lifetime of battery systems. However, the realization of these benefits critically depends on the implementation of sophisticated control algorithms. From fundamental rule-based systems to advanced predictive and intelligent control strategies, the evolution and integration of these algorithms are driven by the need to efficiently manage the power flow, optimize energy utilization and ensure the long-term reliability of hybrid energy storage systems. This study briefly presents (in the form of a mini review) the research in this field and the development directions and application of state-of-the-art control algorithms, also highlighting the needs, challenges and future development directions. Based on the analysis made, it is found that from the point of view of performance vs. ease of implementation and computational resource requirements, fuzzy algorithms are the most suitable for HESS control in the case of common applications. However, when the performance requirements of HESSs relate to special and high-tech applications, HESS control will be achieved by using convolutional neural networks. As electric vehicles continue to evolve, the development of more intelligent, adaptive and robust control algorithms will be essential for achieving the full potential of integrating ultracapacitors into electric mobility.

Keywords: ultracapacitors; battery; electric vehicles; hybrid energy storage systems (HESS); control; algorithms; artificial intelligence

1. Introduction

The contemporary need to reduce the amount of pollutant emissions caused by transport is essential for several important reasons, related to public health, environmental protection and combating climate change. Transport, especially road transport, generates emissions of fine particles (PM 2.5), nitrogen oxides (NOx) and other toxic substances that increase the incidence of respiratory and cardiovascular diseases, the risk of cancer and other chronic conditions caused directly and/or indirectly by them [1,2]. The transport sector is also responsible for a significant part of greenhouse gas emissions (CO_2, methane, etc.), which contributes to the global warming process with immediate effects in increasing the frequency of extreme weather phenomena (periods of prolonged drought, floods, extreme heat waves, etc.), and moreover, it should not be forgotten that pollutant emissions due to transport affect air and water quality, biodiversity and disrupt urban and rural

 https://doi.org/10.3390/batteries11110395

ecosystems. The difference is substantial if we consider the Global Warming Potential parameter values (g CO_2-eq/km) as a reference point, with electric vehicles having an average value of 182.9 while internal combustion engine vehicles have a value of 258.5 [3].

By their widespread introduction in transport, electric vehicles (EVs) play an essential role in reducing polluting emissions from transport, significantly contributing to the transition to a more sustainable mobility system, they directly contribute to the elimination of direct emissions, reduce air pollution in urban environments (where traffic is intense), reduce noise pollution and in the long term have a lower carbon footprint than vehicles with internal combustion engines (even if the industrial processes of battery production involve increased polluting emissions, but also taking into account second-life applications [4]). Mainly all these advantages presented above are due to the superior energy efficiency (40–70%) because the electric motors used for propulsion are much more efficient than those with internal combustion (11–27%) and less energy is lost in the form of heat dissipated in the environment [5].

Thus, it can be said that electric vehicles are at the forefront of progress towards sustainable transport, but under current conditions of technological development, their performance is intrinsically linked to the energy capacities of their battery energy storage systems (BESS), storage systems based mainly on Li Ion technology, a technology that has demonstrated its reliability in operation (Table 1) [6–8]. One of the barriers identified in the massive penetration of EVs on the automotive market is the fact that consumers want a vehicle autonomy almost equal to that of vehicles with thermal engines, and in this context an important parameter is the energy storage capacity in batteries.

Table 1. Comparison between energy density and power density for the main types of batteries used in electric vehicle applications and energy storage systems [6–8].

Energy Storage System	Energy Density (Wh/kg)	Power Density	Life Cycle	Safety	Relative Cost
LPF battery	90–120	High	2.000	Very good	Low
NMC battery	150–220	Average	1.500	Good	Average
LTO battery	50–80	Very high	10.000	Excellent	High
NiMH battery	60–120	Average	500	Good	Low
Ultracapacitors	5–10	Extremely high	100.000	Excellent	High

Specifically, Li-ion battery-based energy storage systems face major challenges in terms of power density [9,10]. This particular characteristic becomes a problem in demanding driving situations, such as aggressive acceleration or rapid energy absorption during regenerative braking. Cold weather further aggravates this problem, as a low ambient temperature: slows down the chemical reactions inside the battery, decreases the conductivity of the electrolyte, and increases the internal resistance of the battery [11]. Also, when the battery is subjected to high or sudden energy demands (e.g., rapid acceleration, starting at low temperatures) or EV operation is carried out under extreme conditions (cold weather, rapid charging or deep discharging, charging/discharging frequency), rapid voltage fluctuations/variations occur, which can put significant stress on the battery, accelerating its degradation, shortening its overall lifespan and limiting the maximum performance and efficiency of the electric vehicle (Table 2) [12–15].

Table 2. Operating temperature ranges of the main types of batteries used in electric vehicle applications and energy storage systems [12–15].

Energy Storage System	Optimal Operating Temperature	Minimum Temperature	Maximum Temperature	Comments
LPF battery	20–45 °C	−20 °C	60 °C	Excellent thermal stability making it ideal for various environments
NMC battery	15–40 °C	0 °C	60 °C	High energy density, but less thermally stable than LFP
LTO battery	−30–55 °C	−40 °C	60 °C	Works well in extreme temperatures, making it ideal for industrial applications
NiMH battery	0–45 °C	−20 °C	60 °C	Average performance given sensitivity to extreme temperatures
Ultracapacitors	−40–65 °C	−40 °C	85 °C	Ideal for applications with high power requirements as they are extremely thermally resistant

Based on the performances presented above, it can be stated that supplementary/complementary energy sources (such as ultracapacitors) are becoming an important choice and application for increasing the performance and improving the long-term viability of electric vehicle technology. Ultracapacitors (UC), also known as supercapacitors or electrochemical capacitors, represent such an ideal complementary energy storage solution due to their unique technical characteristics. Their high-power density allows them to deliver and absorb extremely high currents almost instantaneously. This capacity is measured in kW/kg, with typical values ranging from 1 to 10 kW/kg, significantly higher than the 0.2 to 0.4 kW/kg of a standard lithium-ion battery. This makes UC perfect for handling peak power demands during aggressive acceleration and for capturing rapid energy pulses from regenerative braking [16–20].

The superior power density of ultracapacitors comes from their energy storage mechanism. Unlike batteries, which rely on slow electrochemical reactions, UCs store energy electrostatically by separating charge at the interface between an electrode and an electrolyte. This physical process is incredibly fast and reversible, allowing for rapid charge and discharge cycles. The energy stored (E) is given by the formula $E = CV^2/2$, where C is the capacitance and V is the voltage. In addition, UCs possess a very low equivalent series resistance (ESR), typically in the milliohm range. This low internal resistance (R) translates into high efficiency, often exceeding 95%, as less energy power (P_{loss}) is lost as heat ($P_{loss} = I^2R$) during power transfer (depending on current intensity I).

The non-destructive nature of their physical charge storage mechanism gives UCs exceptional cycle life and makes them a robust and long-lasting component within a hybrid energy storage system. They can withstand millions of charge/discharge cycles with negligible degradation, a stark contrast to lithium-ion batteries, which are limited to a few thousand cycles before significant capacity loss occurs. UCs also have a much wider operating temperature range than batteries. They can operate effectively at temperatures from approximately −40 °C to +65 °C, while lithium-ion batteries suffer significant performance degradation and risk of damage at extreme temperatures, particularly below zero degrees.

This temperature resistance ensures consistent performance in a variety of climates without the need for extensive thermal management systems.

The integration of ultracapacitors into electric vehicle powertrains, typically in a hybrid energy storage system (HESS) configuration alongside batteries, aims to capitalize on these advantages. The primary goal is to discharge the battery due to peak power demands, thereby extending battery life, improving overall system efficiency, enhancing vehicle performance (e.g., faster acceleration, more efficient regenerative braking), and potentially reducing the size and cost of the battery pack (compared to the entire lifetime) [21,22]. Using a HESS for the powertrain of a mining haul truck, it was found that in addition to the advantages related to traction power, under the specific conditions of the research, the lifecycle cost of HESSs can be reduced by as much as 23.94% (compared to the battery-only solution) [23].

Also, the implementation of a HESS (consisting of 7100 LiFePO4 electrochemical cells and 72 UCs) in a plug-in electric vehicle demonstrated, under operating conditions based on the UDDS cycle, a reduction in operating costs by 11.9%, in parallel with an extension of the battery life cycle by 21.7% (also compared to the battery-only solution) [24].

It should be noted that these results directly depend on the way the HESS is composed (the ratio between the number of battery cells and the number of ultracapacitors), which directly influences the total cost and life cost of the HESS.

2. Architectures of Hybrid Energy Storage Systems in Electric Vehicles Propulsion Systems

The effectiveness of utilizing the superior energy performance of the UC in the powertrain of an electric vehicle depends largely on how the UC is physically integrated with the battery and the rest of the powertrain. In general, common HESS architectures have a constructive topology divided into 4 classes (passive, semiactive, active and full active), shown in Figures 1–4.

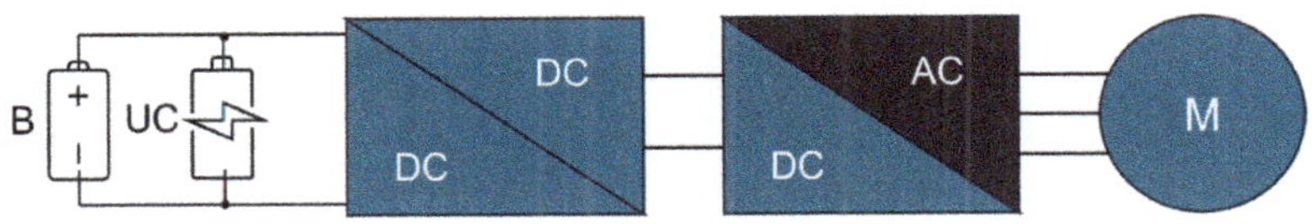

Figure 1. Passive HESS architecture.

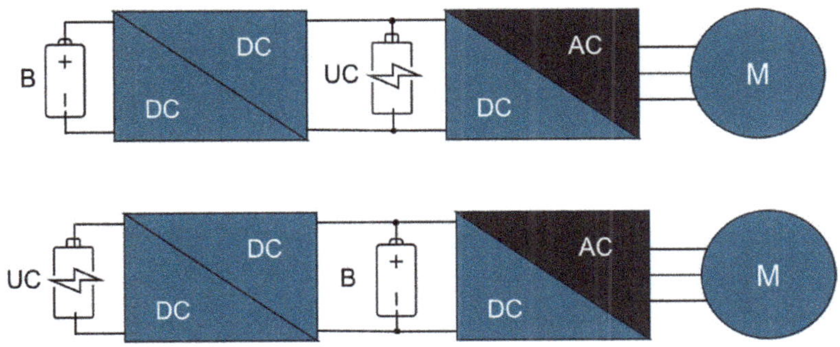

Figure 2. Semiactive HESS architecture.

When selecting various HESS topologies to outfit electric vehicle powertrains, technical and financial performance criteria may be taken into consideration. In this regard, Table 3 is presented which summarizes the comparative CAPEX/OPEX expenses, weight and dimensions, efficiency and suitability for applications for above presented HESS topologies using batteries and UC: passive, semi-active, active and fully active. Additionally, it takes into account the range of requirements for EV powertrain applications where passive topologies are still feasible and active topologies are superior.

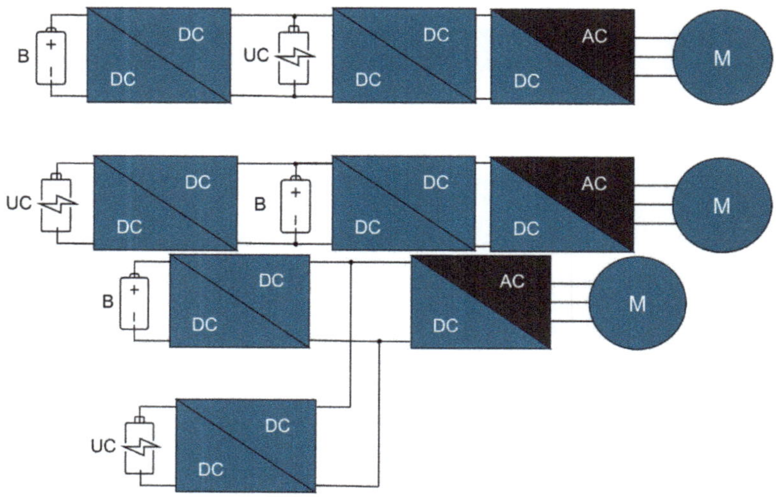

Figure 3. Active HESS architecture.

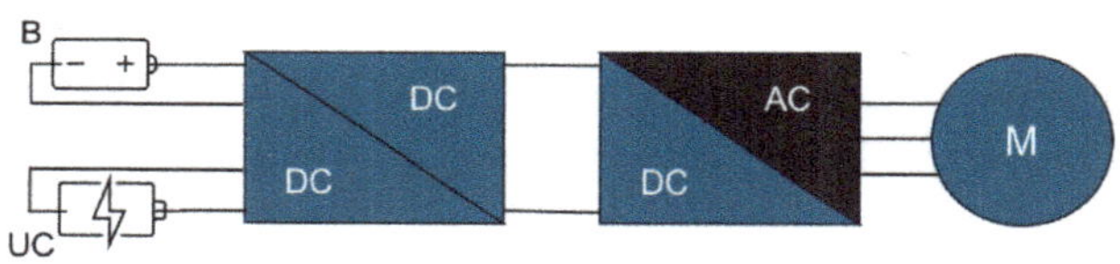

Figure 4. Full active HESS architecture.

Table 3. Comparative analysis of criteria for using a specific HESS topology for EV powertrain [25–28].

Criteria	HESS Topology			
	Passive	Semi-Active	Active	Full-Active
CAPEX	*	**	***	****
OPEX	*	**	**	***
Weight	*	**	**	***
Dimensions	*	**	***	****
Efficiency	**	**	***	****
Control	*	**	***	****
Peak Power Range	<10 kW	10–50 kW	50–200 kW	>200 kW
Peak Frequency Range	<1 Hz	1–10 Hz	10–100 Hz	>100 Hz
SOC Window	±5 . . . 10%	±10 . . . 20%	±20 . . . 40%	±40 . . . 60%
Best for EV class	Small	Mainstream	Luxury	High Performance and Race

* Low; ** Moderate; *** High; **** Very high.

Constructive and interconnection topology is essential to be active type for the future development of HESSs, given the tremendous advancements in artificial intelligence applications, automation and production technologies [21]. However, it must be taken into consideration that active topology currently faces challenges in large-scale applicability and in different electric powertrain configurations (multiple classes and EV models) due to its high costs, complexity, and space requirements.

3. Control Algorithms for Ultracapacitors Integrated in Electric Vehicles' Powertrains

3.1. Challenges and Objectives of Ultracapacitors Integration

The direct integration of ultracapacitors into the electric vehicle powertrain, despite the immediate energy benefits, presents a number of challenges that require complex control systems to optimize the low energy density, voltage variation, thermal management and, last but not least, cost optimization.

Because integrated ultracapacitors store much less energy per unit mass or volume than batteries, they are not capable of sustaining an electric vehicle alone for long periods of time (if energy densities of present lithium-ion batteries are 150–250 Wh/kg, ultracapacitors are typically limited to 5–15 Wh/kg). On the other hand, because UC serves as a high-power buffer for the main battery, absorbing the demands of high-power transient energy events, their incorporation into a hybrid energy storage system is necessary to protect the battery and increase its lifetime.

The voltage of an ultracapacitor decreases linearly with the amount of energy discharged, unlike batteries, whose voltage remains relatively constant during discharge (until near exhaustion) (an ultracapacitor at half its maximum voltage has only 25% of its stored energy left). This significant voltage fluctuation is problematic for the electric vehicle's traction motor inverter, which requires sophisticated control and DC/DC converters to maintain a constant powertrain voltage. Complex control algorithms can be used to command and regulate the DC/DC converters to manage the optimal SOC (State of Charge) of the UC and dynamically change the voltage from the UC to the level required by the DC bus [29–31].

Although they are less susceptible to operating at extremely high or low temperatures than batteries, they can still produce excessive heat when operating at high currents, as might occur during high-speed driving. This heat can then spread or dissipate to other parts of the energy storage system, requiring the inclusion of a specific thermal management system in the storage system design. To prevent component degradation and maintain system reliability, a specialized thermal management system must be considered in the overall design of energy storage systems, even though UCs are robust devices to temperature fluctuations [32,33]. The need to monitor system/component temperatures and, if necessary, limit current to prevent overheating requires the use of a control system and sophisticated command and control algorithms.

The cost-effectiveness of integrating a UC package is not measured in terms of energy capacity, but rather in the value it adds by extending battery life, creating efficiency, and providing adequate dynamic performance. This is because, despite continued cost reductions, UCs can still be relatively expensive to implement in propulsion systems compared to the energy capacity they provide [17]. To optimize these benefits, support the higher initial investment, and increase the economic viability of the HESS or solution, the control algorithms must be sufficiently complex.

The challenges presented above related to the inclusion of UC in energy storage systems, emphasizing the critical role of advanced control algorithms in efficiently managing the power flow and optimizing the performance of electric vehicle propulsion systems integrated in UC. Control algorithms for ultracapacitors in electric vehicle propulsion systems are designed to achieve several interconnected objectives considering basic HESS KPIs: optimizing power sharing, extending battery life, maximizing system efficiency, regulating system voltage, managing battery state of charge (SOC) and UC, improved vehicle dynamic performance, fault tolerance and operational safety (Table 4, Figure 5).

Table 4. HESS main KPIs.

KPI Name	Unit	Description	Method or Source
RMS_{I_B}	A	Root Mean Square of battery current over a drive cycle	Simulation or measurement
I_{peak_B}	A	Peak current drawn from the battery	Simulation or measurement
ΔSOC_B	%	Change in battery State of Charge during operation	SOC model or BMS data
ΔSOC_{UC}	%	Change in ultracapacitor State of Charge	UC voltage-based SOC estimation
Recuperation Share	%	Share of braking energy recovered and stored	Energy flow analysis
System Efficiency	%	Overall energy efficiency of the HESS	$\eta = \frac{E_{out}}{E_{in}} \times 100$
Thermal Load	°C/W	Temperature rise per unit thermal resistance	Thermal model or sensor data
Degradation/Cycle	Ah/cycle	Battery degradation per cycle	Aging model or empirical data
Computational Complexity	-	Algorithmic complexity	Theoretical analysis of control algorithm
Step Time	ms	Time per simulation/control step	Profiling during simulation
Memory Requirements	MB	Memory capacity (and speed) needed for simulation or real-time control	System profiling or estimation

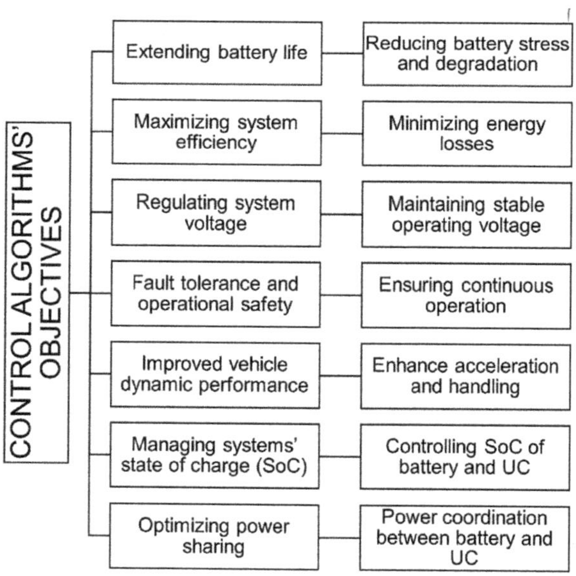

Figure 5. Interconnected main objectives of control algorithms for UC in electric vehicles propulsion systems.

The control objectives presented above are not the only possible ones, they are mainly (or be must) correlated with the architecture of the energy storage system and the operational and performance requirements of the electric vehicle.

3.2. Control Algorithms

Control algorithms for integrating ultracapacitors into electric propulsion systems can be classified according to how they approach energy management and the optimization of the overall system performance. They range from classical methods (such as threshold-based control or PID) to heuristic approaches (which use empirical rules for energy distribution). In recent years, there has been a trend towards intelligent algorithms, which use neural networks, fuzzy logic or model-based predictive control (MPC) to anticipate power requirements and adapt the control strategy in real time. The choice of methods depends on the complexity of the system, the performance requirements and the available processing capacity [34,35]. The integration of these algorithms significantly contributes to increasing energy efficiency, protecting the battery and improving the durability of the propulsion system.

3.2.1. Rule-Based Control (RBC)/Heuristic Control

RBC is the simplest approach because it relies on predefined rules and thresholds (specific to design and operating parameters) to manage the power flow based on the following general principle: power splitting decisions are based on the power demand from the driver, the battery and ultracapacitor SOC, as well as other operational parameters (Figure 6). These rules are usually derived from extensive empirical data, simulations or expert knowledge. The main advantages are simple implementation, low computational burden, easy to understand, but there are also some disadvantages, such as suboptimal performance under variable operating conditions, inadaptability to system degradation or external disturbances, difficulty in defining process of exactly optimal thresholds.

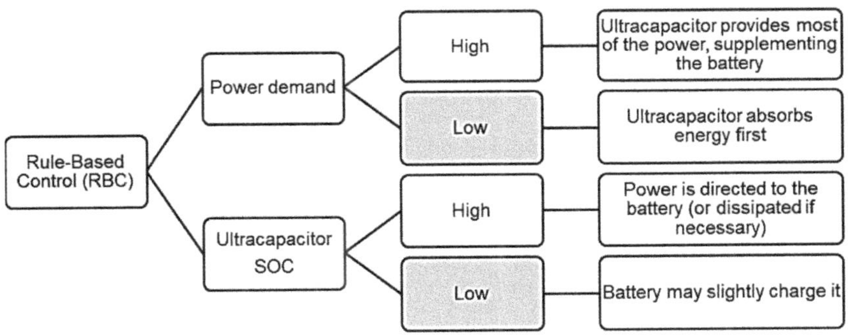

Figure 6. Examples of Rule-Based Control (RBC) considering exploitation conditions.

A multi-layer control approach is used to identify the best energy management strategy for HESSs in electric vehicles [36]. The efficiency of a standard rule-based algorithm, an adaptive rule-based algorithm, and an advanced adaptive rule-based algorithm were considered in the analysis taking into account the following functional properties:

- The standard rule-based algorithm uses fixed parameters that do not take into account the topographic features of the route
- The adaptive rule-based algorithm adjusts the parameters based on the energy flow produced by regenerative braking
- The advanced adaptive rule-based algorithm continuously updates the parameters considering the operating cycle.

Regardless of the driving (operating) cycle considered, the standard rule-based algorithm efficiently distributes the vehicle load current between the battery and the UC, while maintaining the battery current at a predetermined target value (usually the highest). The

HESS manages the energy flow by considering the direction of the energy flow, the total requested load current, and the state of charge of the UC. The operational parameters of the algorithm are established using Equation (1).

$$\text{If } (I_t > 0) \text{ and } (I_t < I_{Bmax}) \text{ then } I_{CO} = 0$$

$$\text{If } (I_t > 0) \text{ and } (I_t > I_{Bmax}) \text{ and } (SOC_{UC} > SOC_{UCmin}) \text{ then } I_{CO} = (I_t - I_{Bmax})$$

$$\text{If } (I_t < 0) \text{ and } (SOC_{UC} < SOC_{UCmax}) \text{ then } Ico = I_t$$

where I_t is total load current of the vehicle, I_{CO} is output current of DC-DC converter, I_B is battery current, V_{CO} is output voltage of DC-DC converter and SOC_{UC} is ultracapacitor state of charge.

When the electric vehicle's total load current is less than the maximum battery current, the HESS can supply current from the battery to the vehicle using a standard rule-based algorithm. Additionally, it limits the battery current to its maximum value during high load driving cycles and uses the UC to capture all regenerative energy during the driving cycle's deceleration.

The adaptive rule-based algorithm dynamically modifies the algorithm coefficients to improve system performance. Both the regenerative current and the total current demand required for a given driving cycle are calculated taking into account variables such as the electric vehicle model parameters, vehicle speed, and road gradient. The energy distribution between the battery and UC is also independently optimized. By estimating the potential regenerative energy and establishing an energy sharing ratio between the battery and the UC, which is modified according to the driving cycle, the energy management system (EMS) ensures that the HESS operates efficiently and that the UC utilizes all the regenerative energy during the driving cycle.

By estimating the regenerative energy from the current driving cycle, the advanced adaptive rule-based algorithm calculates the energy sharing ratio between the battery and the UC. This method guarantees that the supercapacitor manages peak loads and cooperates with the battery to meet the EVs' transient load requirements by adjusting the energy sharing ratio and the maximum permitted battery current in accordance with the amount of regenerative energy available (this method maximizes the reuse of the regenerative energy produced, ensuring sufficient capacity to deliver the powertrain the energy needed during vehicle acceleration).

Simulations were conducted in [36] using three standard driving cycles (UDDS, NYCC, and Japan1015) and varying initial states of charge of the UCs (in the first case 92%, in the second case 51% and in the third case 20%) to evaluate the effectiveness of these control strategies. In the first case, the standard adaptive rule-based algorithm allowed the HESS to complete 29 driving cycles for UDDS conditions, 234 for NYCC conditions and 89 for Japan1015 conditions. In contrast, the advanced rule-based algorithm significantly improved these results, allowing 30 driving cycles for UDDS conditions, 357 for NYCC conditions, and 100 for Japan1015 conditions. Thus, it was proven that the HESS control by advanced adaptive algorithm proved effective for all three considered driving cycles, as it efficiently manages the energy sharing between the battery and the ultracapacitor in the HESS.

The results of the aforementioned study demonstrated that both the adaptive rule-based algorithm and the standard rule-based algorithm successfully mitigate current peaks and reduce battery power consumption (as well as the total power consumption of electric vehicles). Comparing the performance of these two algorithms with the efficiency of the advanced adaptive algorithm, it is observed that in addition to the benefits presented above, an increase in the maximum number of driving cycles that can occur is also achieved.

The rule-based HESS active power system (battery controller and ultracapacitor controller) for a low-power electric vehicle is approached differently in [37]. Since the proposed HESS active power scheme reduces the battery load, it improves the controllability of the system and provides efficient and superior control while an electric vehicle is in operation (Figure 7). Simulation results for different control modes demonstrate that the battery and ultracapacitor can effectively share active power based on the state of charge of the energy storage device, the peak load requirement, the instantaneous fluctuations of the electric vehicle load, and the load voltage adjustments.

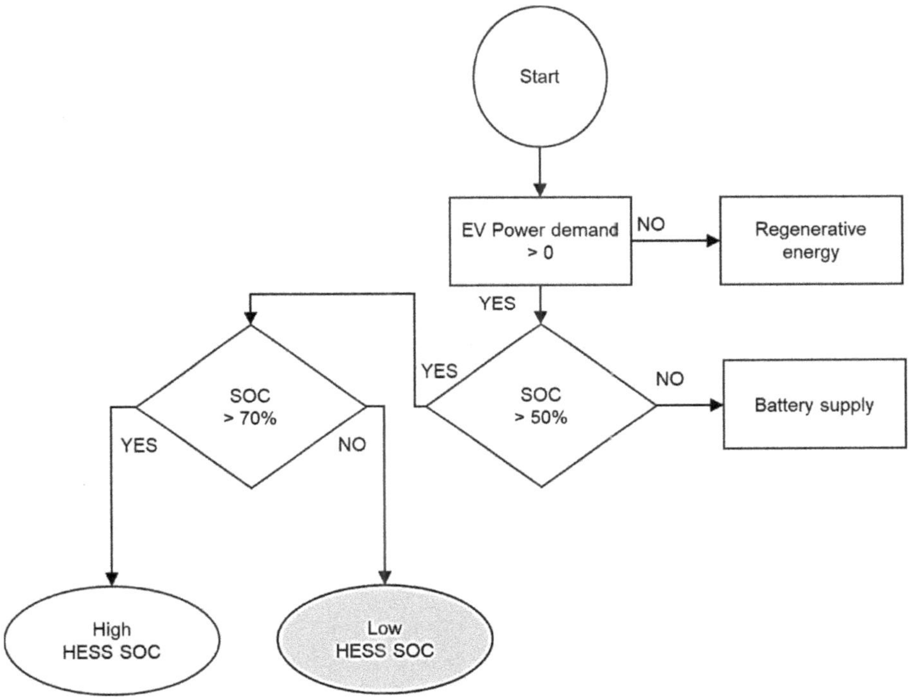

Figure 7. Flowchart of the rule-based algorithm for a HESS (adapted from [37]).

Real-time parameters and suboptimal power splitting, which has nonlinear data and linguistic knowledge, are used in fuzzy rule-based control to determine the best result. With real-time parameters, the primary benefits are adaptability (easy to adjust) and robustness (tolerance to imprecise measurements). Conventional fuzzy strategy, adaptive fuzzy strategy, and predictive fuzzy strategy are its subcategories.

Figure 8 represents a general fuzzy logic-based control algorithm for a HESS consisting of a battery and an ultracapacitor, with a detailed explanation of the algorithm steps. The fuzzy control algorithm for HESS allocates power between a battery and an ultracapacitor based on load demand and SOC levels. Inputs include the load power (P_{load}), its rate of change (dP/dt), and the SOC of both the battery and UC. The controller fuzzifies these inputs, applies a rule base to decide the power sharing, and defuzzifies the outputs to produce two references: P_{B_ref} and P_{UC_ref}. Fast power transients are handled by the UC, the battery provides constant power and charges the UC as needed, and a ramp limiter is used to filter out peaks to prevent battery degradation. The final outputs control the DC-DC converters to maintain DC bus stability and ensure SOC constraints.

INPUTS: P_{load} - Load power demand, SOC_B - Battery State of Charge, SOC_{UC} - UC State of Charge; dP/dt - Rate of change of P_{load}

START OF ALGORITHM

Step 1. Measure P_{load} from the DC bus

Step 2. Calculate dP/dt

Step 3. Measure SOC_B and SOC_{UC}

Step 4. Fuzzification:
Convert inputs (P_{load}, dP/dt, SOC_B, SOC_{UC}) into fuzzy sets:
P_{load} - {Low, Medium, High}
dP/dt - {Negative, Zero, Positive}
SOC_B - {Low, Medium, High}
SOC_{UC} - {Low, Medium, High}

Step 5. Apply Fuzzy Rule Base:

- IF dP/dt is High AND SOC_{UC} is High THEN UC supplies most power
- IF dP/dt is Low AND SOC_B is High THEN Battery supplies most power
- IF SOC_{UC} is Low THEN Battery charges UC

Step 6. Inference:

Combine all matching rules to determine fuzzy outputs for:

- Battery (B) contribution

- Ultracapacitor (UC) contribution

Step 7. Defuzzification:

Convert fuzzy outputs into crisp values:

- P_{B_ref} (Battery power reference)

- P_{UC_ref} (UC power reference)

Step 8. Apply constraints:
- Limit P_{B_ref} ramp rate
- Clamp outputs to respect SOC_B and SOC_{UC} limits

Step 9. Send P_{B_ref} to Battery DC-DC Converter

Step 10. Send P_{UC_ref} to UC DC-DC Converter

Step 11. Ensure power balance:
$P_{B\_ref} + P_{UC\_ref} \approx P_{load}$

END OF ALGORITHM

OUTPUTS: P_{B_ref} - Battery power reference; P_{UC_ref} - Ultracapacitor power reference

Figure 8. General steps of a fuzzy based control algorithm for a HESS.

One of the first and early attempts in this regard was carried out through simulation activities, which were used to verify the energy management strategy within the Urban Dynamometer Driving Schedule (UDDS) dynamic driving cycle by controlling the HESS (battery + UC) with fuzzy algorithms [38]. The results show that the proposed energy management strategy based on fuzzy logic can ensure the operation of the battery pack in a high efficiency range and can present better performances than the traditional control strategy based on logical thresholds. The electricity economy of the HESS was improved by 4.1%, and the negative influences of high-current discharge and charge on the battery pack were avoided.

The real-time implementation of a fuzzy logic energy management strategy was applied to a hybrid battery-ultracapacitor energy storage system and associated with a permanent magnet synchronous motor (PMSM) emulating the traction part of an electric vehicle [39]. Based on the results, it was determined that the fuzzy logic supervisor divides the frequency efficiently and acts intelligently to smoothly permute between the various operating modes. The suggested fuzzy logic supervisor was able to guarantee the proper operation of each energy source (based on the dynamics of the energy requirement) and deliver quick and high performance at various EV speed levels, in addition to the seamless operation of the entire HESS. An ideal power flow to the powertrain was maintained (while maintaining the UC operation within a safe voltage range) as a result of the discrete and fluid regulation of both the DC bus voltages and the UC voltages (regardless of the variations in the vehicle speed profile, according to the operating conditions).

In order to identify the optimal solution for the HESS control strategy of an electric vehicle, the efficiency of fuzzy, GA-Fuzzy and PSO-Fuzzy control algorithms was analyzed under the same optimization conditions (electric vehicle operation in three driving cycles: UDDS, NEDC and China City) [40]. Based on the simulation results, it was observed that, with fuzzy control, the fluctuation of the battery output current is more stable. Under the selected UDDS, NEDC and China City cycles, the peak current of GA-Fuzzy control is lower than PSO-Fuzzy control by 35.6001 A, 19.9046 A, and 46.5270 A, respectively. Also, the total power consumption of GA-Fuzzy control decreased by 2.4489%, 9.0604%, and 2.5332%, respectively, and the system power consumption of PSO-Fuzzy control decreased by 1.0859%, 0.9659%, and 0.2650%, respectively, compared with fuzzy control. Combined with the comparative results of battery operating current, the optimization effect by using GA-Fuzzy control algorithm is the best in terms of battery protection and stability of battery life.

In the examples presented above, it is observed that several driving cycles have been used. It should be noted that there are major differences between them, and normally all research should be related to the WLTC cycle (it has the combination of urban and extra-urban traffic). Based on the profile, it can be said that, for example, WLTC is more dynamic and aggressive than UDDS, with higher acceleration and more frequent transitions. In the case of WLTC fuzzy control, the membership functions will be affected in terms of power demand (wider range, higher granularity), recovery power (more frequent and higher power peaks) and SOC thresholds (faster depletion-the SOC decrease rate is about 1% for UDDS and 2.5% for WLTC). An example of how the outputs in the case of HESS control are affected for different driving cycles is shown in Figure 9.

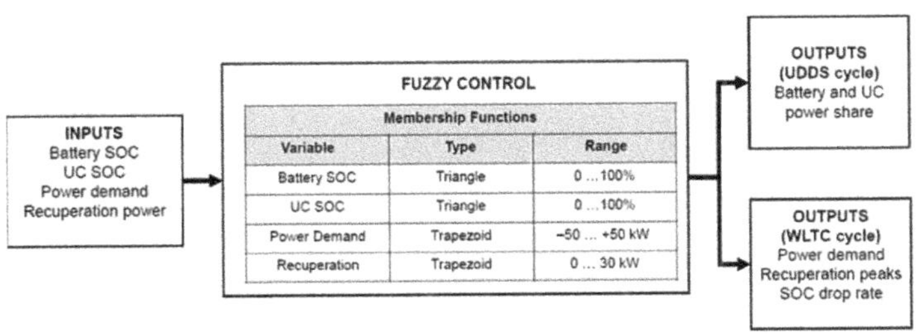

Figure 9. Effects of different driving cycle characteristics on outputs for fuzzy control of HESS.

This method exploits the distinct frequency characteristics of the power demand to distribute power between the battery and the ultracapacitor. The operating principle is based on a low-pass filter (LPF) applied to the total power demand. The low-frequency components (constant, long-term power) are assigned to the battery, while the high-frequency components (transient, peak power) are handled by the ultracapacitor.

The main advantage of the method is that it effectively separates the steady-state power from the transient power, but there is also the important disadvantage that the choice of the filter cutoff frequency is critical and can have a significant impact on performance (a fixed/preset cutoff frequency may not be optimal for all driving/discharging cycles or battery SOC ranges).

One of the energy management strategies for EVs equipped with HESSs is the use of low-pass filter (LPF), a strategy researched and developed in [41]. An iterative approach based on the Ragone graph (called LPF-Iterative) was used to solve the primary problem of LPF (Low Power Factor) based power management, which is figuring out the ideal decoupling frequency. Comparative analyses of this initial method's performance were primarily conducted in relation to the use of an iterative process that was optimized using the Particle Swarm Optimization (LPF-PSO) algorithm. The results obtained showed that LPF-Iterative performed better than LPF-PSO, offering a notable 83.51% increase in computational speed. It was also determined that the control method utilizing the LPF-Iterative algorithm provides a workable solution for enhancing power distribution and battery life in EVs whose power sources are outfitted with HESSs.

In [42], a hybrid HESS (battery + UC) power management approach based on adaptive digital filters (ADFBEMS) was put forth for electric vehicles. The digital filter tracks the instantaneous load spectrum using the sliding discrete fast Fourier transform (SDFFT), and it achieves frequency-based load distribution by using a low-pass filter with an adaptive cutoff frequency. The battery supplies the remaining load component, while UCs handle the high-frequency portion of the load. By contrasting it with the conventional power management approach based on the fixed cutoff frequency filter, hardware-based experiments conducted under the WLTC driving cycle have demonstrated the control's optimization.

The battery module lifetime benefited from the reduction in the high-frequency ratio by 27.46% and the power load stress by 11.06%, respectively.

The moving average filter (MAF) method determines the sampling frequency in HESSs, while the upper limit of the battery charging and discharging power fluctuation frequency is regarded as the cutoff frequency. Additionally, the wavelet transform (WT) is a powerful analysis tool in signal processing that is used in the development of frequency-decoupled algorithms.

In this regard, various power management strategies through MAF and WT control algorithms have been applied in [43], to achieve/analyze the optimal distribution of load power between the battery and the supercapacitor. Through a parameter called "power allocation effect", it was concluded that the WT method has a better power allocation effect than the MAF method (the average values calculated for different driving cycles are 0.9479 and 0.8123, respectively).

In a brief recap of the control methods of HESS hybrid systems presented so far, in Figure 9 the performance of the different control methods in terms of RMS battery current and specific energy savings is compared, depending on 3 standard driving cycles (the differences and influences due to the different dynamics of the driving cycles can be observed). The data presented in Figure 10 were compiled from references [44–47].

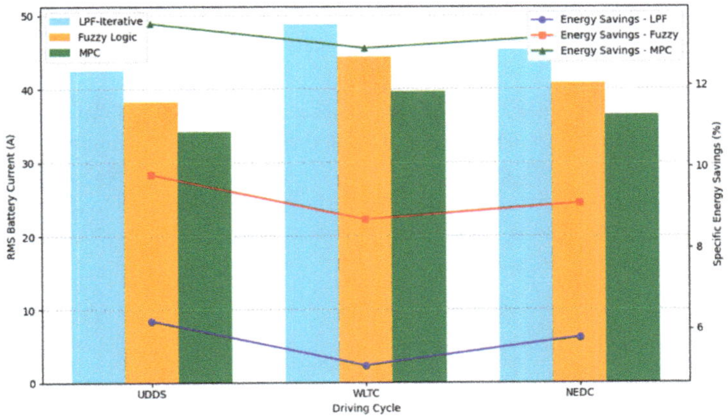

Figure 10. Comparison of LPF-Iterative, Fuzzy Logic, and MPC strategies in a HESS across UDDS, WLTC, and NEDC driving cycles.

3.2.2. Optimization-Based Control

The control approach has the ability to handle complex constraints, adapt to changing circumstances, and incorporate data on future driving habits (or operating conditions, if considered and incorporated into the model). The goal of the algorithms is to minimize an objective function to determine the best power distribution, often taking into account a number of constraints (such as current limits or SOC). However, implementing this approach is difficult because it requires a precise system model, which is very sensitive to the accuracy of the prediction, and requires a significant amount of computing power.

Several control strategies, including the following, can be used to instantly optimize the flow and demand of energy:

- Equivalent Consumption Minimization Strategy (ECMS)
- Dynamic Programming (DP)
- Model Predictive Control (MPC)

Equivalent Consumption Minimization Strategy, or ECMS, is a popular method for hybrid electric vehicles that can be modified for HESS. It attempts to minimize the total equivalent consumption at each moment by converting the energy consumption of the ultracapacitor into an equivalent fuel/energy consumption of the battery. An important component that can be modified online is the "equivalence factor". The possibility of using/implementing the ECMS aging control algorithm in HESS control was investigated in [48]. The Ah flow method was used as the aging term in the ECMS cost function (regardless of the driving cycles considered) and a fixed equivalence factor was taken into account. The energy capacity of the vehicle battery was correlated with the aging coefficient. Based on the results, it was determined that the optimal ECMS aging controller was developed. It aims to preserve the battery life by minimizing the state of charge ripples and optimizes the HESS operation without requiring predictions or knowledge of future driving actions. By integrating operating conditions determined by environmental perception with ECMS, this research [49] proposes an environmental perception-based HESS control architecture to maximize the parallel power system management for hybrid vehicles. The overall control process is divided into two parts: online testing and offline optimization. The offline process involves optimizing the equivalence factors based on varying degrees of environment and training the environmental perception using GCN and attention mechanisms [50]. The lookup of the equivalence factor table in online testing is done based on the environmental level determined by the environmental perception. Consequently, the optimized equivalence factor is used to complete the energy management control, and the simulation results indicated a 7.25% improvement in performance (compared to the traditional ECMS model).

Dynamic Programming (DP) is an offline optimization technique that determines the best global energy management plan for a given driving cycle but cannot be implemented in real time due to its computational complexity. It is extremely useful when creating rules for real-time controllers and for benchmarking. The early research, presented in [51], use dynamic programming (DP) analysis to identify the optimal control mode in terms of both energy saving and battery life extension (Figure 11).

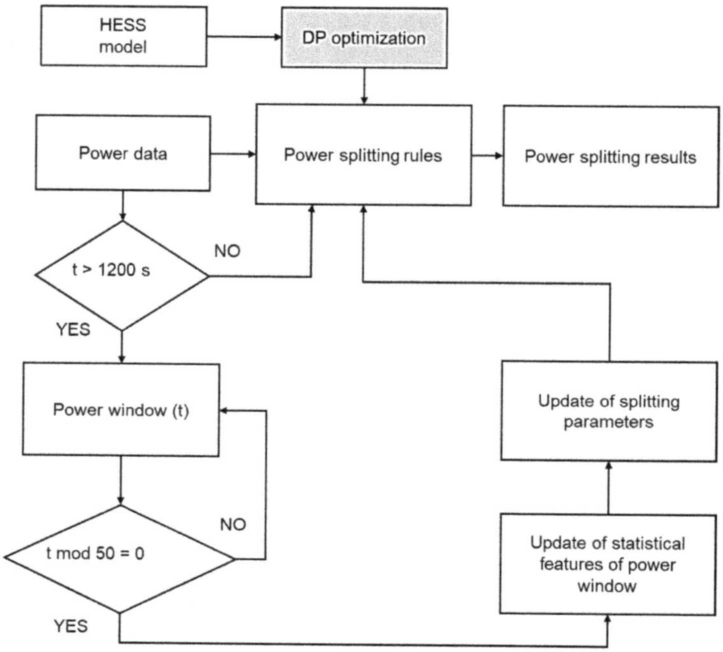

Figure 11. Flowchart of the DP load-adaptive rule-based control (adapted from [47]).

A functional relationship between the power-sharing parameters and charging statistics is established (using a rule-based control strategy adaptive to load variation) by extracting three-segment control rules from the DP results for four distinct types of charging cycles. The results demonstrate that the suggested strategy is more capable of protecting the battery and conserving energy under unknown charging conditions than the rule-based control strategy (the battery discharge in Ah and the total energy loss are reduced by 3.4% to 15.7% and 3.0% to 15.1%, respectively). It is determined that the suggested strategy (DP) can achieve near-optimal real-time energy management with low computational costs, despite the fact that the obtained results are fairly close. However, the authors conclude that more research is required in this area.

There are methods that use DP algorithms to solve the problem of optimizing energy management in HESSs, because the rule-based strategy is empirical and cannot guarantee to obtain the best efficiency optimization solution. According to [52], the HESSs managed by the DP algorithm strategy can reduce energy consumption by 4.8% (for NEDC driving cycle conditions) compared to the rule-based strategy. In addition, the HESS managed by the power sharing strategy derived from the DP approach can reduce consumption by 17.6% compared to a battery-only electric vehicle. The final conclusion of the study demonstrates that the optimized DP strategy provides lower energy consumption and higher efficiency of the energy storage system compared to the rule-based strategy.

The idea behind using Model Predictive Control (MPC) as optimization-based control technique is to predict how a system will behave in the future, over a limited period of time, using a dynamic model of the system (battery, ultracapacitor, and vehicle dynamics). At each time step, it then solves an optimization problem to identify the best course of action to control a cost function (such as energy consumption or battery degradation) while respecting constraints (Figure 12). The procedure is repeated after the first control action is applied.

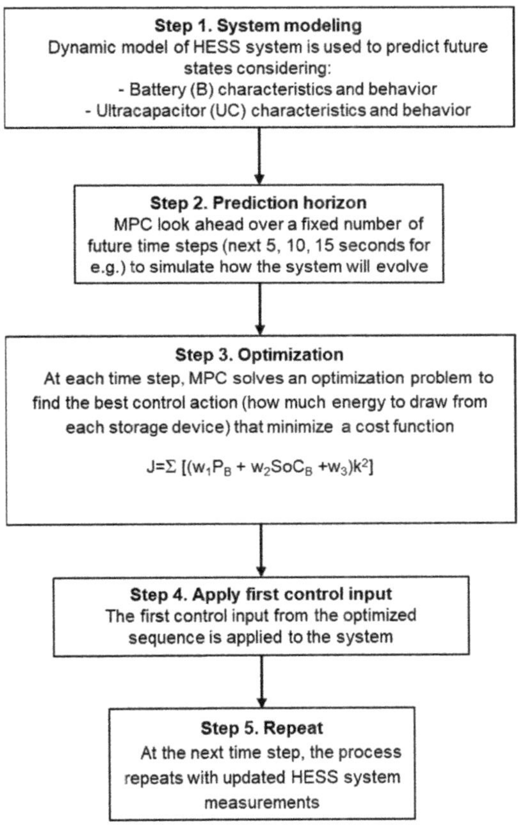

Figure 12. Algorithm's general structure of Model Predictive Control (MPC) (w_1, w_2, w_3 are weights for different objectives, k is deviation from demand and J is the prediction horizon).

With current parametric modeling techniques, model predictive control (MPC)-based power management in electric vehicles with hybrid energy storage systems is susceptible to model accuracy effects and parameter sensitivity. To overcome these shortcomings, a new data-driven hierarchical predictive control-based power management system is proposed in [53], in which the upper layer uses an optimized long-short-term memory (LSTM) network for driving prediction (allowing cost-effective acquisition of load power demands). To maximize power distribution between the UC and the battery (while minimizing battery capacity losses), a data-driven predictive control for HESS is proposed in the lower layer. In contrast to traditional MPC, data-driven predictive control is based on a non-parametric model that is constructed solely from HESS input-output data. This model enables flexible handling of a range of nonlinearities and uncertainties in different tasks and operating environments. When compared to predictive control based on nonlinear models, data-driven predictive control can lower overall operating costs by up to about 23%.

3.2.3. Intelligent Control (AI/ML-Based)

The application of convolutional neural networks, hybrid models combining regression techniques, ensemble learning, reinforcement learning, genetic algorithms, short-term memory algorithms, and graph neural networks for a variety of tasks (classification, regression, optimization, or system management) in the energy sector has grown exponentially in the last several years [54]. In addition to providing a number of opportunities to enhance energy, economic, and environmental performance and thereby support sustainability goals, artificial intelligence is poised to become a crucial component of research in energy conversion systems and the energy sector as a whole [55]. It will also significantly improve benchmark performance for a variety of energy conversion and storage tasks when compared to traditional or non-AI methods. By using machine learning and artificial intelligence techniques, these methods for optimizing energy management in HESSs for electric vehicles can learn the best control strategies from data or adapt to changing conditions, which define how an electric vehicle operates in real-world scenarios. The most common approaches, along with the general operating principle, advantages and disadvantages of intelligent control algorithms are presented in Table 5.

Table 5. General operating principle, advantages and disadvantages of intelligent control algorithms.

Control System	Principle	Advantages	Disadvantages
Fuzzy Logic Control (FLC):	FLC uses linguistic rules and fuzzy sets to map input variables (e.g., power demand, battery and UC SOC) to output control actions (e.g., power split ratio).	Robust to uncertainties. Does not require a precise mathematical model. Intuitive for rule definition.	Requires expert knowledge to define rules and membership functions. Tuning can be challenging.
Neural Networks (NNs)	NNs can learn complex nonlinear relationships between inputs and outputs from training data. Can be used to predict optimal power split or estimate system states.	Can learn highly complex relationships. Adaptive.	Requires large datasets for training. "Black box" nature can make interpretation difficult. High computational cost.
Reinforcement Learning (RL)	An RL agent learns an optimal policy by interacting with the environment (e.g., vehicle powertrain simulation) and receiving rewards or penalties for its actions. The goal is to maximize cumulative rewards over time.	Can learn optimal strategies without prior knowledge of system dynamics. Adaptable to changing environments.	Requires extensive training. High computational intensity. Ensuring stability and safety during real-world deployment is challenging.

Strong machine learning models called neural networks (NNs) can capture intricate, nonlinear relationships between input and output properties. By learning directly from data, NNs are able to model intricate dependencies that are frequently challenging to articulate analytically or through traditional control strategies. Using both historical and real-time data, neural networks provide a scalable and flexible method to enhance the performance, efficiency, and reliability of complex energy systems. There are several ways in which NNs can be used to control electric vehicle HESSs (Table 6) [56]:

- Predicting optimal energy sharing.
- Estimation of system states: NNs can be trained to estimate internal system states, such as battery and UC status, energy consumption trends, or thermal conditions, which are expensive or difficult to measure directly. Predictive maintenance, fault detection, and system monitoring can be improved through these estimates.

Table 6. Common types of general NN architecture with potential in controlling HESS of electric vehicles.

Architecture	Task	Strengths	Limitations	Use Cases for HESS Control
Feedforward Neural Network (FNN)	Static estimation	Simple and fast; good for static input-output mappings	• Limited in handling sequential or time-dependent data	• Battery and UC state estimation • Fault detection
Recurrent Neural Network (RNN)	Complex control policies	Captures temporal dependencies in sequential data	• Prone to vanishing gradients • Less effective for long sequences	• Driving cycle analysis • Power split prediction
Long Short-Term Memory (LSTM)	Time-series prediction	Handles long-term dependencies; robust to vanishing gradients	• More complex and computationally intensive	• Predictive energy management • Vehicle behavior modeling (energy consumption)
Convolutional Neural Network (CNN)	Spatial-temporal data	Good at extracting local patterns; efficient with structured input	• Less suited for purely sequential data unless adapted	• Sensor array analysis
Deep Belief Network (DBN)	Complex control policies (useful in scenarios where labeled data is sparse)	Effective for unsupervised features learning and pre-training	• Training can be complex • Less commonly used in real-time applications	• System health monitoring • Anomaly detection

In the case of optimal energy sharing prediction, neural networks (NNs) can learn to predict the most efficient distribution of energy between energy sources by evaluating real-time input data, including driver driving style, battery and UC charge status, vehicle speed, and environmental conditions. This makes it possible to implement dynamic, data-driven energy management plans that adapt to changing operational circumstances.

For more accurate system state estimation, neural networks (NNs) can be trained to estimate specific internal system states (such as battery and UC status), energy consumption trends, or thermal conditions, which are expensive and/or difficult to measure directly. Also, collaterally, other estimates regarding predictive maintenance, fault detection, and system monitoring can be used to increase the efficiency of energy storage systems.

A reinforcement learning (RL)-based power management (control) strategy for minimizing real-time energy losses is proposed to achieve the optimal energy distribution between the battery and the ultracapacitor (Figure 13), which is a critical issue for the hybrid energy storage system [57]. The Kullback–Leibler divergence rate is used to determine when the power management strategy update is initiated, and the power transition probability matrices are updated according to the new application duty cycle to further minimize the energy losses. The suggested control method has been verified under various circumstances, taking into account influencing factors such as duty cycles, operating states, temperatures, and SOC values (battery and UC). According to a comparison between rule-based energy management and RL-based online energy management, the latter approach can increase the efficiency of energy management (the relative reduction in total power losses can reach 16.8%).

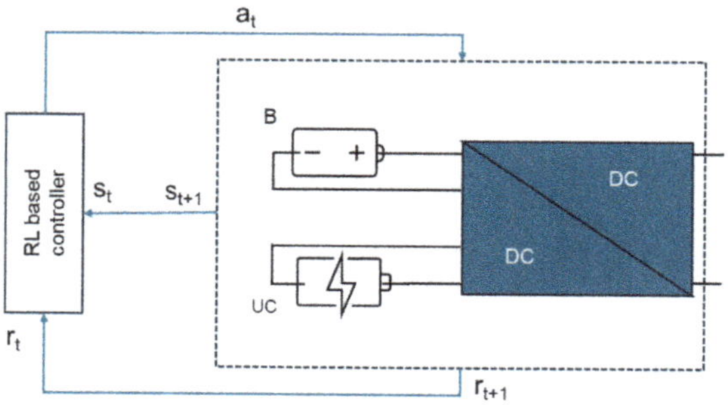

Figure 13. Structure of RL-based power management (control) strategy (a-action variable, r-reward function, s-state variables, t–time, B–battery, UC-ultracapacitor) (adapted from [57]).

Because of its instantaneous optimization and computational simplicity, the indirect optimal control method of EV energy management (EMS) based on the Pontryagin minimum principle (PMP) draws attention in this direction. In order to reduce battery degradation in a hybrid HESS EV (Figure 14), the online hybrid EMS solution is examined in [58] by integrating PMP and deep reinforcement learning (RL). By keeping the UC state of charge within the desired range and minimizing the battery current for various driving profiles, the experimental results demonstrate the efficacy of deep reinforcement learning for optimal cost estimation to satisfy the UC load sustainability. Comparing this EMS control method to the standard EMS performance, a notable improvement of 900 charging cycles is obtained.

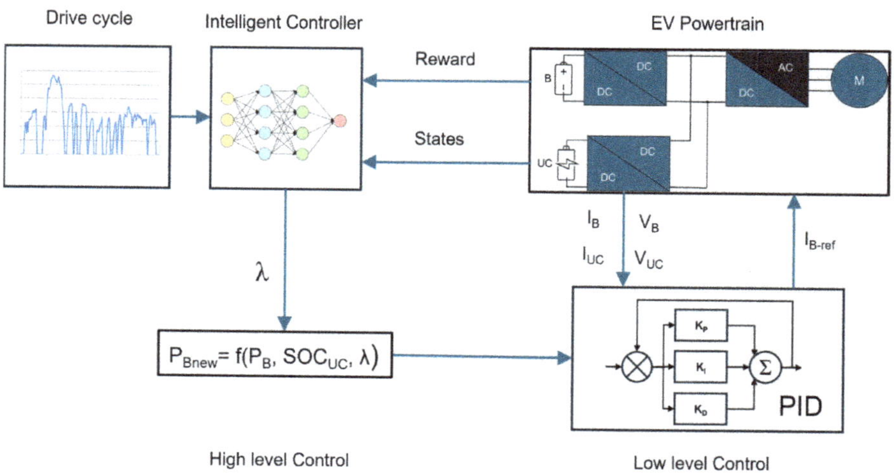

Figure 14. Energy source management framework for hybrid HESS EV based on the Pontryagin minimum principle (PMP) (adapted from [58]).

The application of DL in HESS management can be applied to manage the hybrid energy system of electric vehicles (battery, ultracapacitor, motor) by controlling the energy flow between the energy storage and different consumption modules [59]. The vehicle speed, the desired speed, the electric motor power, SOC of the battery and the ultracapacitor, the terrain topology and the outdoor temperature are the inputs used by the proposed Artificial Neural Network (ANN). The neural network construction contains two hidden layers (each layer having fifty nodes), having the hyperbolic tangent as the activation function and the identity function at the output connections (Figure 15). The input/output power flow of the battery and UC are the outputs of the neural network. The neural network was validated using 20 distinct datasets (16 for training and 4 for testing). According to the results of the neural network training, the system can achieve a 2.68% increase in the travel efficiency for the UDDS driving cycle, which is represented by the autonomy of the electric vehicle (due to the optimization of energy consumption).

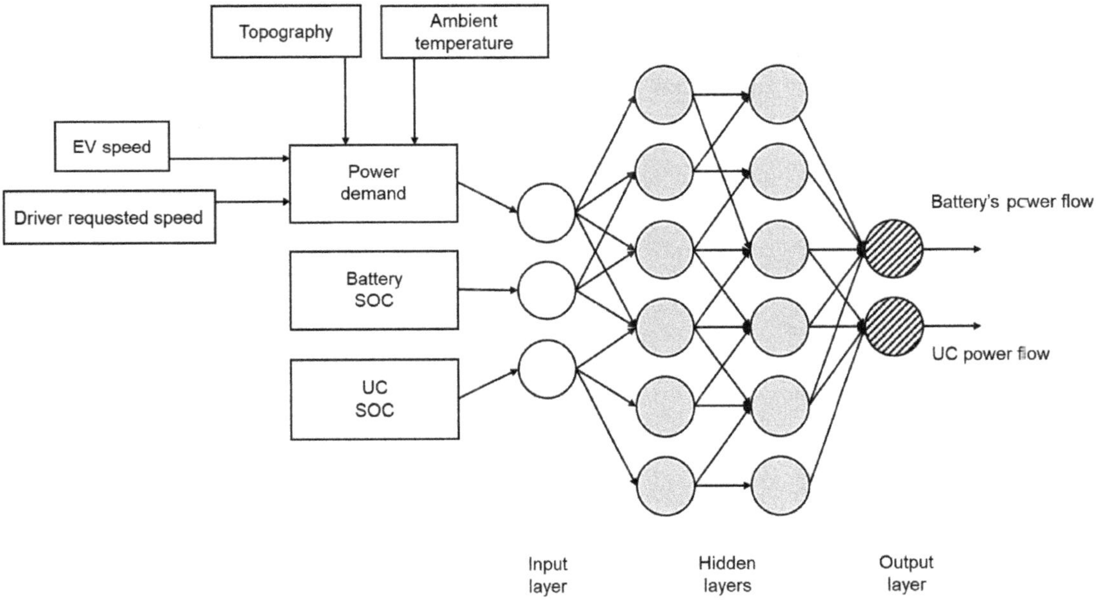

Figure 15. Structure of NN for HESS control (adapted from [59]).

As an advanced energy management strategy between the battery and the energy storage system of an electric vehicle HESS, the DNN control method combined with the SS-IFS technique is used to maximize the benefits of both the battery and the UC storage components while also ensuring the safety and stability of the battery packs (Figure 16) [59]. By recovering energy during deceleration stages and extending battery life through advanced HESS control—a control that guarantees the HESS operates at high efficiency by solving a multi-objective optimization problem—this superior energy management aims to increase vehicle autonomy. A Deep Neural Network (DNN) is constructed using the results of the initial PID controller. The meta-heuristic algorithm SS-IFS is further utilized to generate the best control signals for the HESS. In comparison to the conventional control schemes that use the WOA, PSO, SOA, and SSA algorithms, the adopted SS-IFS model is 74.96%, 79.9%, 40.39%, and 70% better in the rise time analysis. The research also yielded stabilization times that were 74.71%, 79.61%, 39.61%, and 69.72% faster than the current WOA, PSO, SOA, and SSA methods. Furthermore, the suggested DNN + SS-IFS model's computation time is 7390.3 s, compared to 5262.6 s for the conventional WOA, 6259 s for the PSO, 4381 s for the SOA, and 7390.3 s for the SSA.

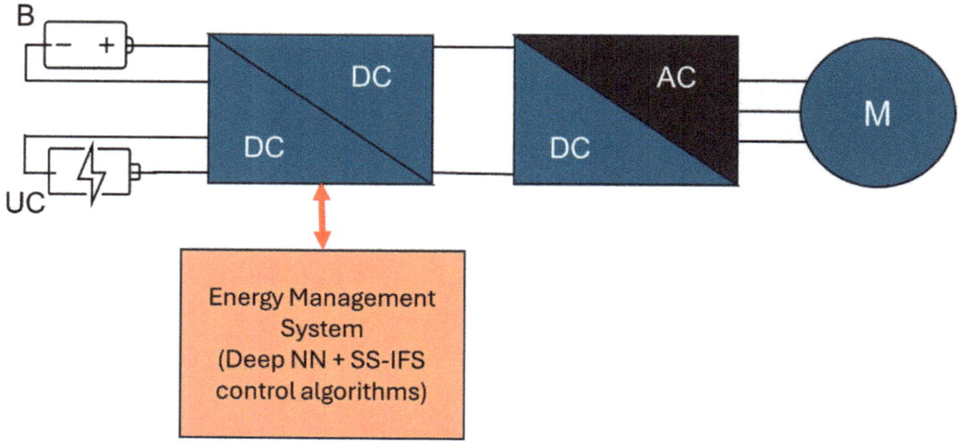

Figure 16. Structure of DNN + SS-IFS control strategy for HESS EV (adapted from [59]).

3.3. Application of Algorithms Depending on the Architectures of Hybrid Energy Storage Systems

The choice of the type/topology of the HESS architecture (passive, semi-active, active or fully active-see Figures 1–4) directly influences the complexity, response speed and flexibility of the control strategies that can be used. Active and fully active architecture, which allows bidirectional energy flow and independent control of each energy storage component (e.g., batteries and UC), offers the greatest flexibility for implementing advanced control algorithms. These architectures allow for dynamic energy management, real-time optimization and predictive control, which are essential for applications that require high performance, efficiency and adaptability (Table 6).

It should be emphasized that the effectiveness of the control algorithms depends largely on how the ultracapacitor is physically integrated with the battery and the rest of the powertrain and on the choice of specific control strategies for the HESS. Specific command and control strategies for optimizing the energy efficiency of storage systems are often combined to form a comprehensive control system.

In the case of ultracapacitor SOC management, multiple approaches include:

- Fenestration: aimed at maintaining UC's SOC within a defined operational window (e.g., 50–90%) to ensure availability for both charging (regenerative braking) and discharging (acceleration).
- Charge/discharge prioritization: during regenerative braking, UC charging is prioritized, and during acceleration, UC discharging is prioritized.
- UC recharging: if the UC's SOC drops too low, the battery can provide a small charge to bring the UC back within the defined operational window.

In the case of managing the current/power limitation delivered to the propulsion system, multiple approaches aim to ensure that the current consumption from or supplied to the battery and ultracapacitor remains within their safe operating limits (to prevent possible fires, component damage, and to extend the lifetime of both the components and the storage system as a whole). Furthermore, to this end, control algorithms can incorporate thermal models to predict and mitigate thermal stress on both the batteries and the ultracapacitor, possibly by adjusting power sharing strategies depending on temperature.

In addition to the HESS topologies that can equip the transmission of an electric vehicle, the specific operating conditions of the electric vehicle must also be taken into account in the research carried out. There are few works that show comparatively the results of applying various driving cycles knowing that there are dynamic differences that can directly influence the way the power (charge) is distributed between the HESS elements (battery and UC). Table 7 presents the HESS adaptations according to dynamic characteristics of two different driving cycles (UDDS and WLTC).

Table 7. Comparative insights of HESS adaptation for different driven cycles [38,41,42,44–47,52].

Parameter	**HESS Adaptation**	
	UDDS Cycle	**WLTC Cycle**
I_{Bmax}	Lower peak current (favors battery protection)	Higher peak current (allows more battery load)
SOC_{UC} Range/Window	30% to 90% (favors UC usage)	10% to 70% (enables early UC dispatch)
UC Dispatch Behavior	More conservative, full dispatch only at high SOC	More aggressive, dispatch starts at low SOC
Drive Cycle Characteristics	Urban traffic conditions, stop-and-go, moderate acceleration	Mixed urban/highway traffic conditions, dynamic acceleration
Battery Stress	Lower	Moderate to High
UC Buffering Effectiveness	High	Moderate
SOC Stability	High	Moderate
Thermal Load	Low	High

The main conclusion from Table 6 is that HESS adaptation for UDDS cycle prioritizes battery longevity and UC buffering during frequent stops and starts, while HESS adaptation for WLTC cycle supports higher power demands and faster UC response, suitable for dynamic driving conditions.

Considering the cost of a HESS, the issue of the overall lifetime of the HESS must also be addressed. This can be achieved through an intelligent degradation control system. Advanced control algorithms (AI-based) can incorporate battery and UCs degradation models to make decisions that not only optimize efficiency but also extend the overall lifetime of the HESS (e.g., by dynamically adjusting power division based on estimated degradation rates).

4. Future Trends and Research Directions

The integration of ultracapacitors into electric vehicle powertrains represents an innovative solution for improving energy performance and component durability. Due to their high-power density and exceptionally fast response to changing energy demands, ultracapacitors are perfect for controlling peak loads and for efficient energy recovery during regenerative braking. Therefore, it is necessary to continue research and development of control algorithms for electric vehicle powertrains with UC, especially in promising areas connected to predictive control algorithms, adaptive hybrid control, artificial intelligence-based methods, and integrated thermal and energy management (considering operational conditions of electric vehicles and related KPIs—Table 8).

Table 8. Application of control algorithms depending on the HESS's architecture.

HESS Architecture	Characteristics	Operational Conditions	Control Algorithms	KPIs
Passive Parallel	Battery and ultracapacitor are directly connected in parallel. No power electronics; energy flow is dictated by natural voltage/current characteristics.	Very limited direct control over power split. Power is shared based on internal resistances and voltage differences. Simplest, but least effective.	• Rule-based control-simple logic based on thresholds (if battery voltage (SOC) < X (%), reduce load). • No active control-system relies on passive balancing and limited algorithmic intervention.	• Low cost • Low efficiency • Limited control
Active Parallel (Semi-Active)	The ultracapacitor is connected to the DC bus via a bi-directional DC/DC converter, while the battery is directly connected. One storage device (usually an ultracapacitor) is actively controlled and the other storage device is passive.	The DC/DC converter regulates the power flow to/from the ultracapacitor. Common and effective architecture, offering good control over UC contribution.	• State-of-Charge (SOC) management-controls supercapacitor charging/discharging based on battery SOC. • Fuzzy logic control-handles uncertainty in load demand and energy flow between devices.	• Moderate battery life • Good transient handling • Moderate control
Full Active	Both the battery and ultracapacitor are connected to the DC bus via separate bi-directional DC/DC converters. Both storage devices are actively controlled via power converters.	Offers the highest degree of control over both energy sources, allowing independent optimization. Most complex and costly but provides maximum flexibility and efficiency.	• Model Predictive Control (MPC)-predicts future states and optimizes control actions over a time horizon. • Dynamic Programming (DP)-optimizes energy flow based on cost functions (e.g., minimizing battery degradation). • Neural Network-based control (NN)-learns optimal energy management strategies from data.	• High efficiency • Extended battery life • Advanced control
Multi-Input Converter	A single converter integrates multiple energy sources. Independent bidirectional control of each device; full flexibility.	Requires complex control of the multi-input converter to manage power flow from both sources simultaneously.	• Real-time optimization-continuously adjusts energy flow based on system state and external conditions. • Reinforcement Learning (RL)-learns optimal policies through interaction with the environment. • Multi-objective control-balances trade-offs between efficiency, lifespan, and performance (e.g., Pareto optimization).	• Predictive control • Long battery life • Maximum performance

First of all, robust and adaptive MPC algorithms that are less susceptible to computational burden and model errors need to be developed. These algorithms should also be able to incorporate real-time parameter estimation. The development of reinforcement learning algorithms for real-time control of energy management systems helps to overcome current limitations related to training stability and computational intensity/load. In addition, this approach can include transfer learning and combining RL models with classical control. Command and control algorithms can also be integrated into prognostic and health management (PHM) systems. Predicting the remaining useful life of batteries and single control units (UC) is a benefit of integrating PHM capabilities into the control system. In addition, control strategies can be chosen or modified based on this information to increase the overall system lifetime.

In the context of contemporary technological development directions, it is also necessary to use cloud computing and big data analytics to collect vast amounts of data about the vehicle and its operating mode (driving style), optimize offline control strategies, and then implement updated/modified algorithms. An immediate illustration of this is the proactive optimization of energy management by using predictive data from navigation and ADAS (such as impending hills, traffic, weather, etc.).

With the development and widespread application of V2X (Vehicle to Everything) technologies, challenges arise related to control strategies, strategies that will need to incorporate the bidirectional power flow capabilities of the UC and batteries for network services, requiring more sophisticated energy management.

Last but not least, one of the challenges and/or trends will be the increasing efforts to standardize communication protocols and control interfaces for HESS components.

5. Conclusions

The main challenges to achieve increased efficiency, reliability, flexibility and profitability of HESS are given by: the compatibility of different types of HESS with the desired application and the compatibility of the components, the optimization of the storage capacity configuration, efficient control strategies, compatible integration into the network architecture, the development of materials with high energy storage performance, the differentiated assessment of the aging mechanisms of the components, the environmental impact and recycling methods [57].

To ensure stable and effective operation, advanced control strategies are necessary for managing the charge–discharge cycles of the various storage components in a hybrid system, forecasting energy demands, and balancing loads. The accuracy and applicability of current control methods are limited by a variety of factors, including model simplifications and parameter tuning, implementation challenges brought on by intricate algorithms and multi-agent coordination, limitations in real-time optimization and dynamic response, and computational cost. A hybrid HESS's control method selection must be in line with all of the aforementioned specifications, weighing the method's efficacy against both implementation and operating expenses (Figure 17).

The integration of ultracapacitors into electric vehicle powertrains offers significant promise for improving performance, improving efficiency, and extending the lifespan of battery systems. However, realizing these benefits critically depends on the implementation of sophisticated control algorithms. From fundamental rule-based systems to advanced predictive and intelligent control strategies, the evolution of these algorithms is driven by the need to efficiently manage energy flow, optimize energy utilization, and ensure the long-term reliability of hybrid energy storage systems. As electric vehicles continue to evolve, the development of more intelligent, adaptive, and robust control algorithms will be essential to unlock the full potential of ultracapacitor technology in electric mobility.

It should be noted that the analysis of research conducted on the algorithmic control of HESSs shows that misinterpretations of the results may occur, due to the characteristics of the driving cycles used by the researchers (UDDS, NEDC, WLTC, Japan 1015, NYCC, etc.). The reproducibility of the results for different operating conditions is distorted/influenced by the fact that these driving cycles have different acceleration dynamics and the number of start-stops (for example) with direct influence on the dynamics of power sharing between the battery and the UC. For this reason, in future, it is necessary that any research conducted on the subject of this article be validated experimentally, in real traffic conditions and real operating and environmental parameters.

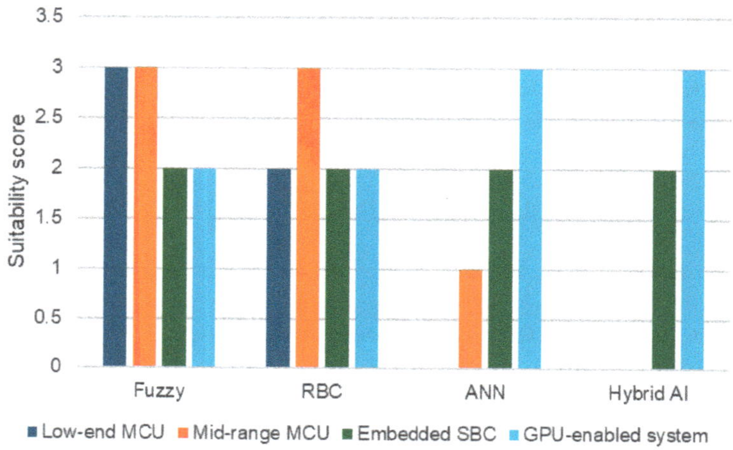

Figure 17. Suitability score for using different control techniques for HESS depending on computational resource (0-not suitable, 1-limited, 2-good, 3-optimal).

Funding: This paper is supported by the European Union's Horizon 2020 research and innovation program under grant agreement no. 101121530, project ZERO-MOVE Zero Emission Mobility (HORIZON-RIA-SGA-NZC).

Data Availability Statement: No new data was created.

Conflicts of Interest: The author declares no conflicts of interest.

Abbreviations

The following abbreviations are used in this manuscript:

AC	Alternative Current
ADAS	Advanced Driver Assistance Systems
ADFBEMS	Adaptive Digital Filter
AI	Artificial Intelligence
ANN	Artificial Neural Network
B	Battery
CAPEX	Capital Expenditure
China City	Chinese Urban Driving Cycle
CNN	Convolutional Neural Network
DBN	Deep Belief Network
DC	Direct Current
DP	Dynamic Programming
DNN	Deep Neural Network
ECMS	Equivalent Consumption Minimization Strategy
EMS	Energy Management System
ESR	Equivalent Series Resistance
EV	Electric Vehicle
GA	Genetic Algorithm
GCN	Graph Convolutional Network
GPU	Graphics Processing Unit
HESS	Hybrid Energy Storage System

Japan1015	Japanese Driving Cycle
LPF	Low-Pass Filter
LSTM	Long-Short-Term Memory
MAF	Moving Average Filter
MCU	Microcontroller Unit
MPC	Model Predictive Control
NEDC	New European Driving Cycle
NN	Neural Network
NYCC	New York City Cycle
OPEX	Operational Expenditure
PID	Proportional–Integral–Derivative
PHM	Prognostic and Health Management System
PMP	Pontryagin Minimum Principle
PMSM	Permanent Magnet Synchronous Motor
PSO	Particle Swarm Optimization
RBC	Rule-Based Control
RL	Reinforcement Learning
SBC	Single Board Computer
SDFFT	Sliding Discrete Fast Fourier Transformation
SOA	Seagull Optimization Algorithm
SOC	State of Charge
SSA	Squirrel Search Algorithm
SS-IFS	Squirrel Search with Improved Food Storage Algorithm
UC	Ultracapacitor
UDDS	Urban Dynamometer Driving Schedule
V2X	Vehicle to Everything
WLTC	Worldwide Harmonized Light Vehicle Test Cycle
WOA	Whale Optimization Algorithm
WT	Wavelet Transformation

References

1. Liu, J.; Brandt, J.; Christensen, J.H.; Ye, Z.; Chen, T.; Dong, S.; Geels, C.; Yuan, Y.; Nenes, A.; Im, U. The recent and future $PM_{2.5}$-related health burden in China apportioned by emission source. *NPJ Clean Air* **2025**, *1*, 7. [CrossRef]
2. Li, C.; Managi, S. Contribution of on-road transportation to $PM_{2.5}$. *Sci. Rep.* **2021**, *11*, 21320. [CrossRef] [PubMed]
3. da Costa, V.B.F.; Bitencourt, L.; Dias, B.H.; Soares, T.; de Andrade, J.V.B.; Bonatto, B.D. Life cycle assessment comparison of electric and internal combustion vehicles: A review on the main challenges and opportunities. *Renew. Sustain. Energy Rev.* **2025**, *208*, 114988. [CrossRef]
4. Kotak, Y.; Marchante Fernández, C.; Canals Casals, L.; Kotak, B.S.; Koch, D.; Geisbauer, C.; Trilla, L.; Gómez-Núñez, A.; Schweiger, H.-G. End of Electric Vehicle Batteries: Reuse vs. Recycle. *Energies* **2021**, *14*, 2217. [CrossRef]
5. Albatayneh, A.; Assaf, M.N.; Alterman, D.; Jaradat, M. Comparison of the Overall Energy Efficiency for Internal Combustion Engine Vehicles and Electric Vehicles. *Environ. Clim. Technol.* **2020**, *24*, 669–680. [CrossRef]
6. Ding, Y.; Cano, Z.P.; Yu, A.; Lu, J.; Chen, Z. Automotive Li-Ion Batteries: Current Status and Future Perspectives. *Electrochem. Energ. Rev.* **2019**, *2*, 1–28. [CrossRef]
7. Liu, Y.; Zhang, R.; Wang, J.; Wang, Y. Current and future lithium-ion battery manufacturing. *iScience* **2021**, *24*, 102332. [CrossRef]
8. Nasajpour-Esfahani, N.; Garmestani, H.; Bagheritabar, M.; Jasim, D.J.; Toghraie, D.; Dadkhah, S.; Firoozeh, H. Comprehensive review of lithium-ion battery materials and development challenges. *Renew. Sustain. Energy Rev.* **2024**, *203*, 114783. [CrossRef]
9. Heo, J.; Gu, H.; Lee, C.; Sung, J.; Kim, D.-H.; Han, J.; Oh, Y.-S.; Ahn, S.; Jeon, I.; Par, J.-W. Recent Advances in Achieving High Energy/Power Density of Lithium–Sulfur Batteries for Current and Near-Future Applications. *Battery Energy* **2025**, *4*, e20240051. [CrossRef]
10. Fang, H. Challenges with the Ultimate Energy Density with Li-ion Batteries. *IOP Conf. Ser. Earth Environ. Sci.* **2021**, *781*, 042023. [CrossRef]
11. Cao, W.; Zhang, J.; Li, H. Batteries with high theoretical energy densities. *Energy Storage Mater.* **2020**, *26*, 46–55. [CrossRef]
12. Wu, X.; Xia, Y.; Guan, S.; Sun, J.; Du, J. Research on influencing mechanism of time gap for fast charging on battery capacity degradation in electric vehicles. *J. Energy Storage* **2024**, *103*, 114425. [CrossRef]

13. Madani, S.S.; Shabeer, Y.; Allard, F.; Fowler, M.; Ziebert, C.; Wang, Z.; Panchal, S.; Chaoui, H.; Mekhilef, S.; Dou, S.X.; et al. A Comprehensive Review on Lithium-Ion Battery Lifetime Prediction and Aging Mechanism Analysis. *Batteries* **2025**, *11*, 127. [CrossRef]
14. Tian, Y.; Wang, L.; Liao, C.; Yan, G. Comprehensive Investigation of the Durability of Lithium-Ion Batteries Under Frequency Regulation Conditions. *Batteries* **2025**, *11*, 75. [CrossRef]
15. Njema, G.G.; Ouma, R.B.O.; Kibet, J.K. A Review on the Recent Advances in Battery Development and Energy Storage Technologies. *J. Renew. Energy* **2024**, *2024*, 2329261. [CrossRef]
16. Olabi, A.G.; Abbas, Q.; Al Makky, A.; Abdelkareem, M.A. Supercapacitors as next generation energy storage devices: Properties and applications. *Energy* **2022**, *248*, 123617. [CrossRef]
17. Dissanayake, K.; Kularatna-Abeywardana, D. A review of supercapacitors: Materials, technology, challenges, and renewable energy applications. *J. Energy Storage* **2024**, *96*, 112563. [CrossRef]
18. Mone, P.; Deore, S.; Balgude, S.; Pandit, V. Fabrication, design and performance evaluation of supercapacitors review. *Mater. Today Proc.* **2022**, *53*, 130–133. [CrossRef]
19. Volfkovich, Y.M. High power supercapacitors. Review. *J. Electroanal. Chem.* **2024**, *963*, 118290. [CrossRef]
20. Jamadar, N.M.; Jadhav, H.T. Effectiveness of supercapacitor during braking operation of electric vehicle. *Mater. Today Proc.* **2022**, *56*, 314–319. [CrossRef]
21. Urooj, A.; Nasir, A. Review of Hybrid Energy Storage Systems for Hybrid Electric Vehicles. *World Electr. Veh. J.* **2024**, *15*, 342. [CrossRef]
22. Mariasiu, F.; Kelemen, E.A. Analysis of the Energy Efficiency of a Hybrid Energy Storage System for an Electric Vehicle. *Batteries* **2023**, *9*, 419. [CrossRef]
23. Liu, Q.; Tong, Y.; Liu, B.; Yang, J.; Feng, Y. Life Cycle Cost-Oriented Optimization of Hybrid Energy Storage System for Mining Haul Truck. In Proceedings of the 19th Annual Conference of China Electrotechnical Society, Xi'an, China, 20–22 September 2024; Lecture Notes in Electrical Engineering. Yang, Q., Bie, Z., Yang, X., Eds.; Springer: Singapore, 2024; Volume 1406. [CrossRef]
24. Lu, X.; Wang, H. Optimal sizing and energy management for cost-effective PEV hybrid energy storage systems. *IEEE Trans. Ind. Inform.* **2020**, *16*, 3407–3415. [CrossRef]
25. Pipicelli, M.; Sessa, B.; De Nola, F.; Gimelli, A.; Di Blasio, G. Assessment of Battery–Supercapacitor Topologies of an Electric Vehicle under Real Driving Conditions. *Vehicles* **2023**, *5*, 424–445. [CrossRef]
26. Dong, Z.; Zhang, Z.; Li, Z.; Li, X.; Qin, J.; Liang, C.; Han, M.; Yin, Y.; Bai, J.; Wang, C.; et al. A Survey of Battery–Supercapacitor Hybrid Energy Storage Systems: Concept, Topology, Control and Application. *Symmetry* **2022**, *14*, 1085. [CrossRef]
27. Kotana, R. Cost & Efficiency analysis of Battery & SC based Hybrid Energy Storage system for Solar OFF Grid applications. *J. Eng. Res. Sci.* **2025**, *4*, 1–7. [CrossRef]
28. Banerjee, A.; Mosier, T.M.; Shafiul Alam, S.M. Impact of Hybrid Energy Storage System (HESS) Topologies on Perfor-mance: Exploration for Hydropower Hybrids. In Proceedings of the 54th Hawaii International Conference on System Sciences, Kauai, HI, USA, 5 January 2021; pp. 3102–3110.
29. Korompili, A.; Monti, A. Review of Modern Control Technologies for Voltage Regulation in DC/DC Converters of DC Microgrids. *Energies* **2023**, *16*, 4563. [CrossRef]
30. Karamanakos, P.; Liegmann, E.; Geyer, T.; Kennel, R. Model predictive control of power electronic systems: Methods, results, and challenges. *IEEE Open J. Ind. Appl.* **2020**, *1*, 95–114. [CrossRef]
31. Karami, Z.; Shafiee, Q.; Sahoo, S.; Yaribeygi, M.; Bevrani, H.; Dragicevic, T. Hybrid model predictive control of DC-DC boost converters with constant power load. *IEEE Trans. Energy Convers.* **2021**, *36*, 1347–1356. [CrossRef]
32. Dhumal, A.R.; Kulkarni, A.P.; Ambhore, N.H. A comprehensive review on thermal management of electronic devices. *J. Eng. Appl. Sci.* **2023**, *70*, 140. [CrossRef]
33. Benisi Ghadim, H.; Godin, A.; Veillere, A.; Duquesne, M.; Haillot, D. Review of thermal management of electronics and phase change materials. *Renew. Sustain. Energy Rev.* **2025**, *208*, 115039. [CrossRef]
34. Folea, R.; Slusanschi, E.-I. A new metric for evaluating the performance and complexity of computer programs: A new approach to the traditional ways of measuring the complexity of algorithms and estimating running times. In Proceedings of the 23rd International Conference on Control Systems and Computer Science (CSCS), Bucharest, Romania, 26–28 May 2021; pp. 157–164. [CrossRef]
35. Yang, X.-S.; He, X.-S.; Fan, Q.-W. Chapter 7—Mathematical framework for algorithm analysis. In *Nature-Inspired Computation and Swarm Intelligence (Algorithms, Theory and Applications)*; Academic Press: Cambridge, MA, USA, 2020; pp. 89–108. ISBN 978-0-12-819714-1. [CrossRef]
36. Wai, C.K.; Sadeq, T.; Hau, L.C. Advanced adaptive rule-based management for hybrid energy storage systems (HESS) to improve electric vehicle range. *Vehicles* **2025**, *7*, 6. [CrossRef]

37. Kachhwaha, A.; Rashed, G.I.; Garg, A.R.; Mahela, O.P.; Khan, B.; Shafik, M.B.; Hussien, M.G. Design and Performance Analysis of Hybrid Battery and Ultracapacitor Energy Storage System for Electrical Vehicle Active Power Management. *Sustainability* **2022**, *14*, 776. [CrossRef]
38. He, H.; Xiong, R.; Zhao, K.; Liu, Z. Energy management strategy research on a hybrid power system by hardware-in-loop experiments. *Appl. Energy* **2013**, *112*, 1311–1317. [CrossRef]
39. Sellali, M.; Abdeddaim, S.; Betka, A.; Djerdir, A.; Drid, S.; Tiar, M. Fuzzy-Super twisting control implementation of battery/super capacitor for electric vehicles. *ISA Trans.* **2019**, *95*, 243–253. [CrossRef]
40. Ye, K.; Li, P.; Li, H. Optimization of Hybrid Energy Storage System Control Strategy for Pure Electric Vehicle Based on Typical Driving Cycle. *Math. Probl. Eng.* **2020**, *2020*, 1365195. [CrossRef]
41. Maghfiroh, H.; Wahyunggoro, O.; Cahyadi, A.I. Novel iterative Ragone plot-based optimization of low pass filter for hybrid power sources electric vehicles. *e-Prime-Adv. Electr. Eng. Electron. Energy* **2024**, *7*, 100389. [CrossRef]
42. Lee, Y.-L.; Lin, C.-H.; Chang, C.-H.; Liu, H.-D.; Chen, C.-C. A Novel Hybrid Energy Storage System with an Adaptive Digital Filter-Based Energy Management Strategy for Electric Vehicles. *IEEE Trans. Transp. Electrif.* **2024**, *10*, 5131–5142. [CrossRef]
43. Chen, F.; Ge, C.; Tang, D.; Ding, S.; Gong, X. Energy management and nonlinear control strategy of hybrid energy storage system for electric vehicle. *Energy Rep.* **2021**, *8*, 11161–11173. [CrossRef]
44. Maghfiroh, H.; Wahyunggoro, O.; Cahyadi, A.I. Optimizing Low Pass Filter Cut-off Frequency for Energy Management in Electric Vehicles with Hybrid Energy Storage Systems. *Math. Model. Eng. Probl.* **2023**, *10*, 1857–1865. [CrossRef]
45. Gunjal, M.; Mathew, L.; Shimi, S.L. Optimizing Battery Efficiency: Design and Implementation of a Fuzzy Logic Based Controller for Hybrid Energy Storage System in Electric Vehicles. In *Data Science and Applications*; ICDSA 2024; Lecture Notes in Networks and Systems; Nanda, S.J., Yadav, R.P., Gandomi, A.H., Saraswat, M., Eds.; Springer: Singapore, 2025; Volume 1237. [CrossRef]
46. Yu, H.; Castelli-Dezza, F.; Cheli, F. Multi-objective Optimal Sizing and Energy Management of Hybrid Energy Storage System for Electric Vehicles. *arXiv* **2018**, arXiv:1801.07183.
47. Yu, H.; Cao, D. Multi-objective Optimal Sizing and Real-time Control of Hybrid Energy Storage Systems for Electric Vehicles. In Proceedings of the 2018 IEEE Intelligent Vehicles Symposium, Changshu, China, 26–30 June 2018; pp. 191–196. [CrossRef]
48. Zhou, B.; Burl, J.B.; Rezaei, A. Equivalent Consumption Minimization Strategy With Consideration of Battery Aging for Parallel Hybrid Electric Vehicles. *IEEE Access* **2020**, *8*, 204770–204781. [CrossRef]
49. Pu, S.; Chu, L.; Hu, J.; Li, S.; Hou, Z. An Equivalent Consumption Minimization Strategy for a Parallel Plug-In Hybrid Electric Vehicle Based on an Environmental Perceiver. *Sensors* **2022**, *22*, 9621. [CrossRef] [PubMed]
50. Chen, M.; Wei, Z.; Huang, Z.; Ding, B.; Li, Y. Simple and Deep Graph Convolutional Networks. *arXiv* **2020**, arXiv:2007.02133. [CrossRef]
51. Liu, C.; Wang, Y.; Wang, L.; Chen, Z. Load-adaptive real-time energy management strategy for battery/ultracapacitor hybrid energy storage system using dynamic programming optimization. *J. Power Sources* **2019**, *438*, 227024. [CrossRef]
52. Pan, C.; Liang, Y.; Chen, L.; Chen, L. Optimal Control for Hybrid Energy Storage Electric Vehicle to Achieve Energy Saving Using Dynamic Programming Approach. *Energies* **2019**, *12*, 588. [CrossRef]
53. Chen, B.; He, G.; Hu, L.; Li, H.; Wang, M.; Zhang, R.; Gao, K. Energy management of electric vehicles based on improved long short term memory network and data-enabled predictive control. *Appl. Energy* **2025**, *384*, 125456. [CrossRef]
54. Chander, B.; Gopalakrishnan, K. Chapter 1—Integration of Artificial Intelligence Techniques for Energy Management. In *Sustainable Management of Electronic Waste*; Abhishek, K., Pramod, S.R., Ashutosh, K.D., Arun, L.S.T., Ananth, K., Vishal, D., Eds.; Scrivener Publishing LLC: Beverly, MA, USA, 2024; pp. 1–46. [CrossRef]
55. Mira, K.; Bugiotti, F.; Morosuk, T. Artificial Intelligence and Machine Learning in Energy Conversion and Management. *Energies* **2023**, *16*, 7773. [CrossRef]
56. Fang, Z.; Shek, J.K.H.; Sun, W. A review of grid-connected hybrid energy storage systems: Sizing configurations, control strategies, and future directions. *J. Energy Storage* **2025**, *118*, 116226. [CrossRef]
57. Xiong, R.; Cao, J.; Yu, Q. Reinforcement learning-based real-time power management for hybrid energy storage system in the plug-in hybrid electric vehicle. *Appl. Energy* **2018**, *211*, 538–548. [CrossRef]
58. Nambisan, P.; Khanra, M. Optimal power-split of hybrid energy storage system using Pontryagin's minimum principle and deep reinforcement learning approach for electric vehicle application. *Eng. Appl. Artif. Intell.* **2024**, *135*, 108769. [CrossRef]
59. Alaoui, C. Hybrid Vehicle Energy Management Using Deep Learning. In Proceedings of the International Conference on Intelligent Systems and Advanced Computing Sciences (ISACS), Taza, Morocco, 26–27 December 2019; pp. 1–5. [CrossRef]

batteries

MDPI

Article

$NiCo_2O_4$ Electrodes Prepared by Inkjet Printing on Kapton Substrates for Flexible Supercapacitor Applications

Angeliki Banti [1,*], Paris Pardalis [1], Eleni Mantsiou [1,2], Michalis Charalampakis [2], Vassilios Binas [1,2], Andronikos C. Balaskas [1] and Sotirios Sotiropoulos [1,*]

1 Physical Chemistry Laboratory, Chemistry Department, Aristotle University of Thessaloniki, 54124 Thessaloniki, Greece; el.mantsiou@iesl.forth.gr (E.M.)

2 Institute of Electronic Structure and Laser, Foundation for Research and Technology-Hellas (FORTH-IESL), 70013 Heraklion, Greece

* Correspondence: ampantic@chem.auth.gr (A.B.); eczss@chem.auth.gr (S.S.)

Abstract

This study explores the fabrication and electrochemical performance of flexible $NiCo_2O_4$-based pseudo-capacitor electrodes, inkjet-printed onto flexible Kapton substrates. To circumvent the insulating nature of Kapton, a thin Au interlayer (20 nm) was introduced, significantly enhancing electrical conductivity. The effect of $NiCo_2O_4$ mass loading, ranging from 0.1 to 0.5 mg cm^{-2}, was investigated. Optimal performance was achieved at a loading of 0.3 mg cm^{-2} on Au/Kapton substrates, yielding a specific capacitance of 520 F g^{-1} at 3.3 A g^{-1} and 90% capacitance retention after 1000 charge–discharge cycles. These results confirm that inkjet-printed $NiCo_2O_4$ electrodes, when combined with a conductive interlayer, exhibit excellent pseudo-capacitive behavior on flexible, non-conductive substrates. This approach demonstrates the feasibility of scalable, low-temperature fabrication techniques for high-performance flexible energy storage devices, suitable for emerging wearable technologies.

Keywords: flexible supercapacitors; $NiCo_2O_4$ electrodes; inkjet printing; Kapton substrate; wearable energy storage devices

1. Introduction

The rapid advancement of portable and wearable electronics has created an urgent demand for flexible energy storage systems capable of delivering high power density, fast charge–discharge rates, and long cycle life. Among the various energy storage technologies, supercapacitors have garnered significant attention due to their ability to meet these requirements either through electrostatic charge storage mechanisms or fast redox reactions in thin films, which entail faster kinetics compared to faradaic processes in batteries [1–3]. Despite their superior power density and cycling stability, supercapacitors typically suffer from lower energy density, which restricts their applicability in scenarios requiring both high energy and power output [4–6].

To address this limitation, recent research has focused on the development of advanced electrode materials that can enhance the energy density of supercapacitors without compromising their inherent advantages. Transition metal oxides, particularly nickel cobaltite ($NiCo_2O_4$), have emerged as promising candidates due to their high theoretical capacitance, excellent electrical conductivity, and robust electrochemical stability [7–10]. $NiCo_2O_4$, a mixed-valence oxide of nickel and cobalt, exhibits pseudo-capacitive behavior, enabling fast and reversible redox reactions that contribute to higher energy storage capacity compared to conventional electric double-layer capacitors (EDLCs) [11–13].

The spinel structure of $NiCo_2O_4$ provides a large number of electrochemically active sites, which enhances specific capacitance. Furthermore, the synergistic interaction between nickel and cobalt ions improves electrical conductivity and structural integrity, which are critical for maintaining performance over prolonged cycling [14–17]. However, the electrochemical performance of $NiCo_2O_4$ electrodes is highly dependent on factors such as morphology, surface area, and the nature of the substrate. Optimizing these parameters is essential for realizing the full potential of $NiCo_2O_4$ in flexible supercapacitor applications [18–21].

Flexible substrates play a pivotal role in the integration of supercapacitors into wearable and portable electronics, where mechanical durability and conformability are essential. Traditional substrates such as carbon cloth and nickel foam offer high conductivity and surface area but often lack the mechanical flexibility required for dynamic applications [22–24]. Kapton, a polyimide film, has emerged as a promising alternative due to its exceptional thermal stability, mechanical flexibility, and chemical resistance [25–27]. When coated with a conductive layer such as gold (Au), Kapton provides a stable and conductive platform for the deposition of active electrode materials like $NiCo_2O_4$, thereby enabling the fabrication of flexible supercapacitors with enhanced performance and durability [28–30]. The selection of Au as an interlayer was based on its excellent chemical inertness, high electrical conductivity, and strong adhesion to Kapton. Unlike Ag, which is prone to oxidation and migration, or carbon paints, which may suffer from poor film uniformity, Au provides a stable and conductive platform for inkjet printing. Although PET/ITO is a common alternative, its brittleness and limited thermal tolerance make it less suitable for flexible and thermally processed devices.

The fabrication of flexible supercapacitors necessitates precise and scalable deposition techniques that are compatible with advanced materials and substrates. Inkjet printing has gained prominence as a versatile and cost-effective method for fabricating supercapacitor electrodes, offering precise control over material deposition and patterning at the microscale [31–34]. This technique allows for the modulation of electrode thickness and morphology, which are critical parameters influencing electrochemical performance. Moreover, inkjet printing operates at relatively low temperatures, preserving the integrity of flexible substrates such as Kapton [35–37]. Compared to conventional methods like electrodeposition, sputtering, and chemical vapor deposition, inkjet printing offers reduced material waste, design flexibility, and scalability for industrial production [38–40].

One of the key factors influencing the performance of $NiCo_2O_4$ electrodes is the thickness of the catalyst layer. Electrode thickness directly affects specific capacitance, energy density, and mechanical flexibility. Recent studies have demonstrated that optimizing the number of printed layers can significantly improve electrochemical performance [1,8,36].

The main aim of this work is to report the proof-of-concept for an inkjet printing fabrication technique for flexible supercapacitor electrodes. Toward this end, $NiCo_2O_4$ was used as a typical electrocatalyst material and Kapton as a typical flexible support. When compared to other flexible substrates such as carbon cloth and nickel foam, a Kapton support offers a distinct advantage in terms of mechanical flexibility and structural integrity. Although carbon cloth and nickel foam coated with $NiCo_2O_4$ typically yield high capacitance values (750–1200 F g^{-1}), their limited flexibility constrains their use in wearable applications [41–44]. In this work, we investigate whether other supports that can be used with inkjet printing techniques, such as the model Au/Kapton system, can result in electrodes with a balanced electrochemical performance and superior mechanical resilience, opening up new avenues for the design of high-performance, flexible energy storage devices tailored for portable and wearable electronics.

2. Materials and Methods

2.1. Materials

Nickel(II) acetate tetrahydrate ($(CH_3COO)_2Ni.4H_2O$), cobalt(II) acetate tetrahydrate ($(CH_3COO)_2Co_4H_2O$), oxalic acid ($HO_2CCO_2H.H_2O$), propylene glycol ($C_3H_8O_2$), and Triton X-100 ($C_{14}H_{22}O(C_2H_4O)_n$) were purchased from Merck (Darmstadt, Germany) and used directly without purification. Distilled water was applied in all synthesis and ink preparation steps. Printing was carried out on Kapton substrates coated with a conductive gold layer (surface resistivity: 10 Ω/sq).

2.2. Synthesis of $NiCo_2O_4$ Nanoparticles

$NiCo_2O_4$ (NCO) nanoparticles were prepared via a hydrothermal process were used as the active material for inkjet printing. A solution containing 5 mmol nickel acetate, 10 mmol cobalt acetate, and 10 mmol oxalic acid in 30 mL distilled water was magnetically stirred for 30 min to form a uniform white suspension. The mixture was sealed in an autoclave and heated at 90 °C for 12 h. The product was centrifuged at 4000 rpm for 20 min, washed twice with ethanol, and dried at 80 °C for 4 h in a vacuum oven. Finally, the obtained powder was annealed in air at 350 °C for 2 h with a heating rate of 5 °C min^{-1}, yielding black $NiCo_2O_4$ nanoparticles.

2.3. Ink Formulation

The NCO powder was ground by ball milling. In detail, 3 g of NCO was dispersed in 25 mL of distilled water and milled in a zirconia container containing 100 g of 0.5 mm zirconia balls. The milling–centrifugation cycle (700 rpm, 30 min) was repeated four times to obtain a stable black nanoparticle-rich dispersion. This dispersion was subsequently mixed with propylene glycol at a 7:3 volume ratio and supplemented with Triton X-100 (0.5 mg mL^{-1}) to improve nanoparticle stability. To avoid clogging of the inkjet nozzles, the final ink was passed through a 0.45 μm nylon syringe filter to remove large agglomerates.

2.4. Conductive Substrate Preparation and Inkjet Printing Procedure

Kapton films (85 μm thick) were cut into 2 × 1 cm^2 pieces and cleaned sequentially with acetone, isopropanol, and deionized water, followed by drying under a nitrogen stream. A 20 nm thick gold layer was deposited onto the Kapton surface using a metal sputter coater ensuring optimal conductivity and adhesion. The Au layer was sputter-deposited on the Kapton substrate at ambient temperature. Although no thermal analysis was performed in this study, Kapton's thermal stability is well documented in the literature, with decomposition temperatures typically above 500 °C [45].

Inkjet printing was performed using a Dimatix DMP-2850 drop-on-demand printer (Fujifilm Dimatix, Santa Clara, CA, USA), equipped with a 10 pL piezoelectric cartridge (DMC-11610, Fujifilm Dimatix, Santa Clara, CA, USA) and a 3 mL ink reservoir. Printing was conducted under ambient conditions, with the cartridge temperature maintained between 25 and 35 °C and a drop spacing of 40 μm. Following deposition, the printed films were annealed in air at 350 °C for 2 h using a controlled heating rate of 5 °C min^{-1} to promote crystallinity and film cohesion.

2.5. Multilayer Electrode Assembly and Thermal Consolidation

To systematically evaluate the impact of film thickness on electrochemical performance, a series of $NiCo_2O_4$ electrodes was fabricated via inkjet printing with 2, 6, and 10 sequential layers. These configurations corresponded to active material loadings of approximately 0.1, 0.3, and 0.5 mg cm^{-2}, respectively. Following the initial high-temperature annealing step at 350 °C, all printed electrodes were subjected to a secondary thermal treatment at 150 °C for 30 min. This post-deposition annealing was implemented to enhance interparticle cohesion and promote strong interfacial adhesion between the $NiCo_2O_4$ film and the underlying Kapton substrate.

The dual-phase thermal protocol—comprising crystallization-driven annealing and low-temperature consolidation—was critical for achieving mechanically stable and electrochemically responsive electrode architectures. Previous studies have demonstrated that annealing temperatures significantly influence the crystallinity, morphology, and electrochemical behavior of $NiCo_2O_4$-based electrodes [46,47]. Moreover, variations in electrode thickness and mass loading have been shown to affect specific capacitance and energy density, underscoring the importance of structural optimization in supercapacitor design [48]. This approach ensured optimal film integrity, minimized interfacial resistance, and facilitated efficient charge transport across the printed layers.

2.6. Characterization Techniques

The crystalline structure of the NCO nanoparticles was examined by X-ray diffraction (XRD) using a Rigaku D/MAX-2000H (Rigaku Corporation, Tokyo, Japan) diffractometer with CuKα radiation (λ = 1.54 Å) and a graphite monochromator. Phase identification and structural analysis were performed with X-Pert Highscore Plus software, version 4.9, PANalytical B.V., Almelo, The Netherlands. The surface morphology was investigated by scanning electron microscopy (SEM) on a JEOL JSM-7000F (JEOL Ltd., Tokyo, Japan) operated at 20 kV. Particle size distribution was determined by dynamic light scattering (DLS) using a Zetasizer Nano ZS90 (Malvern Instruments, Malvern, UK). Atomic force microscopy (AFM) was employed to study the surface topography and film thickness of the electrodes, using a Solver Pro system (NT-MDT, Moscow, Russia).

2.7. Electrochemical Characterization

The electrochemical performance of the inkjet-printed $NiCo_2O_4$ thin film electrodes was evaluated using Cyclic Voltammetry (CV), Galvanostatic Charge–Discharge (GCD), and Electrochemical Impedance Spectroscopy (EIS). All measurements were conducted using a PGSTAT302N electrochemical workstation (Autolab, Metrohm, Herisau, Switzerland) with freshly prepared aqueous 0.1 M KOH as the electrolyte.

CV measurements were performed at potential sweep rates ranging from 5 to 20 mV s^{-1}. EIS analyses were conducted over a frequency range of 10^5 to 0.1 Hz with an AC amplitude of 10 mV. GCD experiments were carried out via chronopotentiometry at constant current densities between 1.0 and 6.0 mA cm^{-2}, with voltage cut-off values defined by the electrochemical window determined from CV profiles. Long-term cycling stability was assessed at a current density of approximately 0.5 mA cm^{-2} over 1000 consecutive charge–discharge cycles.

All electrochemical tests employed a conventional three-electrode configuration, consisting of a platinum wire as the counter electrode, a Saturated Calomel Electrode (SCE) as the reference, and $NiCo_2O_4$/Au/Kapton working electrodes fabricated with 2, 6, and 10 printed layers, corresponding to active material loadings of approximately 0.1, 0.3, and 0.5 mg cm^{-2}, respectively. The active electrode area exposed to the electrolyte was 0.7 cm^2. To minimize iR drop effects, a Luggin capillary was positioned 1 mm from the surface of the working electrode, in accordance with established protocols for reducing solution resistance errors [49].

Finally, to assess the mechanical flexibility and electrochemical resilience of the printed electrodes, CV measurements were performed after mechanically bending the electrode to a 90° angle. Comparative voltammograms recorded before and after bending were used to evaluate any degradation in capacitive behavior or redox activity. This test was designed to simulate mechanical strain conditions relevant to flexible and wearable energy storage devices [50,51].

3. Results

3.1. Structural and Morphological Characterization of Inkjet-Printed $NiCo_2O_4$ Electrodes

The crystalline structure of the inkjet-printed $NiCo_2O_4$ electrode was examined using X-ray Diffraction (XRD), as shown in Figure 1. The diffraction pattern displays prominent peaks indexed to the (111), (220), (311), (222), (400), (331), (420), (422), (511), and (440) planes, consistent with the spinel phase of $NiCo_2O_4$ (JCPDS card No. 40-0276) [1,2]. The sharp and well-defined reflections indicate high crystallinity, with the (311) peak being the most intense, suggesting preferential orientation and phase purity. The absence of secondary phases confirms the successful synthesis of $NiCo_2O_4$ with no detectable impurities, validating the integrity of the inkjet printing process for crystalline oxide deposition [3]. These structural characteristics are essential for achieving efficient pseudo-capacitive behavior, as they provide a stable framework for fast and reversible redox reactions [4]. TEM images of similarly prepared NCO nano-particles used in our previous work in semi-transparent supercapacitor electrodes revealed 5–9 nm diameter nanoparticles, aggregated in 100–300 nm clusters (see Figure 4 in [1]).

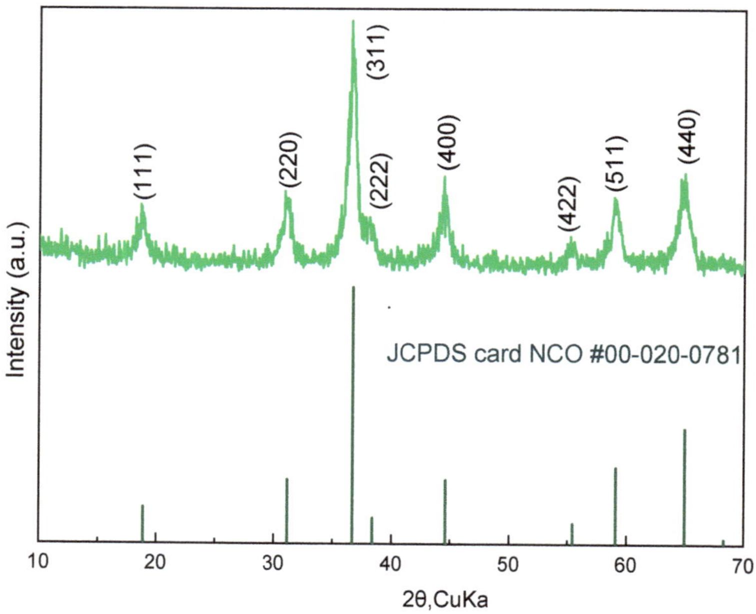

Figure 1. XRD pattern of inkjet-printed $NiCo_2O_4$ showing characteristic spinel peaks.

Figure 2a below shows the top-view photographic image of a $NiCo_2O_4$/Au/Kapton electrode prepared by inkjet printing while Figure 2b depicts bending of the flexible electrode.

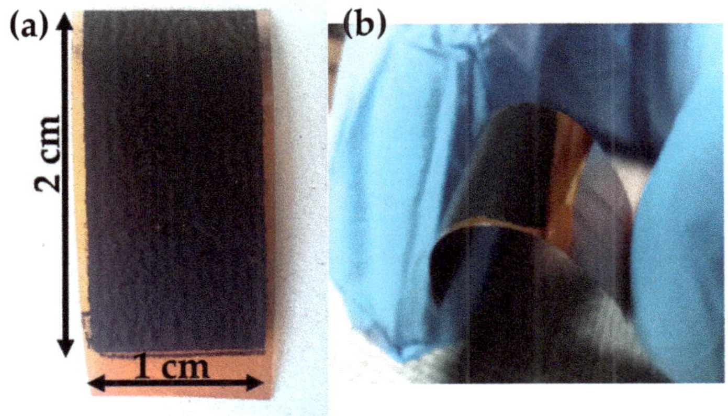

Figure 2. (**a**) Top view of inkjet-printed $NiCo_2O_4$/Au/Kapton electrode and (**b**) demonstration of its mechanical flexibility.

Surface morphology of the electrodes was investigated using Scanning Electron Microscopy (SEM), presented in Figure 3. The micrographs (Figure 3a,b) reveal a rough, granular texture composed of interconnected nanoscale particles, forming a porous network across the Au/Kapton substrate. Figure 3b shows a high-magnification SEM image of the $NiCo_2O_4$ film, revealing its nanoparticulate structure which is beneficial for ion transport and electrochemical activity. Figure 3c presents the associated particle size analysis, confirming that the particle size in the film (average diameter 60 nm) remains almost the same as in the ball-milled powder. The NCO particles exhibit irregular shapes and a porous morphology, with average diameters ranging from 200 to 800 nm. In some regions, agglomerates reach sizes of 1–2 µm, indicating localized clustering. Following mechanical grinding, the NCO nanoparticles achieve a more uniform size distribution, with diameters reduced to below 200 nm. The average particle size decreases from approximately 450 nm to 60 nm, representing a mean reduction of about 385 nm as shown in Figure 3b (high resolution SEM) and 3c (histogram of particle size analysis), in excellent agreement with similar films prepared on glass substrates and presented in [1]. The average particle size decreases from approximately 450 nm to 65 nm, representing a mean reduction of about 385 nm. This refinement enhances surface homogeneity and increases the electrochemically active area, which is beneficial for ion transport and redox activity [8,27,30]. The uniform distribution of particles and absence of cracks or delamination indicate strong adhesion and consistent film formation, both critical for mechanical stability in flexible devices [9,26]. The observed surface features are characteristic of pseudo-capacitive materials and support efficient electrochemical performance [10,28].

Atomic Force Microscopy (AFM) was employed to assess the topography of the printed $NiCo_2O_4$ layer, as shown in Figure 4. The 2D surface profile reveals nanoscale roughness with height variations ranging from 0 to approximately 300 nm. The presence of elevated regions and valleys suggests a textured surface that enhances electrolyte contact and charge storage capacity [30]. The AFM micrograph also reveals a homogeneous and compact film with pronounced nanoroughness, indicating uniform deposition and strong surface cohesion. This topographical complexity complements the SEM findings and confirms the formation of a uniform, high-surface-area film. Together, the XRD, SEM, and AFM analyses demonstrate that inkjet-printed $NiCo_2O_4$ on Au/Kapton exhibits the structural and morphological characteristics necessary for high-performance flexible supercapacitor electrodes [8].

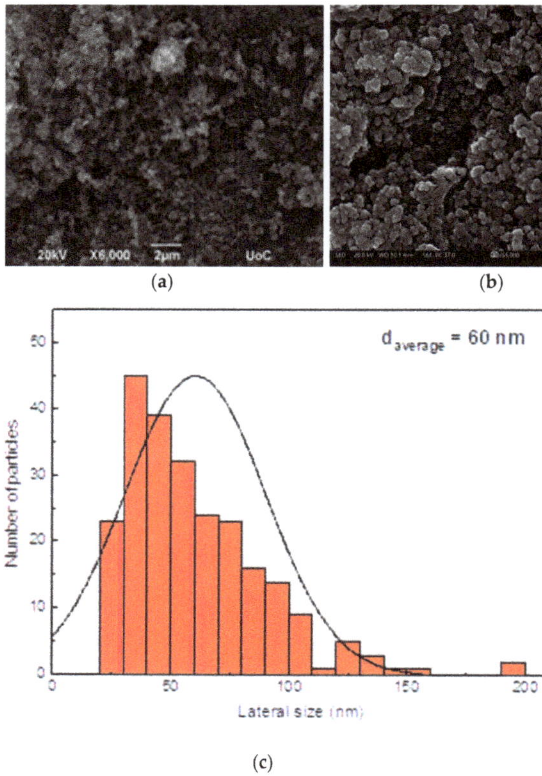

Figure 3. SEM images of $NiCo_2O_4$ films on Au/Kapton substrates, (**a**) Prepared from NCO particles as synthesized (white scale bar of 2 μm) and (**b**) after ball-milling (white scale bar of 200 nm); (**c**) Particle size analysis of the NCO film depicted in (**b**).

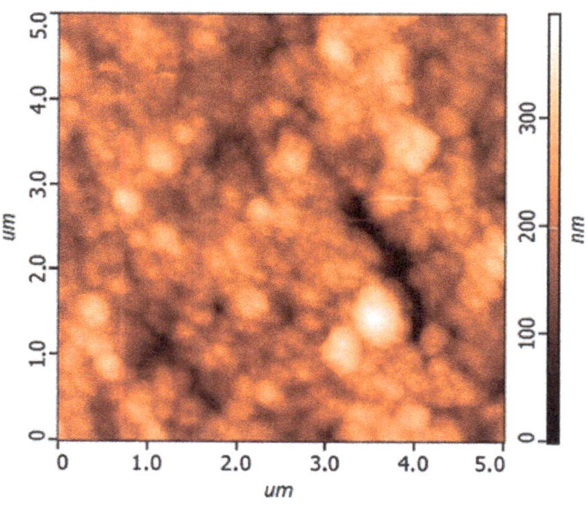

Figure 4. AFM topography of the printed $NiCo_2O_4$ layer showing nanoscale surface roughness.

3.2. Electrochemical Performance

Cyclic voltammetry (CV) was employed to evaluate the capacitive behavior of the inkjet-printed $NiCo_2O_4$ electrodes on Au/Kapton substrates as a function of the number of printed catalyst layers. Figure 5 presents the CV responses of the electrode composed of two, six and ten printed layers (corresponding to 0.1, 0.3 and 0.5 mg cm^{-2} loadings), recorded at scan rates of 5, 10, and 20 mV s^{-1} within the potential window of 0–0.55 V vs. SCE. The voltammograms exhibit quasi-rectangular shapes with distinguishable redox peaks, indicating the coexistence of electric double-layer capacitive and pseudocapacitive behavior. This hybrid charge storage arises from additional reversible faradaic reactions of Ni^{2+}/Ni^{3+} and Co^{2+}/Co^{3-} redox couples. Specifically, the nickel-based redox process follows:

$$NiO + OH^- \rightleftharpoons NiOOH + e^- \quad (1)$$

For cobalt, the redox activity proceeds through a two-step mechanism:

$$Co_2O_3 + H_2O \rightleftharpoons 2CoOOH \quad (2)$$

$$2CoOOH + 2OH^- \rightleftharpoons 2CoO_2 + 2H_2O + 2e^- \quad (3)$$

Alternatively, the overall cobalt reaction can be expressed as:

$$Co_2O_3 + 2OH^- \rightleftharpoons 2CoO_2 + H_2O + 2e^- \quad (4)$$

These reactions contribute to the broad and overlapping redox peaks observed in the voltammograms, confirming the pseudo-capacitive nature of the inkjet-printed $NiCo_2O_4$ electrodes. As the scan rate increases, both anodic and cathodic peak currents increase correspondingly, without significant deformation of the CV shape, suggesting fast surface redox kinetics. The improved current response of 6- and 10-layer electrodes compared to the 2-layer electrode (Figure 5c) confirms the enhancement in electroactive surface area with increasing film thickness. However, the similarity of the current density for the 6- and 10-layer electrodes (Figure 5a,b), indicates that beyond a certain thickness there is a decrease in porosity, limiting the use of the material in the interior of the thick film.

These results suggest that the electrode with six layers (Figure 5b) achieves a balance between sufficient mass loading for effective charge storage and favorable ion diffusion pathways, making it a promising candidate for flexible supercapacitor applications where both performance and mechanical integrity are critical.

EIS measurements (Figure 6a) and subsequent analysis (Figure 6b) were carried out at the open circuit potential of the optimum 6 L electrode (+0.50 V vs. SCE). The recorded arc is associated with a charge transfer resistance (R_{ct}) that may be attributed to Co and Ni redox transformations (change in oxidation state) under kinetic control, followed by a linear tail at low frequencies, indicative of a diffusion-controlled processes (Warburg impedance), most likely corresponding to OH^- diffusing through the pores (reactions corresponding to the chemical Equations (1), (3) and (4)).

Fitting an appropriate electrical circuit to the data $[R_u(R_pCPE_p)([R_{ct}W]CPE_{dl})]$ accounts for the above-mentioned faradaic and diffusion processes second RC circuit in series; $([R_{ct}W]CPE_{dl})$, as well as for ionic migration inside a porous electrode (first RC circuit in series; $R_u(R_pCPE_p)$ in agreement with others appearing in the literature [41,42]. In this notation Rp is the ionic pore resistance, R_{ct} the faradaic reaction charge transfer resistance, CPE_p and CPE_{dl} the corresponding constant phase elements associated with inhomogeneities and double layer capacitance of the electrode pores and surface and R_u the uncompensated solution resistance up to the outer surface of the electrode.

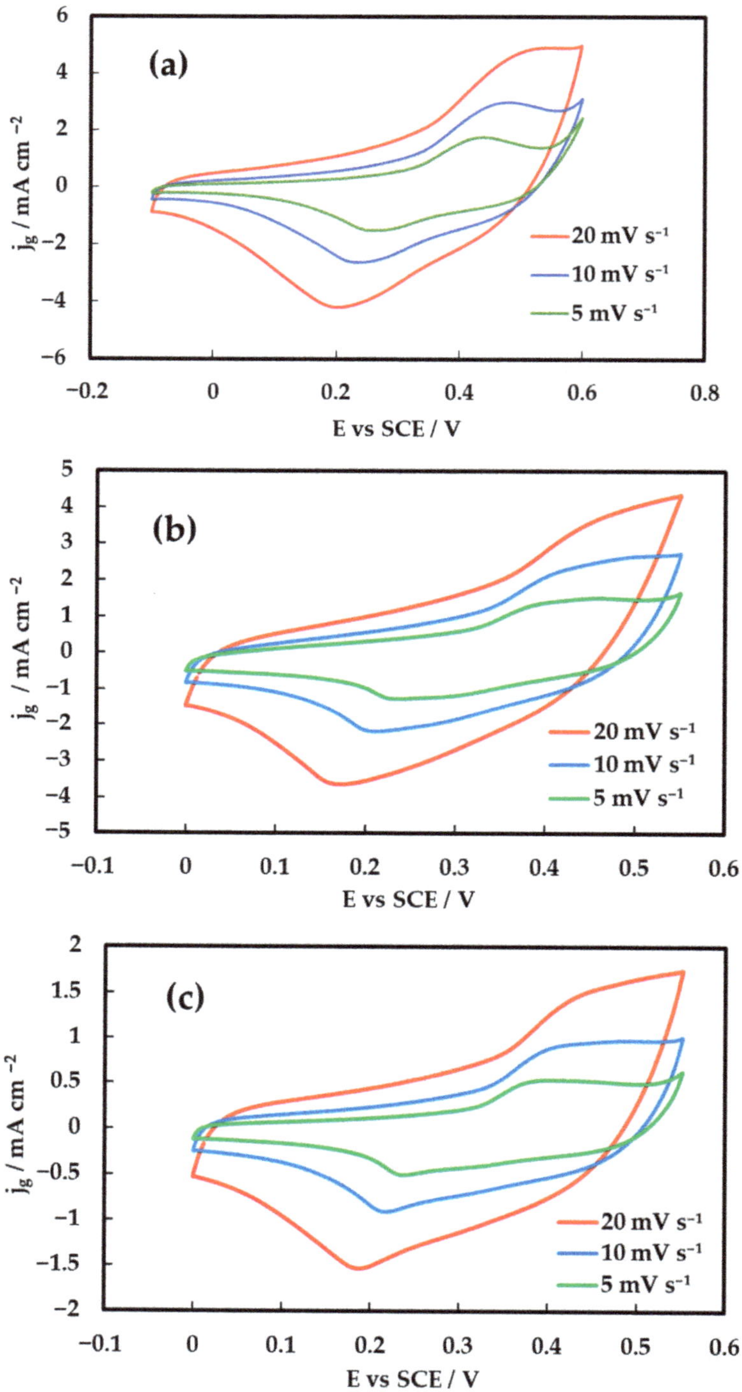

Figure 5. Cyclic voltammetry (CV) curves of $NiCo_2O_4$ electrodes recorded in 1 M KOH within the potential window of 0–0.55 V vs. SCE. (**a**) for 10-layer electrode, (**b**) for 6-layer electrode and (**c**) for 2-layer electrode.

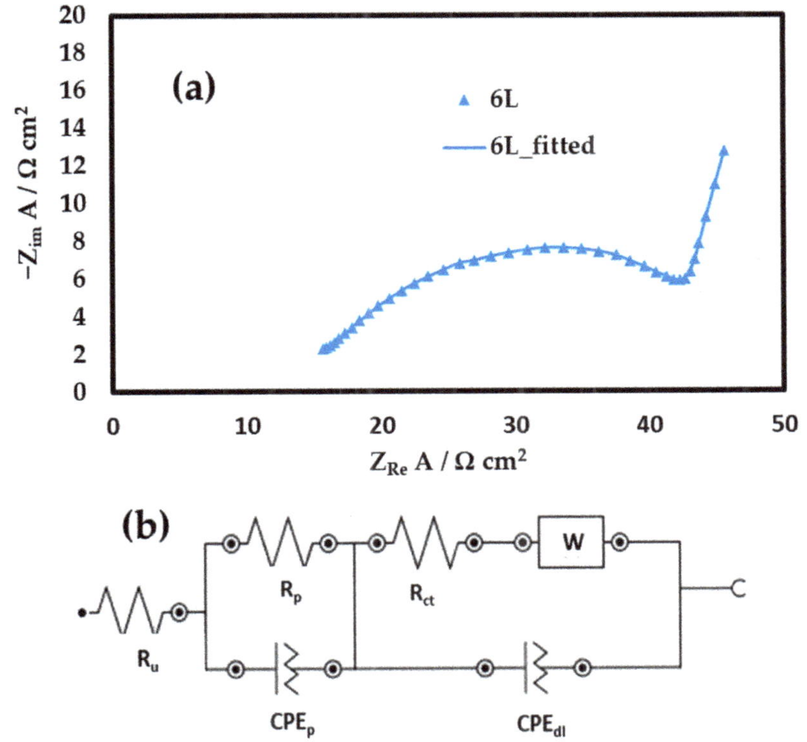

Figure 6. (**a**) Nyquist plots of $NiCo_2O_4$ electrodes with 6 printed layers measured in 0.1 M KOH at +0.55 V and (**b**) corresponding fitted electrical circuit.

The values obtained (Table 1) are comparable to those reported for $NiCo_2O_4$ electrodes inkjet printed on glass substrates [1], indicating that the Kapton/Au does not compromise electrochemical performance. The values of R_u estimated for all electrodes studied here have been used to correct the potential drop during the discharge part of the GCD presented below.

Table 1. Component values of the electrical circuit corresponding to the equivalent model illustrated in Figure 6.

	6L
$R_u/\Omega\ cm^2$	15
$Q_p/\Omega^{-1}\ s^{n1}\ cm^{-2}$	0.0019
n1	0.53
$R_p/\Omega\ cm^2$	33
$Q_{dl}/\Omega^{-1}\ s^{n2}\ cm^{-2}$	0.094
n2	0.95
$R_{ct}/\Omega\ cm^2$	5.6
$W/\Omega\ S^{1/2}\ cm^2$	0.00060
X^2	0.004

Figure 7a presents the galvanostatic charge–discharge (GCD) profiles of inkjet-printed $NiCo_2O_4$ electrodes comprising 2, 6, and 10 printed layers, recorded at a current density of 1.5 mA cm^{-2} in 0.1 M KOH aqueous electrolyte. The voltage–time curves exhibit quasi-linear and symmetric characteristics, with slight deviations from a triangular shape attributed to pseudo-capacitive contributions arising from the reversible faradaic redox reactions involving Ni^{2+}/Ni^{3+} and Co^{2+}/Co^{3+} couples. The hybrid charge storage behavior of $NiCo_2O_4$ electrodes arises from reversible faradaic reactions involving both nickel and cobalt redox couples. During charging/positive-going potential sweep, Ni^{2+} ions are oxidized to Ni^{3+} and Co^{2+} ions to Co^{3+}, while the reverse reaction occurs during discharging/negative-going potential sweep. These transitions are accompanied by hydroxide ion participation and transport from the alkaline electrolyte into the mixed oxide film, accompanied by fast surface electron transfer. The co-existence of Ni and Co red/ox couples provides multiple oxidation states for charge storage, which contributes to the broad and overlapping redox peaks observed in the cyclic voltammograms. This is the origin of the pseudocapacitive nature of the inkjet-printed $NiCo_2O_4$ electrodes and explains their enhanced capacitance and rate capability compared to purely double layer capacitors.

The shape of the discharge curves suggests that the voltage drop during discharge can be reasonably approximated by a straight line connecting the potential limits. This simplification is justified by the fact that the area under the linear approximation deviates by less than 10% from the actual experimental curve. Therefore, for estimating the specific or areal capacitance (C), one can apply the simplified relation commonly used for electric double-layer (EDL) supercapacitors with potential-independent capacitance:

$$C = \frac{j \cdot t_D}{\Delta V_{\text{corrected}}} \tag{5}$$

where j is the discharge current density, t_D is the discharge time and $\Delta V_{\text{corrected}}$ is the voltage drop corrected for ohmic losses.

However, for hybrid supercapacitors—where capacitance varies with potential—a more rigorous approach is required. The average areal capacitance (C_{ave}) should be calculated using the general expression [5,28,37]:

$$C_{ave} = \frac{2j}{\Delta V_{corrected}} \int V\,dt \tag{6}$$

This formulation accounts for the non-linear voltage profile during discharge and provides a more accurate representation of the hybrid system's capacitive behavior.

The electrode with 10 printed layers exhibits the longest discharge duration, yielding the highest areal capacitance of 172 mF cm^{-2}, followed by the 6-layer and 2-layer electrodes with values of 156 mF cm^{-2} and 56 mF cm^{-2}, respectively. The increase in areal capacitance with layer number is ascribed to the greater amount of electroactive material available for charge storage. However, normalization with respect to the active material mass yields a partially reversed trend in mass-specific capacitance, with values of 344 F g^{-1}, 521 F g^{-1}, and 503 F g^{-1} for the 10-, 6-, and 2-layer electrodes, respectively. The superior gravimetric performance of the 6- and 2-layer electrodes suggests more efficient utilization of the active material, likely due to enhanced ion accessibility and reduced diffusion limitations within thinner films. Figure 7b presents the charge–discharge curves of the optimum 6L electrode at various charge–discharge current densities and the same type of behavior is observed at current densities as high as 6 mA cm^{-2}.

On the other hand, the initial potential drop (IR-drop) observed at the onset of discharge more pronounced as the number of layers decreases, indicating an increase in internal resistance due to a decrease in inter-particle contacts in very thin films. These observations are consistent with the cyclic voltammetry results and highlight the trade-off between areal and mass-specific performance. Optimal electrochemical behavior is thus achieved by balancing active material loading with ionic/electronic transport pathways, which is critical for the development of high-performance flexible supercapacitor devices based on inkjet-printed $NiCo_2O_4$ electrodes.

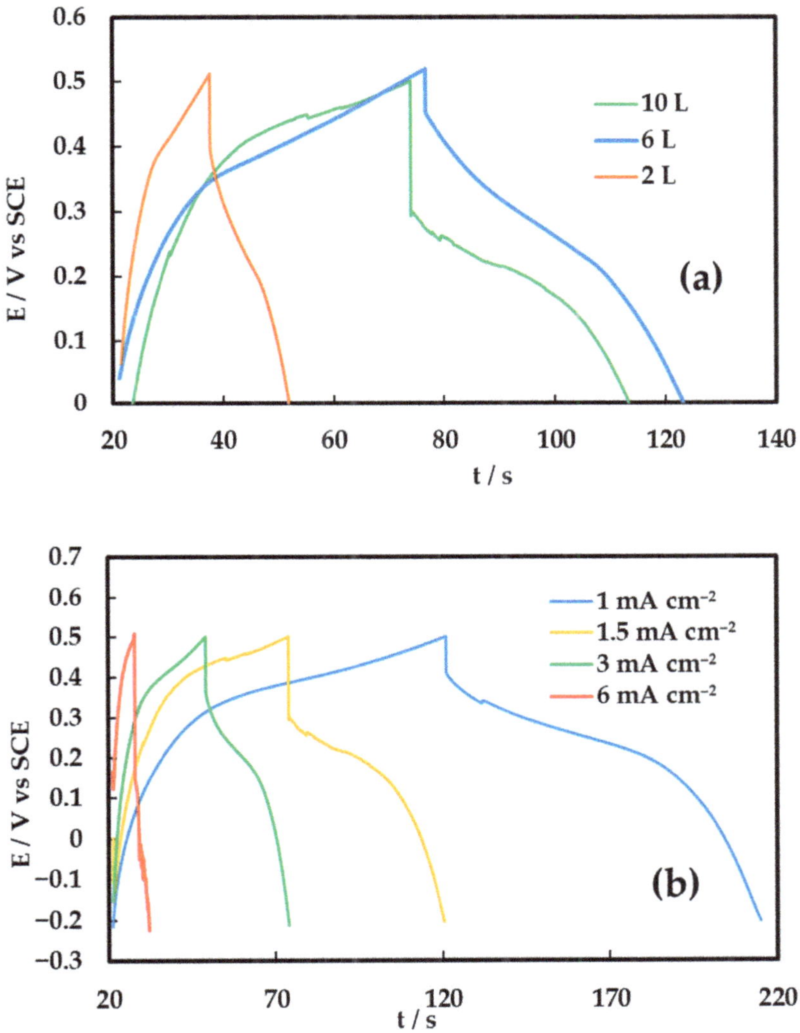

Figure 7. Galvanostatic charge–discharge curves for (**a**) $NiCo_2O_4$ electrodes with 2, 6, and 10 printed layers at 1.5 mA cm^{-2} in 0.1 M KOH and (**b**) the 6 L $NiCo_2O_4$ electrode at various current densities.

Table 2 presents the specific capacitance values of the $NiCo_2O_4$-based electrodes studied in this work. The printed electrodes exhibit specific capacitance values (per substrate geometric area and catalyst mass) in the 50–175 mF cm^{-2} and 345–520 F g^{-1} range, depending on electrode configuration and testing conditions. These values are within the range of those reported for flexible $NiCo_2O_4$ systems in the literature [17,52,53], underscoring the potential of the inkjet printing approach for the development of high-performance, flexible supercapacitor electrodes.

Table 2. Capacitance performance of $NiCo_2O_4$ electrodes with varying layer thickness.

Electrode Material	Loading/ mg cm^{-2}	Areal Capacitance/ mF cm^{-2}	Mass-Specific Capacitance/ F g^{-1}
$NiCo_2O_4$-2 L	0.1	50	504
$NiCo_2O_4$-6 L	0.3	157	520
$NiCo_2O_4$-10 L	0.5	172	345

The long-term electrochemical stability of the $NiCo_2O_4$ electrode with 6 printed layers (NCO–Au–Kapton 6 L) was evaluated over 1000 continuous charge–discharge cycles at a fixed current density (1.0 mA cm^{-2}). As shown in Figure 8, the electrode retains approximately 90% of its initial capacitance after 1000 cycles, indicating moderate cycling stability. While the initial performance is satisfactory, the observed fading highlights the need for further optimization of mechanical adhesion, structural durability, or protective coatings to enhance cycling lifespan in flexible supercapacitor applications.

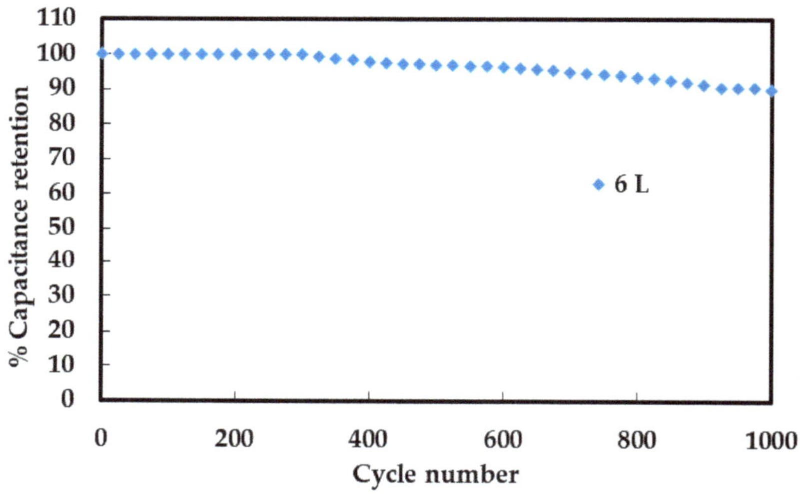

Figure 8. Capacitance retention of $NiCo_2O_4$ (6-layer) electrode on Au/Kapton substrate over 1000 charge–discharge cycles.

3.3. *Electrochemical Performance Under Mechanical Stress*

To assess the mechanical flexibility and electrochemical stability of the inkjet-printed $NiCo_2O_4$ electrodes deposited on Kapton substrates with a gold (Au) interlayer, cyclic voltammetry (CV) measurements were conducted under three mechanical conditions: prior to bending, during bending, and following bending. The CV profiles, recorded at a scan rate of 5 mV s^{-1} with a mass loading of 0.3 mg cm^{-2}, are presented in Figure 9 (while the inset shows a photograph of the 90°-bent electrode). All three curves—green (before bending), orange (during bending), and blue (after bending)—exhibit quasi-rectangular shapes with discernible redox features, indicative of pseudo-capacitive behavior governed by fast and reversible faradaic reactions [8,10].

The small deviation between the CV curves obtained before and after bending demonstrates that the electrode retains its electrochemical integrity following mechanical deformation. A modest reduction in current density observed during bending is attributed to transient strain effects, which do not result in permanent degradation. The recovery of current response post-bending confirms the structural robustness and functional resilience of the $NiCo_2O_4$/Au/Kapton architecture [4,9].

The bending strain, ε, of the bent 6L NCO/Au/Kapton electrode was estimated using two approaches. The first was an approximate one, simply considering the total electrode thickness, $\boldsymbol{d}$, of 85 μm Kapton + 20 nm Au + 1.5 μm NCO = 148.52 μm and the radius, $\boldsymbol{r}$, of the bent structure which is measured as 2.23 cm and the following simple equation [54]:

$$\varepsilon = \frac{d}{2r} \tag{7}$$

The bending strain thus estimated was 0.39%.

The second approach, took into consideration an NCO/Kapton bilayer (assuming that the 20 nm Au interlayer was sufficiently thinner than the other two layers, NCO is 1.5 and Kapton 85 μm, to be ignored as an approximation). Taking the Young's moduli, E, values of NCO and Kapton as 84 and 4 GPa, respectively [54,55], and applying the following equation:

$$\varepsilon = \frac{d}{2r}\left(\frac{1+2\eta+\chi\eta^2}{(1+\eta)(1+\chi\eta)}\right) \tag{8}$$

(where $\eta = d_{NCO}/d_{Kapton}$ and $\chi = E_{NCO}/E_{Kapton}$), the bending strain thus estimated was 0.29%.

These findings validate the mechanical durability of the inkjet-printed $NiCo_2O_4$ electrodes and their ability to maintain stable capacitive performance under dynamic mechanical stress. The integration of a conductive Au interlayer on Kapton not only facilitates efficient charge transport but also supports mechanical compliance, rendering this configuration highly suitable for flexible energy storage systems, particularly in wearable and portable electronic applications where repeated mechanical deformation is anticipated [1,27,28].

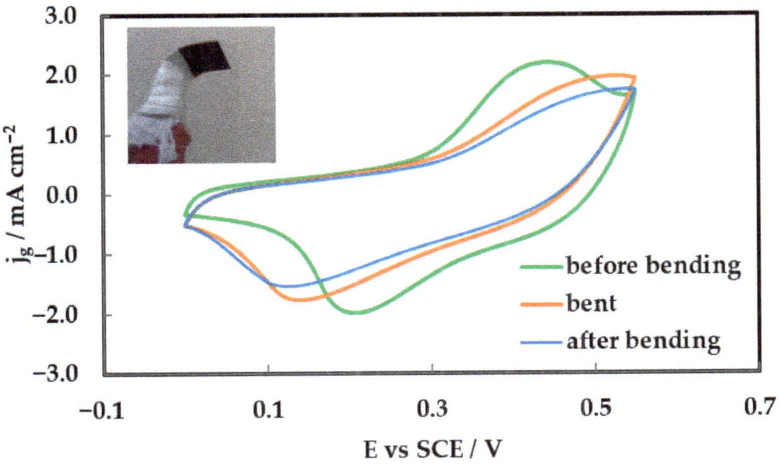

Figure 9. CV curves of inkjet-printed $NiCo_2O_4$ electrodes on Kapton Au (0.3 mg cm^{-2}) in 0.1 M MOH at 5 mV s^{-1}, showing performance before bending (green), during bending (orange), and after bending (blue).

4. Discussion

The comprehensive structural and electrochemical characterization of inkjet-printed $NiCo_2O_4$ electrodes on Kapton/Au substrates highlights their potential for flexible energy storage applications. Each technique—XRD, SEM, AFM, CV, GCD, EIS, and charge–discharge cycling—offers distinct insights into the material's behavior, collectively validating its performance and manufacturability.

X-ray diffraction (XRD) confirmed the formation of a crystalline spinel $NiCo_2O_4$ phase, with sharp reflections corresponding to the (111), (200), (220), (311), and other characteristic planes (Figure 1). The dominance of the (311) peak and absence of secondary phases indicate high phase purity and preferential orientation. This crystallinity, achieved via low-temperature inkjet printing, rivals that of $NiCo_2O_4$ synthesized through hydrothermal or sol–gel methods [5,8], while remaining compatible with flexible substrates.

Scanning Electron Microscopy (SEM) revealed a porous, granular morphology composed of irregularly shaped $NiCo_2O_4$ particles (Figure 3). Particle sizes ranged from 200 to 800 nm, with agglomerates reaching up to 2 μm. After mechanical grinding, the particles exhibited a more uniform distribution below 200 nm, with average diameters reduced from ~450 nm to ~60 nm. This refinement enhances surface homogeneity and increases the electrochemically active area, facilitating ion transport and redox reactions. The porous architecture supports electrolyte penetration and is consistent with pseudo-capacitive behavior.

Atomic Force Microscopy (AFM) provided nanoscale insight into the surface topography of the printed films (Figure 4). The 2D profile revealed a compact, homogeneous film with nanoroughness and height variations up to 300 nm. The presence of elevated regions and valleys enhances electrolyte contact and charge storage capacity. This morphology complements the SEM findings and confirms the formation of a high-surface-area electrode layer, critical for both electrochemical performance and mechanical durability.

Cyclic Voltammetry (CV) was used to probe the charge storage mechanism. The CV profiles (Figure 5) display distinct redox peaks corresponding to Ni^{2+}/Ni^{3+} and Co^{2+}/Co^{3+} transitions, confirming pseudo-capacitive behavior. The increase in peak current with scan rate and layer number reflects enhanced electrochemical activity due to greater active surface area and improved electron transport. These trends align with prior studies on $NiCo_2O_4$ electrodes [5,8], but are here achieved through a scalable inkjet printing process.

Galvanostatic Charge–Discharge (GCD) measurements (Figure 7) quantified the specific capacity. The 10-layer electrode exhibited the highest areal capacitance (345 F g^{-1}), while the 6-layer electrode achieved the highest mass-specific capacitance (520 F g^{-1}), indicating efficient material utilization. This trade-off highlights the challenge of balancing film thickness with ion accessibility, as excessive thickness can lead to underutilization of inner layers due to diffusion limitations.

Electrochemical Impedance Spectroscopy (EIS) (Figure 6) evaluated charge transfer kinetics and internal resistance. Increasing the number of printed layers reduced charge transfer resistance up to an optimal thickness, beyond which transport limitations may arise. This confirms that inkjet-printed $NiCo_2O_4$ films can achieve efficient electron and ion conduction when properly optimized.

Charge–discharge cycling behavior, assessed through repeated GCD cycles (Figure 8), revealed good retention over time. The 6-layer electrode maintained approximately 90% of its initial capacitance after 1000 cycles. This decline may result from mechanical fatigue, structural degradation, or loss of active sites during repeated charge–discharge processes. Enhancements such as surface passivation, flexible binders, or hybrid composites may be necessary to improve long-term reliability.

Mechanical flexibility was evaluated via CV measurements before, during, and after bending (Figure 9). The consistent shape and redox features across all conditions confirm that the $NiCo_2O_4$/Au/Kapton electrodes maintain electrochemical integrity under mechanical stress. A slight reduction in current during bending was reversible, indicating transient strain effects rather than permanent damage. This resilience supports their application in wearable electronics.

In summary, each characterization technique contributes to a holistic understanding of the material: XRD confirms crystallinity, SEM and AFM reveal surface architecture,

CV and GCD quantify charge storage, EIS diagnoses transport efficiency, and charge–discharge cycling tests assess durability. Together, they demonstrate that inkjet-printed $NiCo_2O_4$ on Au/Kapton is a viable platform for flexible supercapacitor electrodes. Future research should focus on improving cycling performance, exploring hybrid or solid-state configurations, and refining ink formulation and substrate engineering to further enhance performance and manufacturability.

5. Conclusions

This study demonstrates the successful fabrication and characterization of inkjet-printed $NiCo_2O_4$ electrodes on Kapton substrates modified with a thin gold interlayer, offering a scalable and substrate-compatible approach for flexible energy storage devices. Structural analyses confirmed the formation of a crystalline spinel phase with a porous, nanostructured morphology and uniform surface topography—features that directly support efficient charge storage.

Electrochemical characterization revealed pseudo-capacitive behavior driven by reversible Ni^{2+}/Ni^{3+} and Co^{2+}/Co^{3+} redox reactions. The six-layer configuration achieved optimal performance, balancing areal and mass-specific capacitance (157 mF cm^{-2} and 520 F g^{-1} catalyst), charge–discharge cycling stability, and mechanical flexibility. Impedance analysis and bending tests further validated the robustness of the electrode architecture under dynamic conditions.

Overall, inkjet printing of $NiCo_2O_4$ onto Au/Kapton substrates presents a viable pathway toward high-performance, flexible supercapacitor electrodes. Future work should focus on enhancing charge–discharge cycling retention, exploring hybrid material systems, and integrating these electrodes into full-cell configurations for wearable and portable energy storage applications.

Author Contributions: Conceptualization, S.S. and V.B.; methodology, A.B.; investigation, P.P., E.M. and M.C.; writing—original draft preparation, A.B.; writing—review and editing, S.S., A.B. and A.C.B. All authors have read and agreed to the published version of the manuscript.

Funding: This research received no external funding.

Data Availability Statement: The original contributions presented in this study are included in the article. Further inquiries can be directed to the corresponding author.

Conflicts of Interest: The authors declare no conflict of interest.

References

1. Banti, A.; Charalampakis, M.; Pardalis, P.; Prochaska, C.; Sotiropoulos, S.; Binas, V. Electrochemical Studies of Inkjet Printed Semi-Transparent $NiCo_2O_4$/ITO Supercapacitor Electrodes. *Catalysts* **2023**, *13*, 1110. [CrossRef]
2. Liang, J.; Jiang, C.; Wu, W. Printed flexible supercapacitor: Ink formulation, printable electrode materials and applications. *Appl. Phys. Rev.* **2021**, *8*, 021319. [CrossRef]
3. Hari Narayanan, K.R.; Kannan, S.; Ramadoss, A. Inkjet Printing Fabrication of Supercapacitors. In *Functionalized Nanomaterials Based Supercapacitor*; Springer: Berlin/Heidelberg, Germany, 2023. [CrossRef]
4. Manjakkal, L.; Dervin, S.; Dahiya, R. Flexible potentiometric pH sensors for wearable systems. *RSC Adv.* **2020**, *10*, 8594–8617. [CrossRef]
5. Wang, Y.; Sun, L.; Wang, C.; Yang, F.; Ren, X.; Zhang, X.; Dong, D.; Hu, W. Organic crystalline materials in flexible electronics. *Chem. Soc. Rev.* **2019**, *489*, 1492–1530. [CrossRef]
6. Shao, Y.; El-Kady, M.F.; Sun, J.; Li, Y.; Zhang, Q.; Zhu, M.; Wang, H.; Dunn, B.; Kaner, R.B. Design and Mechanisms of Asymmetric Supercapacitors. *Chem. Rev.* **2018**, *118*, 9233–9280. [CrossRef]
7. Liu, L.; Niu, Z.; Chen, J. Unconventional supercapacitors from nanocarbon-based electrode materials to device configurations. *Chem. Soc. Rev.* **2016**, *45*, 4340–4363. [CrossRef]
8. Goel, P.; Sundriyal, S.; Shrivastav, V.; Mishra, S.; Dubal, D.P.; Kim, K.H.; Deep, A. Perovskite materials as superior and powerful platforms for energy conversion and storage applications. *Nano Energy* **2021**, *80*, 105552. [CrossRef]

9. Wang, H.; Wang, Y.; Lin, Y.; Huang, X.; García-Tecedor, M.; Peña O'Shea, V.A.; Murrill, C.; Lazarov, V.K.; Oropeza, F.E.; Zhang, K.H.L. Impact of $NiCo_2O_4/SrTiO_3$ p–n Heterojunctions on the Interface of Photoelectrochemical Water Oxidation. *ACS Appl. Mater. Interfaces* **2023**, *15*, 28739–28746. [CrossRef]
10. Chen, S.; Yang, G.; Jia, Y.; Zheng, H. Three-dimensional $NiCo_2O_4$@$NiWO_4$ core–shell nanowire arrays for high performance supercapacitors. *J. Mater. Chem. A* **2017**, *5*, 1028–1034. [CrossRef]
11. Li, X.; Liang, J.; Yang, X.; Adair, K.R.; Wang, C.; Zhao, F.; Sun, X. Progress and perspectives on halide lithium conductors for all-solid-state lithium batteries. *Energy Environ. Sci.* **2020**, *13*, 1429–1461. [CrossRef]
12. Paulsen, B.D.; Wu, R.; Takacs, C.J.; Steinrück, H.-G.; Strzalka, J.; Zhang, Q.; Toney, M.F.; Rivnay, J. Time-Resolved Structural Kinetics of an Organic Mixed Ionic–Electronic Conductor. *Adv. Mater.* **2020**, *32*, 2003404. [CrossRef]
13. Shi, P.; Chen, R.; Hua, L.; Li, L.; Chen, R.; Gong, Y.; Yu, C.; Zhou, J.; Liu, B.; Sun, G.; et al. Highly Concentrated, Ultrathin Nickel Hydroxide Nanosheet Ink for Wearable Energy Storage Devices. *Adv. Mater.* **2017**, *29*, 1703455. [CrossRef]
14. Kim, C.; Marsland, R.; Blick, R.H. The Nanomechanical Bit. *Small* **2020**, *16*, 2001580. [CrossRef]
15. Wang, L.; Fu, X.; He, J.; Shi, X.; Chen, T.; Chen, P.; Wang, B.; Peng, B. Application Challenges in Fiber and Textile Electronics. *Adv. Mater.* **2019**, *32*, 1901971. [CrossRef]
16. Madéo, J.; Man, M.K.L.; Sahoo, C.; Campbell, M.; Pareek, V.; Wong, E.L.; Al-Mahboob, A.; Chan, N.S.; Karmakar, A.; Mariserla, B.M.K.; et al. Directly visualizing the momentum-forbidden dark excitons and their dynamics in atomically thin semiconductors. *Science* **2020**, *370*, 1199–1204. [CrossRef]
17. Bharanitharan, N.T.; Dhinasekaran, D.; Kishore, M.R.A.; Subramanian, B.; Rajendran, A.R. Rational design of $NiCo_2O_4$@carbon hollow spheres as a high-performance electrode material for flexible supercapacitors. *Nanoscale* **2025**, *17*, 2252–2258. [CrossRef] [PubMed]
18. Kim, K.-W.; Park, S.J.; Park, S.-J.; Kim, I.; Park, B.; Kim, S.; Jeong, U.; Jin Kon Kim, K.J.; Yang, C. Deformable micro-supercapacitor fabricated via laser ablation patterning of graphene/liquid metal. *npj Flex. Electron.* **2024**, *8*, 18. [CrossRef]
19. Zhang, G.; Lou, W. General Solution Growth of Mesoporous $NiCo_2O_4$ Nanosheets on Various Conductive Substrates as High-Performance Electrodes for Supercapacitors. *Adv. Mater.* **2013**, *25*, 976–979. [CrossRef]
20. Wang, C.; Zhou, E.; He, W.; Deng, X.; Huang, J.; Ding, M.; Wei, X.; Liu, X.; Xu, X. $NiCo_2O_4$-Based Supercapacitor Nanomaterials. *Nanomaterials* **2017**, *7*, 41. [CrossRef] [PubMed]
21. Zhou, Y.; Li, C.; Li, X.; Huo, P.; Wang, H. $NiCo_2O_4$-based nanostructures for high-performance supercapacitors. *Dalton Trans.* **2021**, *50*, 1097–1105. [CrossRef]
22. Liu, Y.; Ma, Z.; Xin, N.; Ying, Y.; Shi, W. High-performance supercapacitor based on highly active P-doped one-dimension/two-dimension hierarchical $NiCo_2O_4/NiMoO_4$ for efficient energy storage. *J. Colloid Interface Sci.* **2021**, *601*, 793–802. [CrossRef] [PubMed]
23. Secor, E.B. Light scattering measurements to support real-time monitoring and closed-loop control of aerosol jet printing. *Addit. Manuf.* **2021**, *44*, 102028. [CrossRef]
24. Tian, Z.; Wang, D.; Zhang, C.; Meng, F.; Cao, L.; Lin, H. All-solid-state printable supercapacitors based on bimetallic sulfide $NiCo_2S_4$ with in-plane interdigital electrode architecture. *J. Mater. Sci.* **2022**, *57*, 19381–19395. [CrossRef]
25. Zhang, Y.-Z.; Wang, Y.; Cheng, T.; Yao, L.; Li, X.; Wang, H.; Pan, H.; Kim, J.; Wang, L. Printed supercapacitors: Materials, printing and applications. *Chem. Soc. Rev.* **2020**, *49*, 3229–3264. [CrossRef]
26. Zub, K.; Winsberg, J.; Schubert, U.S.; Hoeppene, S. Inkjet-printing of supercapacitors. *ChemistrySelect* **2020**, *5*, 12345–12352. [CrossRef]
27. Yedluri, A.K.; Kim, H.J. Enhanced electrochemical performance of nanoplate $NiCo_2O_4$ for supercapacitor applications. *RSC Adv.* **2020**, *10*, 1034–1042. [CrossRef]
28. Zhang, H.; Xiao, D.; Li, Q.; Ma, Y.; Yuan, S.; Xie, L.; Chen, C.; Lu, C. Porous $NiCo_2O_4$ nanowires supported on carbon cloth for flexible asymmetric supercapacitor with high energy density. *J. Energy Chem.* **2018**, *27*, 195–202. [CrossRef]
29. Peng, X.; Peng, L.; Wu, C.; Xie, Y. Two dimensional nanomaterials for flexible supercapacitors. *Chem. Soc. Rev.* **2020**, *49*, 3303–3323. [CrossRef] [PubMed]
30. Liu, T.; Wei, Y.; Chen, X.; Yu, X.; Chen, L. Hierarchical $NiCo_2O_4$ nanostructures for supercapacitors. *ACS Appl. Mater. Interfaces* **2020**, *12*, 10849–10857. [CrossRef]
31. Wang, X.; Liu, B.; Yu, G.; Hou, X.; Chen, D.; Shen, G. $NiCo_2O_4$ nanowires for flexible supercapacitors. *Nano Lett.* **2020**, *20*, 1931–1938. [CrossRef]
32. Chen, L.; Wang, J.; Yang, Z.; Zhang, J.; Hou, S.; Hao, C.; Zhang, J. Recent advances in flexible supercapacitors. *J. Solid State Electrochem.* **2022**, *26*, 2627–2658. [CrossRef]
33. Li, D.; Raza, F.; Wu, Q.; Zhu, X.; Ju, A. NiCo2O4 Nanosheets on Hollow Carbon Nanofibers for Flexible Solid-State Supercapacitors. ACS Appl. *Nano Mater.* **2022**, *5*, 14630–14638. [CrossRef]
34. Priyadarsini, S.S.; Saxena, S.; Ladole, A.H.; Nehru, D.; Dasgupta, S. Inkjet-Printed Transparent Asymmetric Mi-crosupercapacitors with Mesoporous $NiCo_2O_4$ and Mn_2O_3 Electrodes. *ACS Appl. Energy Mater.* **2024**, *7*, 715–725. [CrossRef]

35. Zhanga, J.-N.; Liua, P.; Jina, C.; Jina, L.-N.; Biana, S.-W.; Zhua, Q.; Wang, B. Flexible three-dimensional carbon cloth/carbon fibers/$NiCo_2O_4$ composite electrode materials for high-performance all-solid-state electrochemical ca-pacitors. *Electrochim. Electrochim. Acta* **2017**, *256*, 90–99. [CrossRef]
36. Zhao, R.-D.; Cui, D.; Sheng, H.-P.; Gammer, C.; Wu, F.-F.; Xiang, J. Core-shell structured $NiCo_2O_4$@$Ni(OH)_2$ nano-materials with high specific capacitance for hybrid capacitors. *Ionics* **2021**, *27*, 1369–1376. [CrossRef]
37. Sun, J.; Wang, w.; Yu, D. $NiCo_2O_4$ Nanosheet-Decorated Carbon Nanofiber Electrodes with High Electrochemical Performance for Flexible Supercapacitors. *J. Electron. Mater.* **2019**, *48*, 3833–3843. [CrossRef]
38. Moghaddam, A.S.; Rahmanian, E.; Naseri, N. nkjet-Printing Technology for Supercapacitor Application: Current State and Perspectives. *ACS Appl. Mater. Interfaces* **2020**, *12*, 34487–34504. [CrossRef]
39. Chen, J.; Ma, T.; Chen, M.; Peng, Z.; Feng, Z.; Pan, C.; Zou, H.; Yang, W.; Chen, S. Porous $NiCo_2O_4$@Ppy core-shell nanowire arrays covered on carbon cloth for flexible all-solid-state hybrid supercapacitors. *J. Energy Storage* **2020**, *32*, 101895. [CrossRef]
40. Li, Y.; Wang, S.; Ni, G.; Li, Q. Facile Synthesis of $NiCo_2O_4$ Nanowire Arrays/Few-Layered Ti_3C_2-MXene Compo-site as Binder-Free Electrode for High-Performance Supercapacitors. *Molecules* **2022**, *27*, 6452. [CrossRef]
41. Zhang, J.; Jiang, D.; Liao, L.; Cui, L.; Zheng, R.; Liu, J. Ti_3C_2Tx MXene based hybrid electrodes for wearable supercapacitors with varied deformation capabilities. *Chem. Eng. J.* **2022**, *429*, 132232. [CrossRef]
42. Liu, Q.; Zhang, X.; Yang, B.; Liu, J.; Li, R.; Zhang, H.; Liu, L.; Wang, J. Construction of Three-Dimensional Homogeneous $NiCo_2O_4$ Core/Shell Nanostructure as High-Performance Electrodes for Supercapacitors. *J. Electrochem. Soc.* **2015**, *162*, E319–E324. [CrossRef]
43. Wang, Q.; Wanga, X.; Xua, J.; Ouyanga, X.; Houa, X.; Chena, D.; Wang, R.; Shen, G. Flexible coaxial-type fiber supercapacitor based on $NiCo_2O_4$ nanosheets electrodes. *Nano Energy* **2014**, *8*, 44–51. [CrossRef]
44. Zhang, D.; Yan, H.; Lu, Y.; Qiu, K.; Wang, C.; Zhang, Y.; Liu, X.; Luo, J.; Luo, Y. NiCo2O4 nanostructure materials: Morphol-ogy control and electrochemical energy storage. *Dalton Trans.* **2014**, *43*, 15887–15897. [CrossRef] [PubMed]
45. Wang, J.; Jiang, A.; Li, Y.; Song, D.; Li, Y.; Cheng, L. Thermal decomposition behavior of polyimide containing flame retardant SiO_2 and $Mg(OH)_2$. *Polymers* **2022**, *14*, 2791. [CrossRef] [PubMed]
46. Thorat, J.P.; Nikam, R.P.; Lokhande, V.C.; Lokhande, C.D. Porous $NiCo_2O_4$ electrodes for high-energy asymmetric supercapacitor: Effect of annealing. *J. Mater. Sci.* **2023**, *58*, 9586–9604. [CrossRef]
47. Lee, S.-H.; Cha, H.-J.; Park, J.; Son, C.-S.; Son, Y.-G.; Hwang, D. Effect of Annealing Temperature on the Structural and Electrochemical Properties of Hydrothermally Synthesized $NiCo_2O_4$ Electrodes. *Nanomaterials* **2024**, *14*, 79. [CrossRef] [PubMed]
48. Tsay, K.-C.; Zhang, L.; Zhang, J. Effects of electrode layer composition/thickness and electrolyte concentration on both specific capacitance and energy density of supercapacitor. *Electrochim. Acta* **2012**, *60*, 428–436. [CrossRef]
49. Bard, A.J.; Faulkner, L.R. *Electrochemical Methods: Fundamentals and Applications*, 2nd ed.; Wiley: New York, NY, USA, 2001.
50. Xu, Y.; Lin, Z.; Zhong, X.; Huang, X.; Weiss, N.O.; Huang, Y.; Duan, X. Flexible solid-state supercapacitors based on three-dimensional graphene hydrogel films. *ACS Nano* **2013**, *7*, 4042–4049. [CrossRef]
51. Chen, T.; Dai, L. Flexible supercapacitors based on carbon nanomaterials. *J. Mater. Chem. A* **2014**, *2*, 10756–10775. [CrossRef]
52. Wang, D.-W.; Li, F.; Liu, M.; Lu, G.Q.; Cheng, H.-M. 3D Aperiodic Hierarchical Porous Graphitic Carbon Material for High-Rate Electrochemical Capacitive Energy Storage. *Angew. Chem. Int. Ed.* **2009**, *48*, 1515–1526. [CrossRef]
53. Chen, X.; Li, H.; Xu, J.; Jaber, F.; Musharavati, F.; Zalnezhad, E.; Bae, S.; Hui, K.S.; Hui, K.N.; Liu, J. Synthesis and Characterization of a $NiCo_2O_4$@$NiCo_2O_4$ Hierarchical Mesoporous Nanoflake Electrode for Supercapacitor Applications. *Nanomaterials* **2020**, *10*, 1292. [CrossRef] [PubMed]
54. Calvo, R.; Fuierer, P. Mechanical integrity of ceramic coatings on Kapton made by a dry aerosol deposition of lunar mare stimulant. *Int. J. Appl. Ceram. Technol.* **2023**, *20*, 395–409. [CrossRef]
55. Farhan, G.K.; Taha, H. A facile method of deriving solar selective nickel-cobalt oxide thin films via spraying process. *Semicond. Sci. Technol.* **2024**, *39*, 115017. [CrossRef]

batteries MDPI

Review

Carbon-Based Anode Materials for Metal-Ion Batteries: Current Status, Challenges, and Future Directions

Salim Hussain, Adeniyi Oyebade, Md Riyad Hossain, Fatima Abbas and Noureen Siraj *

School of Physical Sciences, University of Arkansas at Little Rock, Little Rock, AR 72204, USA; aoyebade@ualr.edu (A.O.); mhossain3@ualr.edu (M.R.H.)
* Correspondence: nxsiraj@ualr.edu

Abstract

The demand for effective, economical, and sustainable anode materials for metal-ion batteries (MIBs) has increased significantly due to the rapid growth of energy storage technologies. Among various candidates, carbon-based materials have emerged as highly promising due to their abundance, structural versatility, and favorable electrochemical properties. This review highlights the current status and future directions of carbon-based anode materials in MIBs, with a particular focus on graphite, hard carbon, carbon nanotubes, heteroatom-doped carbons, carbon-based composites, and other related structures. Various synthesis strategies for these materials are presented, along with discussions on their physicochemical characteristics, including structural features that influence electrochemical performance. Furthermore, we provided an overview on the performance of newly developed carbon-based anode materials in lithium-, sodium-, potassium-, and other emerging metal-ion battery systems to assess the impact of different synthesis approaches. Special attention is given to surface engineering, heteroatom doping, and composite design that can address intrinsic challenges such as limited ion diffusion, low reversible capacity, and poor cycling stability in MIBs. This review does not cover any carbon materials which have been used as an additive. In addition, the review explores emerging opportunities enabled by advanced characterization techniques, computational modeling, and artificial intelligence for optimizing the design of next-generation carbon anode. Finally, this article provides future perspectives and insights into the design principles of novel carbon-based anode materials that can accelerate the development of high-performance, durable, and sustainable MIB technologies.

Keywords: carbon-based anode; metal-ion batteries (MIBs); electrochemical performance

1. Introduction

Today, energy production and energy storage strategies are widely discussed worldwide as the global population is projected to reach 9.8 billion by 2050 [1]. This unprecedented growth is expected to significantly increase energy consumption, placing immense pressure on global resources. To meet this rising demand, non-renewable fossil fuels, including coal, natural gas, and oil, continue to be exploited at unsustainable rates. However, these energy reserves are finite and are expected to be exhausted in the near future [2]. In addition to resource depletion, fossil fuels are a major source of environmental pollution, responsible for nearly 65% of global CO_2 emissions since 2010 [3]. The transport sector alone has experienced a 2% annual increase in emissions, while urban contributions rose from 62% to 70% of the global share between 2010 and 2020 [4].

These alarming figures highlight the urgent need for a transformation of the current energy model in order to mitigate climate change, ensure environmental sustainability, and support future societal development. To achieve this, the promotion of clean fuels and renewable energy technologies must be accompanied by the development of advanced energy storage systems capable of stabilizing power supplies and meeting diverse energy demands [5].

2. Batteries

Among renewable energy technologies, electrochemical energy storage systems such as batteries, fuel cells, and supercapacitors are among the most promising technologies because of their high efficiency, scalability, and environmental compatibility [6]. Unlike conventional energy storage solutions such as pumped hydro or compressed air storage, electrochemical devices can be easily miniaturized or scaled up, making them adaptable for both portable electronics and large-scale grid applications. Among these technologies, metal-ion batteries (MIBs), including lithium-ion (LIBs), sodium-ion (SIBs), potassium-ion (KIBs), etc., have attracted particular attention because of their high energy density, relatively low cost, and increasingly mature industrial infrastructure [7]. LIBs already dominate the commercial market, powering consumer electronics and electric vehicles, while alternative batteries such as SIBs and KIBs are being investigated for large-scale storage due to the abundance and low cost of sodium and potassium resources.

A typical MIB consists of three main components: anode, cathode, and electrolyte. During charge and discharge processes, metal ions shuttle reversibly between the electrodes through the electrolyte, while electrons are transported through the external circuit, thereby delivering usable electrical energy. MIBs exhibit several advantageous features, including fast charging capability, negligible self-discharge, wireless charging compatibility, high round-trip efficiency, and long operational life. These attributes enable their use across a wide range of applications, from portable consumer electronics (mobile phones, laptops, cameras) and renewable energy integration (solar and wind storage) to electric vehicles as well as advanced technologies such as robotics, aerospace satellites, submarines, and military systems [8–10]. Nevertheless, the performance of MIBs depends strongly on the nature of their electrode materials, especially the anode, which directly affects the energy density, power delivery, cycling stability, and safety of the system.

3. Anode in MIBS

Over the years, various classes of anode materials have been explored, including transition-metal oxides, alloy-type anode, silicon-based materials, and carbonaceous anode. Each material class offers distinct advantages but also faces inherent challenges. For instance, silicon (Si) has garnered considerable attention due to its exceptionally high theoretical specific capacity (~4200 mAh/g for Li^+ storage), nearly 11 times higher than that of graphite. However, its practical deployment is severely limited by drastic volume expansion (>300%) during lithiation/delithiation, which leads to mechanical pulverization, unstable solid–electrolyte interphase (SEI) formation, and rapid capacity fading [11]. Similarly, alloy-type anode such as tin (Sn), antimony (Sb), and bismuth (Bi) undergo large volumetric changes during ion insertion/extraction, causing poor cycling stability and limited lifespan [12]. Transition-metal oxides offer relatively high capacity and stability but suffer from intrinsically low conductivity, sluggish ion diffusion kinetics, and poor coulombic efficiency [13].

4. Carbon-Based Anode Materials

Carbon-based anode have emerged as highly attractive alternatives owing to their unique combination of favorable properties. Carbon is one of the most abundant and cost-effective elements on earth, making it economically sustainable for large-scale production. Carbon materials can be fabricated into various dimensionalities, including 0D nanodots, 1D nanotubes and nanofibers, 2D graphene nanosheets, and 3D porous frameworks, since each form offers distinct advantages for ion storage and transport [14]. For instance, low-dimensional carbons such as graphene and carbon nanotubes (CNT) provide high surface areas and fast electronic conductivity, while 3D porous carbons can facilitate rapid ion diffusion and accommodate structural strain during cycling [15,16]. Typically, carbon-based anode materials deliver specific capacities in the range of 200–600 mAh/g, depending on their morphology, crystallinity, and chemical modification strategies [17]. More importantly, they exhibit high structural reversibility, long cycle life, and relatively stable SEI formation, ensuring reliable performance across extended charge–discharge cycles. Their outstanding electrical conductivity supports rapid electron transport, while their mechanical resilience helps withstand repeated ion intercalation/extraction processes. Moreover, carbon surfaces can be engineered via heteroatom doping (e.g., N, B, P, S) to enhance electronic conductivity, create additional active sites, and improve electrolyte wettability. Hybridization with metals, oxides, or polymers further expands their potential by combining the strengths of multiple components [18,19].

Another key advantage of carbon-based anode lies in their broad compatibility with different types of MIBs. While graphite remains the commercial anode for LIBs, expanded or modified carbons have shown promise for hosting larger ions such as Na^+ and K^+, which face difficulties intercalating into the small interlayer spacing of graphite. A schematic illustrating the different types of carbon-based anode employed across various MIBs is presented in Figure 1. Thus, carbon materials are not only vital to current LIB technology, but central to the development of next-generation sodium and potassium-based batteries [20–22].

Considering the effective tactics used by carbonaceous anode in batteries, several researchers have tried to identify carbons with elevated storage capacities by reducing particle size. The small particle sizes or porous structures invariably result in an increased specific surface area of the carbons, which leads to low Initial Coulombic Efficiency (ICE), thereby significantly decreasing the energy density of the battery and causing suboptimal cycling performance in full cell configurations [23]. Consequently, converting materials to nanoscale dimensions is not a comprehensive answer to attain high-performance carbonaceous anode in batteries. Conversely, the metal ion storage process using carbonaceous hosts, referred to as the "house of cards" paradigm by Dahn and colleagues [24], resembles that of Li^+ ion storage [25]. To enable the insertion and extraction of massive metal ions in comparison to Li^+ ions, a substantial interlayer gap is required. Qie et al. [26] concluded from experimental data and theoretical calculations that carbon materials with interlayer distances above 0.37 nm might function as anode for Na^+ ion insertion. An extended graphite with an increased interlayer lattice distance of 4.3 Å was synthesized by Wen et al. [27] using a two-step oxidation-reduction method, demonstrating exceptional cycle stability, thereby qualifying it as an effective anode material for SIBs. Consequently, the interlayer lattice spacing for other MIBs must be considerably larger as discussed earlier.

Taken together, carbon-based anode offer an unparalleled balance of cost-effectiveness, conductivity, structural adaptability, and electrochemical stability, making them ideal candidates for high-performance MIBs. Their versatility, coupled with the vast library of modification strategies, positions them at the forefront of research in sustainable energy storage. Therefore, this review aims to provide a comprehensive overview of the design, synthesis, and application of carbon-based anode across different MIB chemistries, emphasizing recent progress, structure–performance relationships, and future challenges for large-scale commercialization.

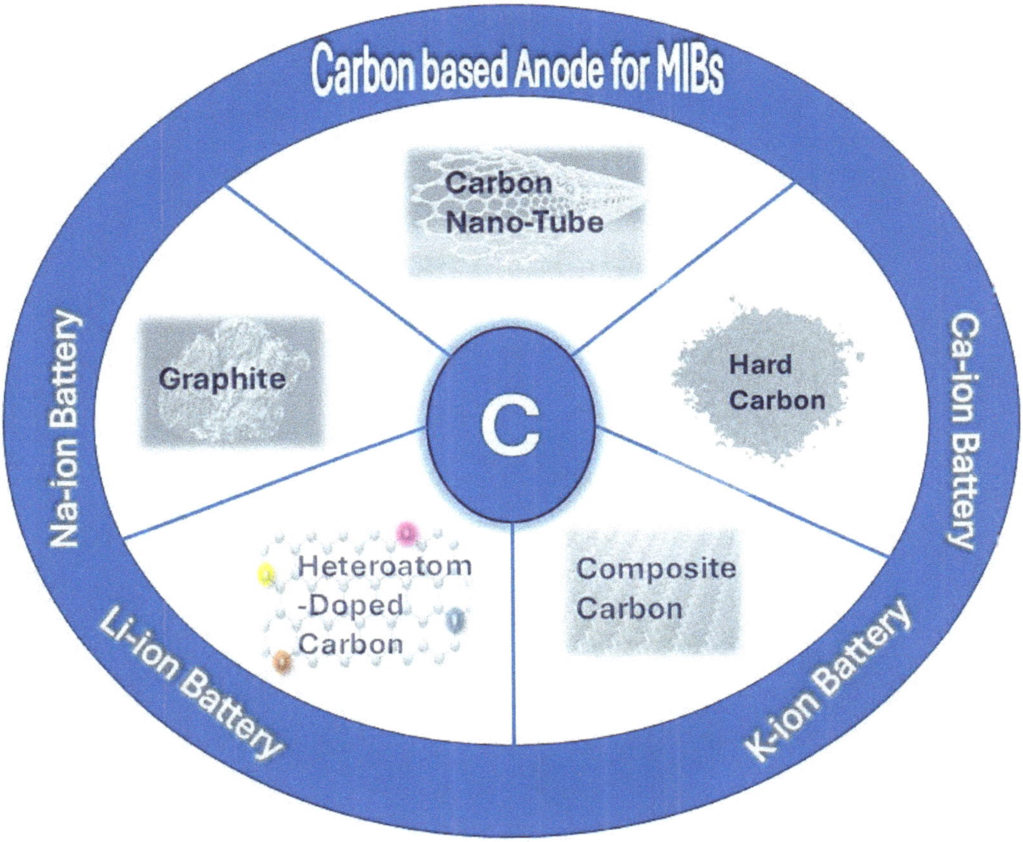

Figure 1. Illustration of carbon-based anode materials for MIB application.

5. Graphite-Based Materials as Anode for MIBS

Graphite became the first commercially viable anode material for LIBs, thanks to the pioneering work of Akira Yoshino and Sony in 1991 [28]. Although graphite's ability to intercalate lithium was known since 1970s, its practical use was limited by instability issues with organic electrolytes. This problem was solved in the late 1980s with the development of a stable carbon material and a compatible electrolyte system, establishing graphite as the standard anode material. A breakthrough occurred in 1993 with the introduction of a mixed electrolyte of ethyl carbonate and dimethyl carbonate for graphite anode, which enabled the stable and reliable use of graphite in LIBs. Since then, graphite has remained the primary choice for commercial anode, with ongoing improvements after 2000 focused on increasing capacity, enabling fast charging, and enhancing overall safety [29,30].

The success of graphite is attributed to its unique structural and electrochemical properties. Its well-ordered layered structure of sp^2-hybridized carbon atoms enables efficient intercalation and deintercalation of Li^+ ions between adjacent graphene layers [31,32]. This interlayer-insertion mechanism is highly favorable compared to pore-filling storage, which is effective only at low potentials, and conversion-type storage, which is often accompanied by large volume expansion during cycling. As a result, graphite delivers stable and reversible lithium storage and ensures long cycle life. Moreover, its natural abundance, low operating potential, high electrical conductivity, and environmental friendliness further enhance its suitability as an electrode material for electrochemical energy storage devices [33,34].

The relatively small interlayer spacing of graphite (~3.35 Å) limits its ability to host larger ions like Na^+ and K^+, leading to slow charge/discharge kinetics and low storage capacity [35]. To overcome this issue, extensive research has focused on structural engineering strategies to develop graphite-based advanced materials, including expanded graphite (EG), graphite intercalation compounds (GICs), and porous graphite, all aiming to improve energy storage performance in MIBs [35–37].

5.1. *Synthesis of Graphitic Materials*

Graphite-based material can be prepared using several synthesis methods, including (i) microwave-assisted, (ii) ultrasonication, (iii) furnace-based, (iv) Hummer's method, and (v) wet chemical oxidation, as illustrated in the schematic below (Figure 2).

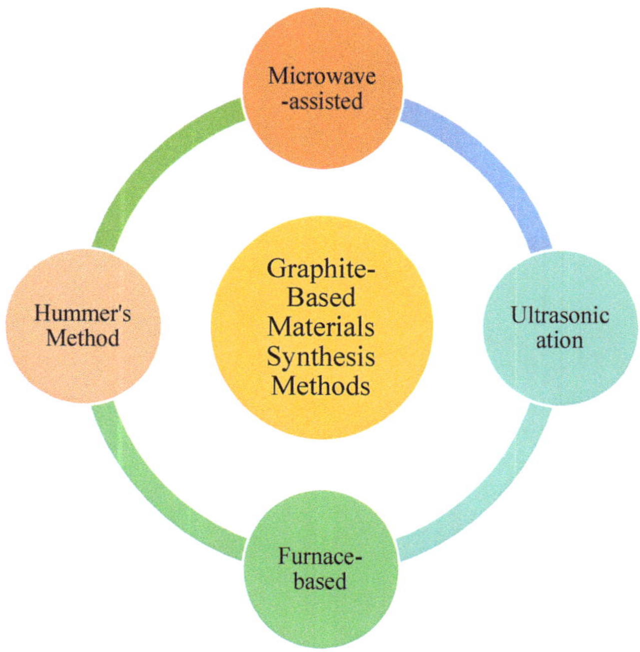

Figure 2. Schematic representation of the primary synthesis approaches used for graphite-based materials.

5.1.1. Microwave-Assisted Method

Microwave-assisted methods for the synthesis of graphite-based materials such as expanded graphite (EG), graphite intercalation compound (GIC) using microwave irradiation with frequencies ranging from 0.3 to 300 GHz to rapidly heat the reaction mixture, enabling uniform energy distribution and fast reaction rates [38]. This technique promotes efficient exfoliation or reduction of graphite oxide into graphene or other graphite-based materials. Compared to conventional heating, it offers shorter synthesis time, lower energy consumption, and better control over material morphology and structure [39].

Thanks to the layered structure property of graphite, various guest species can be inserted between its layers through a process called intercalation forming GICs [40]. The physiochemical properties of GICs depend on both the quality of the graphite and the type of the intercalant used (e.g., alkali metals, metal chlorides, oxides). GIC containing easily vaporized intercalants show excellent expandability and are often used as precursors for EG, while metal chloride-based GICs offer high electrical conductivity, making them useful for energy storage devices [41].

The microwave-assisted method offers several advantages for synthesizing graphite-based materials. It enables rapid and energy-efficient reactions due to uniform volumetric heating, which significantly shortens synthesis time compared to conventional thermal or chemical methods. The process also promotes enhanced intercalation and exfoliation, as microwave-induced plasma or localized heating accelerates ion diffusion between graphite layers. Moreover, it typically requires fewer oxidizing agents and acids, making it a more environmentally friendly and cost-effective approach. The resulting materials often exhibit high purity, uniform morphology, and controlled structural features, which are desirable for energy storage and catalytic applications [42,43].

However, the method also presents certain limitations. Scaling up remains a challenge because uniform microwave penetration is difficult to achieve in larger batches, leading to non-uniform heating or hot spots that can cause localized overheating and incomplete reactions. Additionally, specialized microwave reactors are relatively expensive and may involve complex operation and maintenance. Controlling reaction parameters such as temperature, plasma formation, and exposure time can also be difficult, which affects reproducibility. Furthermore, excessive microwave irradiation may introduce structural defects or partial decomposition of graphite layers. Despite these drawbacks, the microwave-assisted method remains a promising and efficient route for the rapid synthesis and modification of graphite-based materials [44].

Overall, the microwave-assisted method represents a highly promising approach for the efficient and sustainable synthesis of graphite-based materials, particularly GICs and EG. This technique combines the advantages of rapid reaction rates, reduced chemical consumption, and improved energy efficiency while maintaining excellent structural and chemical control. The microwave-induced plasma effect significantly enhances the intercalation and exfoliation kinetics, leading to the formation of high-quality GICs and EG with superior electrochemical and physical properties. Moreover, when utilizing spent graphite, this method not only minimizes environmental impact but also supports the recycling and valorization of graphite waste, making it both economically and ecologically beneficial. Despite challenges in scaling up and ensuring uniform heating, continuous advancements in microwave reactor design and process optimization are expected to overcome these limitations. Overall, the microwave-assisted synthesis route provides an efficient, eco-friendly, and scalable strategy for developing advanced graphite-based materials for energy storage, catalysis, and other functional applications.

5.1.2. Ultrasonication Method

The ultrasonication (mechanical agitation) method for synthesizing graphite-based materials involves dispersing graphite in a suitable liquid medium and applying high-frequency ultrasonic waves to induce acoustic cavitation [45]. The collapse of cavitation bubbles generates intense micro-jets and shock waves, which provide sufficient mechanical energy to exfoliate graphite layers into graphene or graphene oxide (GO)sheets. The choice of liquid medium such as an organic solvent, surfactant solution, or aqueous system plays a crucial role in stabilizing the exfoliated layers and controlling the structural and chemical characteristics of the final product [46]. Viculis and co-workers [47] developed graphite nanoplatelets (GNPs) using mechanical agitation applied to EG, which effectively broke down the material into thinner platelets. GNPs are more cost-effective than CNTs but remain more expensive than carbon black. Xu et al. [48] proposed a novel ultrasonic-assisted wheat straw pulping method was demonstrated at room temperature and atmospheric pressure. Interestingly, graphite material was unexpectedly detected in the ultrasonic-assisted pulp (UP). The ash from the UP contained both inorganic and organic components, with a total inorganic content of 81.9%, primarily composed of SiO_2 (71.9%). The graphitization mechanism is attributed to the ultrasonic cavitation effect, where collapsing cavitation bubbles create localized extreme conditions temperatures around 4000 K and pressures near 100 MPa. The process primarily involves three stages: lignin degradation, graphene formation, and subsequent graphitization.

The ultrasonication-assisted method offers several advantages for synthesizing graphite-based materials. It is a simple, efficient, and environmentally friendly technique that operates under ambient temperature and pressure, eliminating the need for harsh chemicals or high-temperature treatments. The localized high-energy cavitation provides strong mechanical forces capable of exfoliating graphite into thin layers, resulting in high-quality graphene or GO sheets with minimal structural damage. Moreover, the method allows for scalable and controllable exfoliation, as parameters such as ultrasonic power, duration, and solvent type can be tuned to tailor the morphology and thickness of the products. Additionally, this technique is compatible with a wide range of liquid media, making it adaptable for various precursor materials including natural graphite, EG, and even biomass-derived carbon sources [49,50].

Despite its advantages, the ultrasonication-assisted method also has some limitations. The low exfoliation yield and long processing times can restrict its large-scale industrial application. Continuous exposure to strong ultrasonic energy may lead to structural defects or partial oxidation of the exfoliated graphene sheets, which can affect their electrical and mechanical properties. Furthermore, energy consumption associated with prolonged ultrasonication is relatively high. Another limitation is the difficulty in achieving uniform dispersion, as re-aggregation of exfoliated layers may occur after the cessation of sonication, especially in the absence of stabilizing agents or surfactants [51].

In summary, the ultrasonication-assisted method is a promising and green approach for producing graphite-based materials such as GNPs, GICs, and EG. It combines simplicity and versatility, enabling the conversion of graphite or waste carbon sources into high-value materials with desirable structural and functional properties. Although improvements are still needed in terms of yield, scalability, and energy efficiency, ongoing advancements in ultrasonic system design and process optimization continue to enhance its potential for sustainable and large-scale production of graphite-based materials for applications in energy storage, catalysis, and composite development.

5.1.3. Furnace-Based Method

Furnace-based methods for graphite synthesis rely on heating carbon-based precursors at elevated temperatures using induction or resistance heating under controlled atmospheric conditions [52]. These techniques enable the transformation of disordered carbon materials into highly crystalline graphite or carbon nanostructures. Among them, induction graphitization is a widely used process in which carbon parts are heated to temperatures between 1000 °C and 2800 °C through induction heating. This method enhances the crystallinity and purity of the resulting graphite, making it suitable for applications such as brake components, clutch facings, and mechanical seals [53]. In contrast, high-frequency furnaces employ inductive heating of a graphite body, or susceptor, to synthesize advanced carbon nanomaterials, including fullerenes, CNTs, and single-walled carbon nanotubes (SWCNTs) [54–56]. These furnaces typically consist of a graphite susceptor, a carbon shield, and a high-frequency power supply, enabling localized high-temperature zones for nanostructure formation. Another notable technique is resistance heating (Joule heating), in which a graphite crucible is directly heated by passing a large electric current through it. This method allows for extremely rapid heating and cooling cycles, achieving temperatures up to 3200 °C, and is particularly effective for producing graphite films and other materials requiring precise thermal control [57]. Collectively, furnace-based synthesis methods offer high-temperature capability, process flexibility, and tunable structural outcomes, making them integral to the production of both bulk and nanostructured graphite materials.

Liu et al. [58] investigated the effect of heat treatment temperature on the degree of graphitization using biomass-derived activated carbon (AC) as a carbon precursor. The AC was first acid-washed with HCl, dried, and subsequently impregnated overnight with a 1 M acetone solution of $Ni(NO_3)_2$ to introduce a catalytic effect. After vacuum filtration, the treated samples were heat-treated in a tube furnace under nitrogen flow at temperatures up to 1000 °C. XRD analysis revealed that no distinct graphite peak appeared at lower temperatures, while a noticeable (002) diffraction peak near $2\theta = 26^\circ$ emerged at 950 °C. At 1000 °C, this peak became significantly sharper, indicating a higher degree of graphitization. The results demonstrated that the combination of chemical pretreatment and elevated thermal processing effectively converted the activated carbon into a highly porous, ultrathin graphitic carbon material.

Furnace-based synthesis methods offer several advantages that make them highly effective for producing graphite and related carbon materials. One of the key benefits is their ability to achieve extremely high operating temperatures, often reaching up to 3000 °C, which is essential for promoting graphitization and attaining the desired crystalline structure. These furnaces also provide precise temperature control, enabling researchers and manufacturers to tailor material properties and ensure consistent, reproducible results. Furthermore, induction-heating graphitization furnaces are specifically designed to produce high-purity graphite, minimizing contamination during synthesis. The use of graphite-based hot zones creates near-ideal blackbody conditions, resulting in highly uniform heating throughout the chamber. In addition, furnace systems are highly versatile, supporting a wide range of processes such as sintering, graphitization, and pyrolysis [59].

Despite these advantages, furnace-based methods also present several limitations. They are often energy-intensive, particularly in large-scale setups such as Acheson furnaces, leading to high operational costs. Long cycle times are another drawback, as the cooling stages for some furnaces can extend over several days, reducing throughput. Moreover, graphite is susceptible to oxidation at elevated temperatures in the presence of oxygen, necessitating the use of inert or vacuum atmospheres to preserve material integrity. In certain high-temperature environments, especially those involving metal alloys such as nickel and chromium, there is a risk of carbon pickup, where carbon diffuses from the graphite into the metallic components. Additionally, maintenance challenges arise due to the brittleness and limited durability of graphite parts, which may require frequent replacement or careful handling to maintain furnace performance [60].

5.1.4. Hummer's Method

Hummer's method is a well-established chemical process for synthesizing GO/graphite oxide from graphite, which can subsequently be reduced to form reduced graphene oxide (rGO) [61]. This method involves the oxidation of graphite using strong acids typically concentrated sulfuric acid and oxidizing agents such as potassium permanganate and sodium nitrate [62]. In the oxidation step, graphite is dispersed in concentrated sulfuric acid with sodium nitrate, followed by the slow addition of potassium permanganate under controlled low-temperature conditions to prevent overheating. The mixture forms a thick paste that is later heated and diluted with water, initiating vigorous effervescence as the oxidation reaction completes. To remove residual oxidants, hydrogen peroxide (H_2O_2) is added, reducing any remaining permanganate and manganese dioxide, typically producing a brown-colored suspension. The purification process then involves repeated washing and filtration, often using hydrochloric acid and deionized water, to eliminate metal ions and excess acid. Subsequently, exfoliation is achieved through ultrasonication in a polar solvent such as water, yielding single or few-layer GO sheets. The final step, reduction, restores electrical conductivity by removing oxygen-containing functional groups through thermal, chemical, or electrochemical reduction, producing rGO. Hummer's method and its modified variants are widely employed due to their reliability, scalability, and cost-effectiveness, making them one of the most common "top-down" approaches for producing high-quality GO and rGO from bulk graphite [63].

For example, Bannov et al. [64]. investigated the sequential stages of GO formation during the modified Hummers' method by employing a stepwise sampling approach. The modification involved adding the reaction mixture into ice and using excess H_2O_2, which enhanced the hydrolysis of GICs and increased oxygen release. High-purity nipple graphite was used as the precursor to ensure accurate observation of the oxidation process. As reported by Inagaki et al. [65], oxidation of graphite in concentrated sulfuric acid with potassium permanganate and nitric acid first forms unstable GICs, which convert to graphite oxide upon contact with water. In this study, samples were collected after the hydrolysis stage following water washing, allowing precise examination of the transformation from GICs to graphite oxide under the modified synthesis conditions.

Hummers' method is a widely used and efficient technique for synthesizing graphite oxide, valued for its speed, safety, and high yield, making it suitable for large-scale production [66]. The process avoids hazardous reagents such as fuming nitric acid and chlorates, significantly reducing explosion risks, while achieving a high degree of oxidation that enhances the functional properties of the product. The required materials are inexpensive and readily available, further contributing to the method's practicality and scalability [67].

However, the method also presents notable limitations. It produces toxic gases such as NO_2 and N_2O_4, necessitating strict safety and environmental controls. The purification process to remove residual metal and nitrate ions is laborious, and the oxidation step can introduce structural defects that degrade the electrical and mechanical performance of GO. In some cases, incomplete oxidation leads to mixtures of graphite and graphite oxide. Despite these drawbacks, Hummers' method remains a cornerstone for graphite oxide synthesis due to its balance of efficiency and reliability [68].

5.2. Graphite-Based Anode for LIBs

Although graphite has already been widely employed in commercial LIBs, efforts are ongoing to enhance its lithium storage capability. One effective strategy involves preparing EG with enlarged, long-range-ordered interlayer spacing. For example, Bai et al. [69] synthesized EG with an interlayer distance of 0.359 nm by rapidly heating graphite oxide in a preheated muffle furnace at 1050 °C under air. The resulting EG exhibited a capacity of 413 mAh/g at a current density of 0.2 mA cm^{-2} and retained 99% of its capacity after 30 cycles, significantly higher than the 322 mAh/g obtained for natural graphite under the same conditions of 1 M $LiPF_6$/EC-DMC electrolyte in LIBs.

Wang et al. [70] explored $FeCl_3$-graphite intercalation compounds ($FeCl_3$-GICs) prepared by a melt-salt method at 600 °C for 3 h. As shown in Figure 3a, $FeCl_3$ was pre-intercalated within the graphite layers, forming a mixed-stage structure dominated by stage 7 along with stages 3 and 5 (Figure 3b). Upon discharge, Li^+ intercalated into the graphite interlayers to form Li_xC, while $FeCl_3$ reacted with Li^+ to generate LiCl and Fe according to Equation (1):

$$FeCl_3 + 3Li^+ + 3e^- \rightarrow Fe + 3LiCl \quad (1)$$

$FeCl_3$-GICs delivered a high reversible capacity of 506 mAh/g at 0.1 C, well above the theoretical capacity of pristine graphite (372 mAh/g) in LIBs. Remarkable rate performance was also demonstrated, with discharge capacities of 300 and 220 mAh/g at 5 C and 20 C, respectively (Figure 3c). The total capacity could be deconvoluted into three contributions: ~200 mAh/g from conventional Li^+ intercalation into graphite, ~91 mAh/g from reactions involving $FeCl_3$, and the remainder from Li^+ adsorption/desorption on graphene sheet surfaces flanking the $FeCl_3$ intercalation layer. Moreover, the material showed excellent cycling stability, retaining 480 mAh/g after 400 cycles at 100 mA/g (Figure 3d).

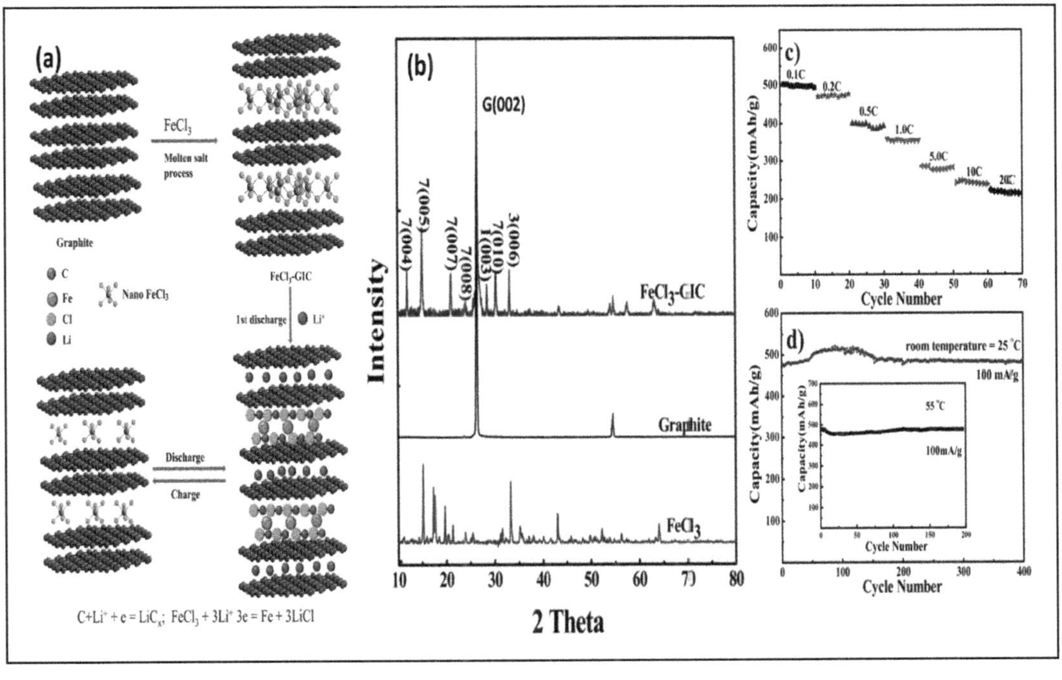

Figure 3. (**a**) Schematic representation of the structure and lithium storage mechanism of $FeCl_3$-GICs. (**b**) Powder XRD pattern of $FeCl_3$-GICs. (**c**) Discharge capacity as a function of cycle number and (**d**) long-term cycling stability of $FeCl_3$-GICs [70].

In addition to interlayer engineering, pore creation in graphite-based electrodes has been proposed as another effective strategy to enhance electrochemical performance. Porous structures can shorten Li^+ diffusion paths, provide abundant active sites, and mitigate volume changes during the cycling process [71]. First-principles calculations by Yang et al. [72] predicted that porous graphene with a hole density of 35% could achieve a theoretical Li^+ storage capacity of 1516 mAh/g about four times higher than that of pristine graphite. The Li^+ diffusion barrier through the extraplanar channels created by the pores was estimated to be only one-fifth that of conventional graphite, suggesting the potential for much faster charge/discharge kinetics.

Recently, porous graphite has attracted considerable attention as an electrode material for multivalent-ion batteries. Various synthetic strategies have been employed, including siliconization, MoO_x-catalyzed gasification, KOH etching and air treatment [73–75]. For instance, Shim et al. [76] demonstrated that KOH-etched porous graphite, treated for 24 h at 80 °C, delivered significantly improved rate capability as an anode for LIBs. This anode material exhibited ~160 mAh/g at 2.5 C with 96.7% capacity retention after 100 cycles in 1.15 M $LiPF_6$/EC-DMC-EMC, compared to ~75 mAh/g for natural graphite in LIBs. Although the BET surface area increased modestly (6.7 → 7.6 m^2/g), but SEM analysis revealed numerous nanoscale pores formed during etching, which facilitated Li^+ access into graphite layers and accounted for the enhanced kinetics.

5.3. Graphite-Based Anode for SIBs

Na-based batteries have served as a cornerstone of "beyond-lithium" energy storage technology since 1985, following the successful development of high-temperature Na/$NiCl_2$ and Na/S batteries utilizing a Na^+-β-alumina ceramic electrolyte [77,78]. Analogous to LIBs, SIBs comprise a cathode made of sodium intercalation materials and an anode, separated by an electrolyte that facilitates sodium-ion transport between the electrodes during charge–discharge cycles.

Anode materials play a crucial role in determining the overall performance of SIBs, with carbon-based materials being widely employed due to their excellent electrical conductivity, structural stability, and tunable electrochemical properties [79]. Among them, graphite stands out as an abundant and low-cost resource, making it a highly promising candidate for large-scale applications in SIB technology. In this regard, Wen and co-workers [27] synthesized EG that has emerged as a highly promising anode material for SIBs. Derived from graphite through a two-step oxidation–reduction process, EG retains the long-range layered structure of graphite while exhibiting an enlarged and tunable interlayer spacing (~0.43 nm), which facilitates efficient Na^+ intercalation and deintercalation. In situ high resolution transmission electron microscopy (HRTEM) has confirmed the reversible sodiation and desodiation behavior of EG, accompanied by stable structural evolution during cycling. Electrochemical measurements demonstrated a high reversible capacity of 284 mAh/g at 20 mA/g, 184 mAh/g at 100 mA/g, and excellent cycling stability with 73.9% capacity retention after 2000 cycles. Owing to its low-cost synthesis, structural tunability, and superior electrochemical performance, EG represents a strong candidate for next-generation, large-scale SIB anode applications.

5.4. Graphite-Based Anode for KIBs

SIBs have gained attention for their chemical similarity to LIBs. However, their relatively high standard reduction potential ($Na/Na^+ = -2.71$ V vs. SHE) limits energy density [80]. To overcome this, KIBs have been proposed, offering a lower reduction potential ($K/K^+ = -2.93$ V vs. SHE), closer to that of Li/Li^+ (−3.04 V vs. SHE) [81]. Additionally, the smaller Stokes radius of K^+ (3.6 Å) compared to Na^+ (4.6 Å) and Li^+ (4.8 Å) in propylene carbonate leads to higher ion mobility and conductivity, making KIBs a promising alternative for high-energy-density storage systems. However, the larger ionic radius of K^+ (1.38 Å) compared to Li^+ (0.76 Å) and Na^+ (0.97 Å) often causes significant structural strain and damage to anode materials during the potassiation–depotassiation process, leading to rapid capacity fading over repeated cycles [82]. Consequently, developing robust, high-performance anode materials is crucial for advancing KIB technology.

A wide range of anode materials, including metals, oxides, sulfides, and phosphides, have been explored for KIBs, demonstrating promising potassium storage capabilities [83–86]. However, challenges such as limited cycling stability and relatively high voltage plateaus continue to constrain the overall energy density of KIB full cells. Considering factors such as cost, stability, and operating voltage, carbon-based materials have emerged as preferred anode for KIBs [87]. Among them, graphite, a classical carbon material extensively used in LIBs, stands out due to its abundance, conductivity, and structural reversibility [88]. Interestingly, while graphite exhibits poor Na^+-ion storage capability in SIBs because of its low reactivity toward Na^+ intercalation [89], it shows a distinct K^+-ion intercalation behavior in KIBs, highlighting its potential as an efficient and stable anode material for next-generation potassium-based energy storage systems.

5.5. Graphite-Based Anode for Calcium-Ion Batteries (CIBs)

CIBs have emerged as a promising divalent-ion system due to their relatively high standard reduction potential (−2.87 V for Ca/Ca^{2+}) and faster charge-transfer kinetics associated with the high charge density of Ca^{2+} (0.49 e Å^{-3}) [90]. Despite these advantages, CIB development is hindered by the irreversible plating/stripping of calcium metal at room temperature. Although partial dissolution of Ca^{2+} from metallic calcium is possible [91], calcium deposition whether on Ca metal or noble-metal substrates is severely restricted by the rapid formation of a passivating surface layer [92]. Increasing the operating temperature to approximately 100 °C and reducing ion-pairing in the electrolyte can partially facilitate Ca^{2+} migration through this surface film [93]; however, substantial overpotentials (0.5–0.9 V) persist, indicating limited practicality for CIBs. Recent studies suggest that the in situ formation of CaH_2 in $Ca(BH_4)_2$/THF electrolytes can mitigate surface passivation, but its slow formation kinetics and the limited anodic stability of the electrolyte (≈3 V) remain significant barriers [94]. These challenges collectively underscore the difficulty of realizing efficient reversible calcium storage in graphite-based anode systems and highlight the need for alternative carbon structures or electrolyte innovations to enable practical CIB performance. Recent progress has demonstrated that graphite can serve as a viable anode material for CIBs through solvent-assisted co-intercalation mechanisms. Richard Prabakar et al. [95] reported the reversible co-intercalation of Ca^{2+} with tetraglyme (G_4) into graphite, a behavior distinctly different from the irreversible reactions typically observed in conventional carbonate- or ether-based electrolytes. In the presence of G_4, graphite exhibits highly stable cycling, maintaining excellent reversibility for up to 2000 charge–discharge cycles at a rate of 1 A/g without noticeable degradation. Using a combination of analytical characterization and computational modeling, the study elucidates the stepwise intercalation process of Ca^{2+}–G_4 complexes within graphite layers. Furthermore, the feasibility of graphite as an anode in a Ca^{2+}-shuttling full-cell configuration was demonstrated, underscoring the potential of solvent-cointercalation strategies to overcome the intrinsic challenges of calcium storage in layered carbon materials.

6. CNT-Based Materials as an Anode for MIBs

The increasing global demand for sustainable and high-performance energy storage systems has stimulated extensive research into advanced electrode materials for rechargeable MIBs. Among various candidates, CNTs have emerged as a promising class of nanostructured carbon materials. CNT was first reported by the Japanese scientist Sumio Iijima in 1991 [96,97]. As novel members of the carbonaceous material family, CNTs exhibit remarkable physicochemical properties, including a large aspect ratio, high electrical conductivity, extensive specific surface area, and low density. Reported electrical conductivity of up to 10^6 S m^{-1} for single-walled CNTs (SWCNTs) and 10^5 S m^{-1} for multi-walled CNTs (MWCNTs). Furthermore, CNTs possess exceptional mechanical strength, with tensile strength values as high as 60 GPa [98,99]. In the context of anode development, CNTs offer significant advantages, including efficient electron transport pathways, large specific surface area for ion adsorption, and the ability to buffer mechanical stress induced by ion insertion/extraction processes. These properties make CNTs highly attractive for a broad range of MIB chemistries, including lithium-ion, sodium-ion, and potassium-ion systems [100–103].

6.1. Synthesis of CNT-Based Materials

CNTs can be synthesized through a variety of techniques; the most commonly employed approaches are shown in Figure 4.

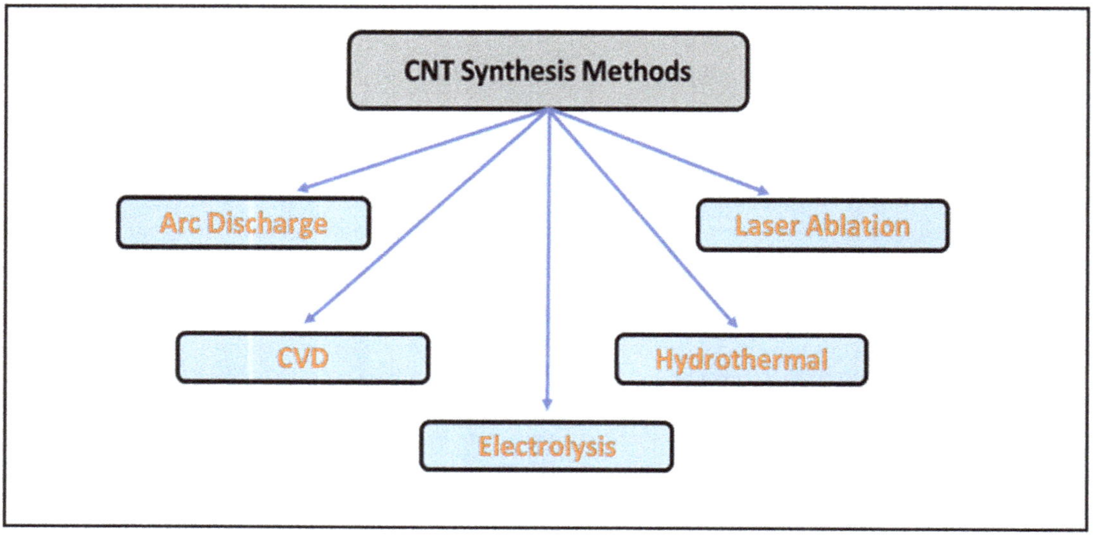

Figure 4. Schematic representation of the major synthesis techniques for carbon nanotubes (CNTs), including arc discharge, laser ablation, CVD, hydrothermal synthesis, and electrolysis.

6.1.1. Arc Discharge Method

The arc discharge method, first employed by Iijima in 1991 for the synthesis of highly crystalline MWCNTs, remains one of the earliest and most established techniques for CNT production [96]. This high-temperature process, also widely used for fullerene synthesis, involves striking an arc between graphite electrodes in an inert gas atmosphere (typically He or Ar), generating plasma temperatures exceeding 3000 °C, sufficient to vaporize carbon. Carbon subsequently deposits on the cathode, leading to nanotube formation. The parameters, such as arc stability, current density, inert gas pressure, and electrode cooling, strongly influence CNT yield, crystallinity, and morphology.

MWCNTs are generally synthesized without catalysts, while transition metals (e.g., Fe, Co, Ni, Mo, Y) are essential for SWCNT growth. Early studies demonstrated successful large-scale synthesis of MWCNTs in helium due to its high ionization potential, while variations such as pulsed arc discharge and liquid-phase arc discharge (e.g., in water, liquid N_2, or salt solutions) expanded synthesis possibilities. Notably, boron- and nitrogen-doped MWCNTs have been produced by arc discharge, and the method has also been adapted to fabricate double-walled CNTs (DWCNTs) [104].

For SWCNTs, catalyst-embedded graphite electrodes have been widely explored, with mixtures such as Ni-Y-graphite and Fe-Co-Ni-graphite proving particularly effective in producing uniform tubes with diameters around 1–2 nm. Process optimization, such as hydrogen-assisted DC arc discharge or pulsed arc discharge, has been shown to improve yield and purity. Key parameters affecting synthesis include chamber temperature, pressure, catalyst type and its composition, carbon precursor, and electrode design.

The primary advantage of arc discharge lies in its ability to produce CNTs with high crystallinity and structural quality. However, the method offers limited control over chirality, requires expensive high-purity graphite and catalysts, and often necessitates additional purification steps, especially for SWCNTs. Despite these limitations, arc discharge remains a benchmark for CNT synthesis due to its ability to produce high-quality nanotubes in substantial quantities [105].

6.1.2. Laser Ablation Method

The laser ablation technique emerged as an alternative to overcome the limitations of the arc discharge method, particularly in achieving uniform and high-purity SWCNTs. First introduced by Guo et al. [106], laser ablation demonstrated remarkable efficiency, producing nearly 500 mg of SWCNTs within five minutes with up to 90% purity [107]. Unlike arc discharge, laser ablation employs a high-energy pulsed light source, typically a Nd: YAG (neodymium-doped yttrium aluminum garnet) laser, to ablate a graphite target.

The process is relatively simple: a quartz tube reactor (25 mm diameter, 1–1.5 m length) placed in a tube furnace at ~1200 °C houses the graphite target, which can be either pure graphite (for MWCNTs) or graphite doped with transition metals such as Fe, Co, or Ni (for SWCNTs). Carrier gases such as Ar, He, or their mixtures are introduced at controlled pressures and flow rates. The Nd: YAG laser beam, operating at 1064 or 532 nm wavelength, ~300 mJ pulse energy, 10 Hz repetition rate, and pulse width <10 ns, is focused on the target surface with a beam diameter of 3–8 mm. The intense irradiation vaporizes the target, forming a plume of carbon species that are swept by the gas stream and condensed on a downstream water-cooled collector [106].

This method not only enables the synthesis of high-purity CNTs but also provides improved structural control compared to arc discharge. However, optimization of experimental parameters such as laser energy, target composition, and gas environment remains essential for maximizing yield and ensuring consistent CNT quality.

6.1.3. CVD Method

CVD is another promising alternative to arc discharge, suitable for controlled nanotube growth, offering scalability and tunability in nanotube structure [108]. The method is based on the catalytic decomposition of hydrocarbon or carbon monoxide gases over transition metal catalysts, a concept long known for filament formation, but first demonstrated for CNT synthesis by Yacamàn et al. [109] Today, catalytic CVD (CCVD) is considered the most economically viable route for large-scale CNT production and for their integration into functional devices [110].

In a typical setup, the process is performed in a flow furnace under atmospheric pressure, with horizontal and vertical configurations. The horizontal design, most widely used, employs a quartz or ceramic boat containing the catalyst, exposed to a hydrocarbon/inert gas mixture at 500–1100 °C. In contrast, vertical furnaces are better suited for continuous CNT or carbon fiber production, where catalysts and feed gases are injected from the top, and CNTs are collected at the bottom. A modified version, the fluidized bed reactor, extends catalyst residence time by using an upward gas flow, enhancing growth efficiency [110].

The general CNT growth mechanism involves hydrocarbon dissociation on the catalyst surface, carbon atom dissolution into the metal nanoparticle, and subsequent precipitation to form sp^2-bonded tubular structures. The resulting CNT characteristics, including diameter, length, crystallinity, and yield, depend strongly on parameters such as reaction temperature, pressure, gas composition, catalyst type, size, support material, and pretreatment. The diameter of CNT can be controlled by tuning the catalyst particle size, while tube length is primarily governed by reaction time [111].

6.1.4. Hydrothermal Method

Hydrothermal techniques have proven effective for preparing a wide variety of carbon-based nanostructures, including nano-onions, nanorods, nanowires, nanobelts, and MWCNTs [112,113]. These approaches offer several advantages over conventional synthesis routes: (i) they rely on easily accessible and stable precursors under ambient conditions, (ii) they operate at relatively low temperatures (150–180 °C), and (iii) they eliminate the need for hydrocarbons or carrier gases. "Hydrothermal" does not always mean very low absolute temperatures. A study reports synthesis at 700–800 °C under high pressure (60–100 MPa) in water + polyethylene mixtures [114]. This resulted in products having both open- and closed-end nanotubes, with wall thicknesses spanning several to over 100 carbon layers, and inner core diameters ranging from 20 to 800 nm [114,115]. Compared to other high-energy processes (arc discharge, some CVD methods), the hydrothermal approach can potentially reduce the total energy costs because of the use of a liquid-phase environment and lower relative thermal gradients [116,117].

Similarly, graphitic CNTs were obtained by employing ethylene glycol as the carbon source under comparable temperature and pressure conditions, also in the presence of Ni catalysts. Transmission Electron Microscopy (TEM) analysis confirmed that these CNTs possessed wide internal channels and Ni inclusions at the tube tips. Typically, hydrothermal nanotubes exhibit wall thicknesses of 7–25 nm and outer diameters of 50–150 nm, while thin-walled tubes with internal diameters from 10 to 1000 nm have also been reported [115]. The growth process often involves infiltration of a supercritical fluid mixture (CO, CO_2, H_2O, H_2, and CH_4) into the developing nanotube cavity.

Using a combined sonochemical/hydrothermal route, Manafi et al. [118] successfully produced CNTs at 150–160 °C for 24 h, starting from a 5 mol/L NaOH aqueous solution containing dichloromethane, metallic lithium, and cobalt chloride. The resulting nanotubes had diameters of ~60 nm and lengths of 2–5 μm. Scanning Electron Microscopy (SEM) revealed uniformly dispersed catalyst nanoparticles, attributed to the ultrasonic pre-treatment of the precursor solution.

Interestingly, MWCNTs and nanocells have also been synthesized in hydrothermal fluids from amorphous carbon without metal catalysts, at temperatures below 800 °C. In these conditions, carbon nanocells formed through the interconnection of multi-walled graphitic layers at ~600 °C, appearing macroscopically as disordered bulk carbon. These nanocells typically exhibit diameters below 100 nm, with internal cavities of 10–80 nm. Additionally, nanotubes produced in such systems usually display diameters of several tens of nanometers and lengths extending to hundreds of nanometers [119].

Notably, the spontaneous growth of short nanotubes and nano-onions has been observed from nanoporous carbon in the presence of elemental cesium at temperatures as low as 50 °C. Microscopic studies demonstrated that the degree of structural ordering and abundance of carbon nanoparticles increased significantly upon heating to 350–500 °C. Remarkably, even at 50 °C, the formation of carbon nanopolyhedra, nanotubes, and onions was observed [120].

6.1.5. Electrolysis Method

Electrolysis is a relatively uncommon approach for synthesizing CNTs, first reported by Hsu et al. [121] in 1995. This process is based on the electrowinning of alkali metals (Li, Na, K) or alkaline-earth metals (Mg, Ca) from their molten chloride salts at a graphite cathode. The deposited metals react with the cathode to form CNTs. Each electrolyte composition requires an optimal operating temperature for CNT formation, with product purity decreasing sharply when deviating from this temperature. For example, in NaCl and LiCl electrolytes, the best results were obtained at temperatures slightly above the melting point of the salts. Following electrolysis, the carbonaceous products are isolated by dissolving the salts in distilled water and filtering the suspension. During this process, the graphite cathode undergoes erosion, producing a mixture that typically contains CNTs, amorphous carbon, carbon-encapsulated metal nanoparticles, spherical carbon particles, and filaments. MWCNTs are the dominant product, though Bai et al. [122] successfully synthesized SWCNTs via electrolytic conversion of graphite in molten NaCl at 810 °C under argon, yielding tubes with diameters of 1.3–1.6 nm comparable to SWCNTs obtained by other methods. Small additions (<1 wt%) of metals or low-melting-point salts such as $SnCl_2$, $PbCl_2$, Bi, or Pb promote the formation of metal nanowires and filled CNTs. They produced MWCNTs typically have diameters of 10–20 nm, lengths exceeding 500 nm, and consist of 10–15 graphitic walls, often aggregated into entangled bundles with amorphous carbon and encapsulated particles.

A distinct advantage of electrolysis is its operation in the condensed phase that directly utilizes graphite as a carbon source at relatively low temperatures. This approach offers several benefits, namely: (i) simple apparatus design; (ii) controllable synthesis through electrolysis parameters; (iii) use of inexpensive raw materials; (iv) relatively low energy requirements; and (v) tunability of CNT morphology, structural features, and heteroatom doping through optimization of both composition and electrolysis conditions [123,124].

Recent progress has expanded the electrochemical route. Novoselova et al. [125] developed a new electrolytic synthesis based on cathodic reduction of CO_2 dissolved in molten salts, leading to in situ carbon deposition on metallic electrodes. Predominantly curved and bundled MWCNTs were obtained by using a ternary chloride melt (NaCl–KCl–CsCl) as an electrolyte, with outer diameters ranging from 5 to 250 nm and inner diameters from 2 to 140 nm, sometimes partially filled with salts. A higher current density favored smaller-diameter CNTs, improved carbon yield, and increased CNT fraction in the product. In another study, nanotube growth was achieved at temperatures as low as −40 °C in liquid ammonia using acetylene as the carbon source, notably without a metal catalyst. Acetylene, generated by CaC_2 hydrolysis and purified through sequential treatments, was dissolved in liquid NH_3 for electrolysis. The resulting deposits formed a porous layer (1–2 μm thick) on the cathode surface, primarily consisting of amorphous, graphitic, and turbostratic carbon, along with MWCNTs. These nanotubes, with an average diameter of ~15 nm, exhibited irregular curvature, bundled morphologies, and exceptionally high aspect ratios (length-to-diameter ratio > 1000).

6.2. CNT-Based Anode for LIBs

LIBs have attracted much attention over the years due to their high theoretical gravimetric capacity, light weight, and long lifespan. Lithium has a high theoretical specific capacity of 3862 mAh/g and a highly negative standard electrode potential (−3.040 V vs. SHE), making it an ideal anode material for LIBs, with an effective ionic radius of 0.76 Å, which promotes fast ion transfer kinetics. While graphite remains the most commonly used anode material for commercial LIBs due to its low cost, its application potential is limited by low energy density (less than 250 Wh kg^{-1}) and a low theoretical specific capacity of 372 mAh/g [97]. Other concerns are the slow diffusion rate of Li^+ ions in graphite, dendrite formation at low voltages, and safety concerns from the non-aqueous electrolyte system (Figure 5) [98].

Recent studies have focused on the use of nanomaterials to overcome these issues, since these materials have large surface areas, which ensure an improved Li-ion diffusion coefficient. These materials include fullerenes (0D), CNT (1D), graphene (2D), and their composites (3D) [1]. Among the various nanocarbon materials, CNTs have become one of the most promising candidates for next-generation LIB anode. Their 1D tubular shape provides continuous electron transport pathways, high electrical conductivity, and many active sites for Li^+ intercalation and adsorption [99]. Additionally, their high mechanical strength and flexibility help accommodate the significant volume changes that often happen during lithiation and delithiation, reducing pulverization and improving cycling stability [126]. Recent research indicates that pristine CNTs can deliver higher reversible capacities than graphite, with improved rate performance resulting from faster Li^+ diffusion along the nanotube channels [127–129]. However, pristine CNTs alone still show relatively low specific capacities (~300–400 mAh/g), which limits their competitiveness with high-capacity alloying materials like Silicon.

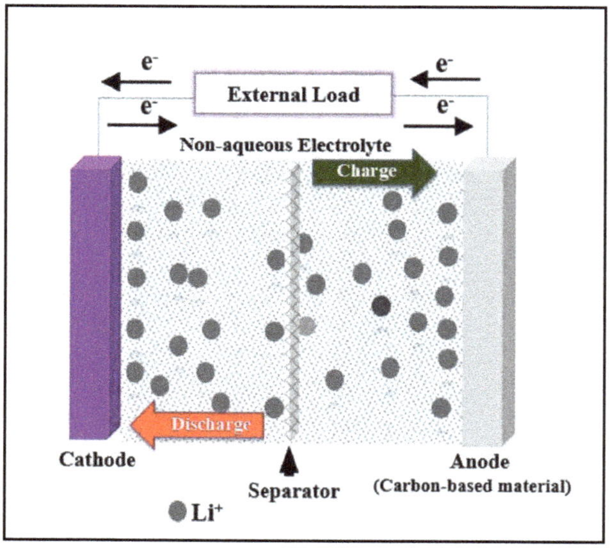

Figure 5. Schematic diagram for LIBs.

To address these challenges, various modification strategies, such as heteroatom doping, have been employed to enhance the electrochemical performance of the anode [97]. Heteroatom doping, where nitrogen, boron, or phosphorus atoms are introduced into the CNT lattice, can create defects and additional Li^+ adsorption sites. This method has been shown to improve conductivity and storage capacity [126]. Another widely studied direction is the development of CNT–silicon composites. In these systems, CNT networks serve as highly conductive scaffolds that buffer the extreme volume expansion of Silicon during lithiation. CNT-Silicon anode have reported superior performance with a high specific capacity of 3250 mAh/g at 0.2 C, a capacity retention of 99.8%, and a coulombic efficiency of 100% after more than 700 cycles [129].

Furthermore, CNT–graphene hybrids are being designed to exploit the complementary advantages of both materials. The combination produces 3D conductive frameworks with high surface areas and rapid electron/ion transport, while also suppressing agglomeration of the carbon nanostructures. Such hybrids have been reported to achieve both high capacity and excellent rate performance [130,131]. Beyond compositing, surface functionalization and defect engineering are also employed to enhance performance. Mild oxidation or plasma treatments can introduce oxygen-containing functional groups that improve electrolyte wettability and facilitate stable solid–electrolyte interphase (SEI) formation. In parallel, controlled defect engineering has been shown to increase Li^+ storage capacity without severely compromising conductivity [132]. Another promising trend is the fabrication of binder-free CNT anode. Direct growth of CNTs onto current collectors, such as copper foils, eliminates the need for polymer binders and conductive additives. This process reduces inactive mass and increases the gravimetric energy density of the electrode. Moreover, this architecture improves structural robustness during cycling, enabling long-term stability at high current densities [133–135].

In summary, CNT-based anode are positioned at the forefront of advanced LIB research. Although their intrinsic capacity is lower than that of Silicon, their superior conductivity, structural flexibility, and chemical tunability make them attractive, either as standalone high-rate anode or as conductive scaffolds in composite systems. Future directions will likely focus on scalable synthesis methods, optimized doping strategies, and hybridization with high-capacity active materials to bridge the gap between laboratory-scale demonstrations and industrial-scale applications [136,137].

6.3. CNT-Based Anode for SIBs

SIBs are considered promising battery systems and viable alternatives to LIBs due to their low costs and the wide distribution of sodium in the Earth's crust, as well as their similar electrochemical mechanisms to LIBs. However, SIBs still face some challenges that have hampered their commercial applications. One challenge is the low operating voltage ($\sim$0.3 V lower than Li) of SIBs. When comparing the metal ions in Group I, the larger effective ionic radii of Na^+ ions (1.02 Å) compared to Li^+ ions can lead to sluggish ion diffusion and reaction kinetics and structural stress in electrodes in SIBs. However, the lower Lewis acidity of Na^+ ions results in their solvated ions being smaller compared to those of Li^+ ions [97]. Therefore, finding appropriate electrode materials with internal spaces large enough to host Na^+ ions should be thoroughly considered.

Recent research efforts have focused on developing better anode materials to enhance the overall performance of SIBs. CNTs have emerged as promising anode materials for SIBs due to their excellent electrical conductivity, mechanical strength, and ability to accommodate volume changes during cycling. CNTs have been integrated into various SIB anode architectures to mitigate the intrinsic challenges of sluggish Na^+ diffusion and substantial volume expansion during cycling. Pristine CNTs have been explored as standalone anode materials for SIBs. However, their limited Na^+ storage capacity, typically below 100 mAh/g, is attributed to weak van der Waals interactions and the absence of active sites for Na^+ intercalation. Therefore, recent studies have focused on modifying pristine CNTs to enhance their electrochemical performance. It has been reported that pristine carbon materials, including CNTs, suffer from low Na^+ intercalation unless doped or hybridized [138]. While pristine CNTs may be structurally robust, the need for surface engineering or doping to achieve practical capacities for SIBs has been emphasized [139]. The foregoing highlights the role of defect engineering and heteroatom doping in improving Na^+ intercalation and affinity. When comparing pristine CNTs with biochar-based hard carbons, it is noted that CNTs offer superior conductivity but lag in Na^+ storage unless functionalized [140]. The limitations of pristine carbon materials were also identified, and advanced nanomaterial tailoring, such as creating oxygen-rich or nitrogen-doped CNTs, was proposed to overcome the low capacity and poor cycling stability [141].

CNTs are increasingly exploited to construct 3D conductive frameworks that simultaneously support efficient electron transport and mechanical stability in SIB electrodes. This process seeks to address the sluggish Na^+ diffusion kinetics and large ionic radius of sodium, which impose significant structural stress on conventional anode. In the 3D CNT frameworks, interconnected conductive networks provide continuous electron pathways, while also accommodating volumetric changes during cycling. The electrochemical advantages of 3D CNT-based anode are increasingly evident in recent studies. Liu et al. [142] demonstrated that CNT/Fe_2O_3 hollow nanostructures anchored on a 3D CNT framework achieved a reversible capacity of 410 mAh/g after 500 cycles, benefitting from both conversion reaction activity and mechanical buffering of CNTs [143]. Similarly, nitrogen-doped CNT aerogels exhibited superior rate capability and mechanical flexibility, maintaining stable sodium storage at high current densities [144]. Other approaches exploit CNT/graphene hybrid aerogels with hierarchical porosity, which delivered improved ion transport and long cycle life compared to single-component carbons [144]. Moreover, CNT–SnS_2 composites achieved high-rate performance due to conductive scaffolding that suppressed pulverization of the active material [145]. Together, these studies highlight the critical role of CNT-based architecture in enabling robust, high-performance electrodes for next-generation MIBs.

Despite promising progress, challenges remain in translating 3D CNT frameworks into practical SIB anode. Large-scale, cost-effective synthesis of CNT networks with controlled porosity and heteroatom doping remains non-trivial. Binder-free electrode designs, while advantageous to attain high energy density, may face limitations in mechanical integrity under industrial processing. Moreover, the intrinsic low-redox activity of pure CNT frameworks necessitates hybridization with other materials to achieve competitive capacities. Future directions include rational defect engineering, hierarchical structure control, and the integration of CNT scaffolds with high-capacity alloying materials such as Sn, Sb, or P to balance kinetics and storage capacity. Given the rapid advances in CNT chemistry and scalable aerogel synthesis, 3D CNT frameworks are poised to play a significant role in advancing SIBs toward commercial applications [144].

6.4. CNT-Based Anode for KIBs

Compared to LIBs and SIBs, the low standard electrode potential of the K/K redox couple results in lower cutoff potential for the most available negative electrode materials, which helps avoid metallic potassium deposition during cycling. Thus, KIBs have the potential for higher-voltage operation and can function over a wider electrochemical window compared to LIBs and SIBs. Moreover, KIBs have high-power densities based on the fast diffusion rate of K^+ ions, due to the weak coulombic interactions of K^+ ions with surrounding anions. These unique advantages of K^+ ions make KIBs a possible alternative to LIBs. However, the large ionic radius of K^+ (1.38 Å) compared to that of Na^+ ions creates even more serious issues with transport kinetics and cycling stability for KIBs [146]. Whereas, the high chemical activity of potassium hindered the KIB use in commercial applications [147].

Comparative analyses across Li-, Na-, and K-ion systems further highlight the unique challenges of KIBs. Hard carbon, while widely used in lithium and sodium storage, shows lower ICE and capacity in potassium systems, underscoring the need for new structural designs for high-performance electrodes that deliver high specific capacity and excellent cycle performance [103,148]. Recent reviews have emphasized that the design of carbon frameworks, including CNTs, plays a crucial role in mitigating volume expansion and enabling stable cycling [149,150]. The effect of the conductivity of a graphene-based anode material on the diffusion kinetics of K^+ ions has also been reported [139]. Several other optimization strategies for CNT anode have also been identified. For instance, heteroatom doping (e.g., N, P, S) can modulate electronic states and create active sites for enhanced K^+ adsorption, while hierarchical CNT architectures facilitate rapid ion transport and shorten diffusion pathways [151]. In addition, coupling CNTs with amorphous or graphitic hard carbon domains can synergistically improve both reversible capacity and cycling life [152]. Binder-free CNT films and sponges have also been explored as flexible anode, offering lightweight electrodes with reduced inactive mass, though mechanical robustness under industrial processing remains a challenge [153]. Future directions include rational defect engineering, scalable aerogel synthesis, and integration of CNT scaffolds with high-capacity alloying materials such as Sn, Sb, or P to balance kinetics and storage capacity [149].

7. Hard Carbon Material as an Anode for MIBs

Hard carbon, also called nongraphitizable carbon, is a type of disordered carbon that cannot be transformed into perfect graphite even at high temperatures [154]. By pyrolyzing organic precursors at temperatures below 2000 °C, hard carbon can be produced while maintaining its turbostratic graphitic domains and amorphous areas. Hard carbon is one of the most appealing anode materials for rechargeable batteries because of these characteristics as well as intrinsic nanopores and flaws. It is useful not only for LIBs but also for sodium-ion systems, potassium-ion systems, calcium-ion systems, magnesium-ion systems, and zinc-ion systems due to its capacity to store a large number of ions reversibly, function at low potentials, and accept ions of different sizes.

Hard carbon's porous structure and increased interlayer spacing facilitate the effective insertion and extraction of bigger ions like Na^+ and K^+, increasing capacity and rate capability [154,155]. Additionally, hard carbon has received more interest recently for large-scale energy storage applications due to its adaptability, cost-effectiveness, and abundance of precursors.

7.1. Synthesis of Hard Carbon Materials

A schematic is presented in Figure 6, showing different methods that have been used to prepare hard carbon.

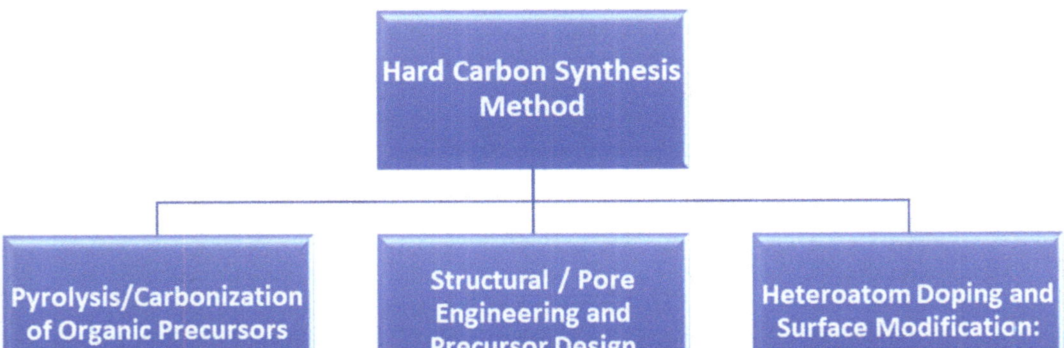

Figure 6. Hard carbon synthesis methods.

7.1.1. Pyrolysis/Carbonization of Organic Precursors

Thermal carbonization of organic precursors in inert or oxygen-deficient environments is the most frequent method of producing hard carbon, which cannot be graphitized even at high temperatures. For instance, heating biomass, polymers, or tiny organic molecules to high temperatures (such as 900–1500 °C) produces hard carbon with a disordered carbon framework that has flaws and closed pores that improve ion-storage capacity. Hard carbon is one of the most attractive anode materials for SIB because of its plentiful supplies and advantageous low-voltage performance, as Tan et al. noted [156]. Hard carbon materials are superior to other anode electrode materials for SIB due to their high capacity, low cost, and low operating voltage [156,157].

Hard carbon is a carbon-based substance that can be made from a variety of precursors, including asphalt, biomass, and resin. At temperatures between 250 and 500 °C , the precursor's carbon atoms are catalyzed, cracked, and reorganized, first establishing the carbon layer structure and then progressively ordered in the heat treatment process that follows, ultimately forming hard carbon [158,159]. Hard carbon's structure and characteristics are directly impacted by the carbonization process's parameters. Two significant factors that influence the structure and characteristics of hard carbon are the temperature and heating rate of the carbonization process. In order to produce bioderived hard carbon, Dahbi et al. [160] carbonized argan shells at various temperatures (800, 1000, 1200, and 1300 °C).

The findings demonstrated that the carbon layer spacing inside hard carbon decreased as the carbonization temperature increased (from 0.4 nm at 800 °C to 0.388 nm at 1300 °C). Both the material's specific surface area (from 99 m^2/g at 800 °C to 2.6 m^2/g at 1300 °C) and the integrated intensity ratio (ID/IG) of the D band to the G band in the Raman spectrum (from 2.66 at 1300 °C to 2.06 at 800 °C) continue to decline, indicating a decrease in defect content. According to the electrochemical test, hard carbon produced by carbonization at 1300 °C performs the best, with an ICE of 83.9% at 25 mA/g and a specific capacity of 300 mAh/g. In order to create sucrose-based hard carbon microspheres, Xiao et al. [161] used a hydrothermally treated sucrose precursor and heated it to 1300 °C at various speeds (0.5, 1, 2, and 5 °C/min). It was discovered that the ICE of hard carbon could be improved by lowering the heating rate, which also had a slight increase in capacity.

The conventional carbonization process uses a lot of energy and takes a long time. Zhen et al. [162] used a novel and effective approach. The pre-carbonized carbon matrix was carbonized at a high temperature using a multifield controlled spark plasma sintering (SPS) technique. SPS uses an extremely quick heating rate (100–500 °C/min) to finish plasma sintering in less than a minute. Compared to conventional carbonization techniques, SPS-prepared hard carbon contains fewer flaws, less porosity, and less oxygen. In another study Guo et al. [163] prepared a bio mass hard carbon anode using the carbonization method with relatively lower heating rate. The synthesized anode displays a high ICE of 82.8% and exhibited decreased initial irreversible capacity loss. The initial charge capacity of 324.6 mAh/g and promising cycle stability with 90.0% capacity retention after 200 cycles at 50 mA/g was achieved.

This technique makes it possible to fine-tune pore structure, defect density, and interlayer spacing all of which are crucial for affecting Na^+ or K^+ insertion behavior. In reality, customized micropore/mesopore volumes, closed-pore architecture, and graphitic domain size can result from controlling the heating rate, final temperature, precursor type, and environment. For instance, Qin et al. report that after 200 cycles at 0.05 A/g, N, P co-doped hard carbons made from maize stover delivered discharge capacities close to ~300 mAh/g [164,165]. These findings highlight the need for ideal pyrolysis and carbonization conditions in order to produce hard carbon with a high reversible capacity.

7.1.2. Structural/Pore Engineering and Precursor Design

To maximize hard carbon performance, structural engineering and meticulous precursor design are frequently used in addition to straightforward carbonization. Hierarchical structures with better storage properties can be created by selecting a precursor (biomass vs. synthetic polymer), using templates or activation, and controlling the distribution of pore sizes and interlayer spacing. Tan et al., for example, point out that essential strategies to improve hard carbon performance include enhancing pore shape, heteroatom doping, and electrolyte design [156].

The precursor is carbonized to create hard carbon, which further reflects performance while retaining some of the precursor's structural features. As a result, the performance can be managed through the design or screening of appropriate antecedents. Biomass-based precursors, polymer-based precursors, asphalt-based precursors, and coal-based precursors are the four primary groups of hard carbon precursors. Phenolic resin is a typical and highly pure precursor for polymers. By controlling the precursor's functional group and degree of crosslinking, it can control the structure and dope with elements [166,167].

Maleic anhydride was used to cure epoxy phenolic resin, which was used as a precursor by Fan et al. [168]. The epoxy rings completely reacted with maleic anhydride, producing a large number of hydroxyl and ester functional groups, according to analysis using Fourier transform infrared (FTIR) spectroscopy.

Although polymer precursors have some advantages, their expensive cost limits their widespread application. Asphalt-based precursors, on the other hand, come from plentiful and reliable sources. Additionally, they yield more carbon than precursors made of biomass or resin. However, the hard carbon created by direct carbonization frequently performs less than ideal since asphalt tends to graphitize at high temperatures. Therefore, pretreatment procedures are frequently used prior to carbonization. Lu et al. [169], for instance, employed asphalt as the precursor and added a pre-oxidation step before carbonization. During this low-temperature pre-oxidation, adding functional groups containing oxygen improves the precursor's crosslinking and prevents the asphalt from melting or undergoing ordered structural rearrangement during high-temperature carbonization.

It has been demonstrated that the capacity of the plateau region is significantly influenced by the pore structure of hard carbon [170]. When assessing ICE and total capacity of hard carbon materials, increasing the number of micropores in the material improves the plateau capacity [171]. Zhang et al. [172] found molecular sieve carbon as an anode material with superior sodium-storage performance for the first time by comparing the microporous architectures and sodium-storage behaviors of commercial activated carbon, molecular sieve carbon, and graphite.

They found that activated carbon with extremely high porosity behaved poorly when storing sodium. Its low ICE (22.1%) was caused by its wide pore apertures, which produced an unduly high specific surface area [172].

By thermally heating a freeze-dried combination of magnesium gluconate and glucose, Kamiyama et al. [173] improved the MgO templating technique to create porous hard carbon. Nanoscale MgO particles were consistently produced within the carbon matrix by a preliminary heat treatment at 600 °C. Reversible capacity was maximized by further acid leaching and high-temperature carbonization at 1500 °C, which further enhanced structural ordering while maintaining a high density of nanoscale micropores. In the first cycle at 25 mA/g, the resultant material showed an ICE of 88% and a reversible capacity of 478 mAh/g.

Yin et al. [174] used phenolic resin as the precursor and added nanoscale ZnO during the precursor synthesis stage, resulting in a one-step carbonization method that produced porous hard carbon. ZnO and carbon combine at high temperatures to produce gaseous zinc and carbon monoxide. Using glucose as the carbon precursor and bis(cyclopentadienyl)nickel as the coating carbon source, Cheng et al. [175] combined surface-coating and templating techniques. They also incorporated a silica template by hydrothermal pretreatment and liquid-phase processes. Following carbonization, the resultant porous carbon spheres coated with an ultrathin carbon layer exhibited high ICE.

Increasing the volume of closed pores and maximizing the mesopore-to-micropore ratio can improve sodium-ion storage's low-voltage plateau capacity. To further improve electrochemical performance, pore engineering is often combined with heteroatom doping and surface functionalization. Ion migration can be accelerated by microstructural changes, as demonstrated by Yin et al. [176], and these effects can be efficiently adjusted by adding dopants to the carbon framework. When taken as a whole, these design approaches provide a thorough road map for creating hard carbons of the future.

7.1.3. Heteroatom Doping and Surface Modification

Hard carbon's electrical structure, conductivity, and interlayer spacing can all be altered by heteroatom doping, which includes N, S, B, P, and other elements. These heteroatoms can typically be added during the carbonization process with the aid of precursors that contain specific elements, and occasionally even through post-synthetic processes including chemical vapor deposition, gas-phase doping, and solution immersion. For instance, Yin et al. showed that doping carbon materials with heteroatoms like N, S, and B effectively modifies microstructures and increases ionic migration rates [176].

According to Wang et al., P-doping hard carbon with phosphoric acid as a precursor boosted reversible capacity from 240.3 mAh/g to 359.9 mAh/g, increased interlayer gap, and promoted Na^+ transport. In comparison to hard carbon without doping, the charge–discharge rate was discovered to be 0.05 C [177]. These findings suggest that the electrochemical performance and structural stability of hard carbon anode are directly impacted by careful dopant selection and doping concentration management. In order to improve rate performance in LIBs, SIBs, and KIBs, sulfur and boron dopants have also been investigated. Sulfur increases surface polarity, while boron increases electronic conductivity.

7.2. *Hard Carbon-Based Anode for LIBs*

Carbon-based materials are still among the most popular and dependable anode for LIBs, because they provide a good mix between stability, affordability, and electrochemical performance. This material has demonstrated exceptional cycling stability, improved low-temperature performance, and fast-charging capabilities [178,179]. Hard carbon is one of the most attractive candidates for high-energy LIB applications. Both intercalation and adsorption-based storage methods are supported by its unique microcrystalline structure, which has disordered graphene layers, nanopores, and a large number of defect sites. These structural benefits directly result in increased high-energy performance and better lithium storage capacity.

Hard carbon nevertheless confronts significant obstacles in its practical application, despite these advantages. Its commercial potential is still constrained by its high voltage hysteresis, significant irreversible capacity loss during the first cycle, and low ICE. Structural engineering, surface changes, improved precursors, and novel synthesis techniques have all been the focus of substantial research efforts over the last ten years to address these problems.

Hard carbon as an anode material for LIBs is thoroughly and currently examined. Xie et al. explained all the principles of lithium storage, classified various hard carbon kinds according to their microstructure and synthesis pathways, and identified the main obstacles that still need to be overcome. In order to provide insight into how hard carbons might develop to satisfy the requirements of next-generation, high-energy battery systems, possible solutions and new research avenues are also examined [178].

7.3. Hard Carbon-Based Anode for SIBs

According to Liu et al. [180], hard carbon stands out among the several anode possibilities for SIBs due to its great compatibility with current industrial manufacturing and economic advantages. Hard carbon's unique microstructure is essential for improving its electrochemical behavior during Na^+ storage [154,181]. Subsequent research has demonstrated that this intricate structural configuration enables hard carbon to attain a sodium-storage capacity considerably greater than that of graphite [182]. As a result, hard carbon has gained more attention as a potential anode material for SIBs [183]. A significant advancement in SIB creation was made when Dahn and his team showed that glucose-derived hard carbon could provide a reversible Na^+ storage capacity of about 300 mAh/g [24]. Hard carbon has emerged as a top contender for upcoming commercial SIB anode in recent years [184].

When cycling at moderate current densities, the hard carbon anode in SIBs typically exhibit reversible capacities in the range of around 250–350 mAh/g. For example, He et al. claimed a capacity gain from 232 mAh/g to 307 mAh/g [185], although one performance benchmark reports ~305 mAh/g. Despite these successes, there are still issues, primarily related to the huge interior surface area and irreversible ion loss during SEI formation, which lead to low ICE and poor rate capability. However, the performance gap for commercial SIB deployment for hard carbon anode has gradually decreased with appropriate pore design, interface tuning, and structural control [186]. Current industrial-scale demonstrations are increasingly validating these improvements, highlighting the commercial readiness of hard carbon for SIBs.

7.4. Hard Carbon-Based Anode for KIBs

Hard carbon's disordered structure and larger interlayer spacing make it suitable for accommodating larger ions such as K^+ (~1.38 Å). Qiu et al., in their recent work, outline that hard carbon materials exhibit excellent rate performance and cycling stability in KIBs when appropriately modified [187].

Larger ions like K^+ (~1.38°) can be accommodated by hard carbon due to its disordered structure and wider interlayer gap. When properly adapted, hard carbon materials show outstanding rate performance and cycling stability in KIBs, according to a recent study by Qiu et al. [187].

Since K^+ shares similar physical and chemical characteristics with Li^+ and Na^+, hard carbon has also emerged as a promising candidate for potassium-ion storage. Early research on carbon-based anode for K^+ was limited, largely due to concerns that the larger ionic radius of K^+ would hinder effective accommodation within carbon structures [184]. This view shifted after a pivotal 2016 study by Jian et al., which demonstrated successful electrochemical intercalation of K^+ in hard carbon, achieving a reversible capacity of 262 mAh/g along with impressive rate performance [184].

The structural engineering of hard carbon and the underlying mechanisms of K^+ storage have been extensively studied as a result of this significant finding [188–190]. Current research continues to concentrate on creating better hard carbon materials for KIBs, as the complex microstructure of hard carbon offers new opportunities for accommodating K^+ ions.

Although direct application of hard carbon in zinc-ion, magnesium-ion, or CIB systems is less developed, hard carbon's architectural advantages make it an obvious choice for these new multivalent systems. When all of these factors are taken into account, hard carbon's adaptability stems from its broad precursor selection, tunable microstructure, and tunability across various ion chemistries. As a result, it becomes one of the primary anode platforms for batteries of the future.

8. Heteroatom-Doped Carbon Materials as an Anode

As mentioned earlier, a variety of carbonaceous materials have been investigated for use as anode, including hard carbon, soft carbon, graphene, EG, etc. Among the various anode materials, hard carbon has garnered significant attention owing to its plentiful availability, stability, and non-toxic nature. Nevertheless, there is an agreement that hard carbon exhibits relatively subpar performance regarding its rate capability, reversible capacity, and ICE. To address these deficiencies, numerous investigations have concentrated on the development of distinctive architectures, such as micro-spherules, nanowires, nanofibers, and porous structures, or the incorporation of heteroatoms to enhance their electrochemical characteristics [191]. The concept of surface functional modification involves the introduction of functional groups through heteroatom doping, which enhances the electronic structure, micro-chemical environment, and surface characteristics of carbon materials [192,193]. Recent reports have focused on doped heteroatoms in carbon, primarily highlighting nitrogen, phosphorus, sulfur and boron as shown in Figure 7 [193].

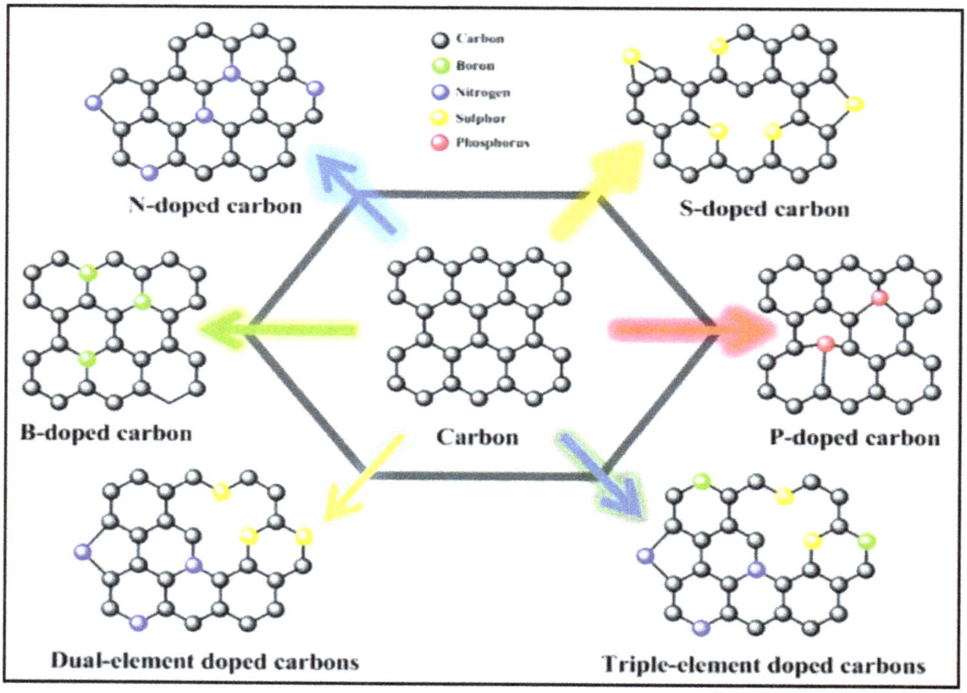

Figure 7. Schematic representation illustrating the architecture of different heteroatom-doped carbon materials [193].

8.1. Synthesis of Heteroatom-Doped Carbon-Based Materials

8.1.1. Pyrolysis Method

The heat-treatment (pyrolysis) process is a prevalent method utilized for the preparation of various N-containing carbon materials. It involves introducing nitrogen into carbonaceous structures such as graphite and graphene. One of the procedures for preparing nitrogen-doped carbon (NC) anode involves the pyrolysis of precursors containing both nitrogen and carbon. For instance, $C_{10}H_{14}N_2Na_2O_8{\cdot}2H_2O$ (ethylenediaminetetraacetic acid disodium salt dihydrate) [194] or $C_9H_{17}NO_5$ (Vitamin B5) [10,195] is pyrolyzed under an argon atmosphere, with the final temperature ranging from 500 to 1000 °C for a duration of two hours. After pyrolysis, the resulting products are washed with 37% HCl to eliminate inorganic impurities and dried in a vacuum oven at 80 °C. Though the procedure is straightforward, it yields highly efficient NC materials suitable for use as anode.

Similarly, boron-containing precursors, such as B_2O_3 or H_3BO_3, can be directly incorporated into carbon precursors, followed by pyrolysis, to yield boron-doped carbonaceous materials [193]. Pitch coke and boron oxide powders are combined in the most straightforward method, which is then baked and heated in an argon environment before being ground up and sieved to produce boron-doped graphite [196].

The pyrolysis method also applies to the direct carbonization of sulfur-containing precursors, such as dodecylbenzene sulfonic acid or poly(3,4-ethylene dioxythiophene) (PEDOT), at high temperatures (700–850 °C), producing sulfur-doped carbon with varying sulfur contents (2–15 wt%) [26].

For phosphorus doping, a related pyrolysis approach can be applied using red phosphorus (RP) or phosphorus-organic compounds as precursors. P-doped GO can be prepared by mixing with triphenyl phosphate (TPP) in alcoholic solvent, drying in the vacuum condition and afterwards annealing in pure argon environment [197].

8.1.2. Carbonization of Mixed Precursors

This method involves combining carbon and heteroatom precursors prior to thermal treatment. For nitrogen-doped porous carbon (NPC), a mixture of sodium citrate and urea is calcined to form a porous structure with uniformly distributed nitrogen atoms.

For sulfur-doped carbon (SC), this approach is widely used due to its ability to achieve high sulfur contents (exceeding 20%). Jiang et al. [16] synthesized sulfur-doped disordered carbon (DC-S) by combining the small molecule 1,4,5,8-naphthalene tetramethylene anhydride (NTCDA) with sulfur powder at 500 °C, resulting in 26.91 wt% sulfur. Solid sulfur sources such as sodium dodecyl sulfate, thiourea, and phenyl disulfide have also been employed [198], although in solid-state mixing, if the interfacial contact is not enough it can cause non-uniform doping.

The choice of carbon precursor significantly affects the resulting structure and morphology. Precursors can include small organic molecules, polymers, and biomass sources such as scallion peel, garlic peel, elm samara, lotus leaf, and bagasse. Uniform doping depends upon more multiple factors such as, the interplay between precursor type, heteroatom source, mixture ratio, and carbonization conditions.

8.1.3. CVD Method

CVD is one of the most used methods for synthesizing heteroatom-doped carbon anode. It is extensively employed for the incorporation of nitrogen and boron into carbon materials. Its significance lies in the fact that it provides highly controlled incorporation of dopants in the carbon lattice and results in homogeneously distributed active sites and tunable electronic properties. CVD with optimized porosity enhance charge transport, cycling stability, and overall electrochemical performance.

N-doped graphene sheets and high-concentration nitrogen-doped carbon nanotubes (HN-CNTs) have been synthesized via ammonia-assisted or CVD-based methods [199,200]. Boron doping in atomic form is also commonly achieved via CVD using boron-containing alkanes or diborane (B_2H_6) diluted in Ar gas. The plasma arc torch variant employs acetylene and diborane (1% B_2H_6 in Ar) as precursors, allowing in situ doping during growth.

8.1.4. Electrospinning Method

Electrospinning represents one of the effective methods to prepare heteroatom-doped carbon materials as it enables precise fabrication of flexible nanofibers with homogenous distribution of dopants. This method leads to controlled diameter, composition, and morphology. After stabilization and carbonization, these dopants are integrated into the carbon framework, generating highly conductive and structurally stable carbon nanofibers. Electrospinning in polymer matrix is particularly used for phosphorus incorporation into carbon nanostructures. H_3PO_4 serves as a phosphorus source and is mixed into a polymer solution (e.g., polyacrylonitrile in N,N-dimethylformamide) to form a spinning solution. Electrospinning produces P-doped 1D macroporous nanofibers, which are subsequently stabilized through thermal treatment in an Ar atmosphere [191].

The high P–C binding energy, PO_x species generation, and oxygen sensitivity of phosphorus can restrict the doping efficiency in this approach by reducing the number of active sites accessible for ion storage [201,202].

8.1.5. Post-Doping Method

Post-doping techniques are used for heteroatom modification of pre-formed carbon structures, particularly for sulfur. This method generally modifies the surface or near-surface regions, causing small changes to the pore structure or morphology [198,203]. In practice, the process often starts with polymerization of thiophene or PEDOT, followed by incorporation of oxidants, dopants, or catalysts, and subsequent heat treatment to achieve stable heteroatom incorporation.

For phosphorus, a novel oxygen-free post-doping approach uses phosphorus trichloride (PCl_3) and cyclohexane (C_6H_{12}) as phosphorus and carbon sources, respectively. By introducing these liquids as mixed gases through N_2 bubbling, carbonization and in situ phosphorus doping are made possible at high temperatures, reaching ultrahigh phosphorus doping values of up to 30 wt% [201].

8.2. NC as Anode

A highly promising approach to enhance the utilization of non-graphitic carbonaceous materials as anode involves the incorporation of N heteroatoms within their structure. N is the most extensively studied heteroatom dopants [204] because N possesses certain advantages compared to other dopants [205]. The introduction of N into carbon materials is straightforward and manageable. Consequently, N can modify the characteristics of the carbon host material while preserving its fundamental structure [206]. Furthermore, N atoms modify the electron count within the structure. Simultaneously, the comparable size of C and N atoms means that the introduction of N does not cause a significant lattice mismatch in the host carbon material. Numerous studies have indicated that N incorporation into the structure of carbonaceous anode active materials improves the electrochemical performance of these anode, especially in terms of their specific capacity [207,208].

The synthesized NC materials were evaluated for their electrochemical behavior in LIBs and KIBs. For LIBs, the C-700 anode synthesized from $C_{10}H_{14}N_2Na_2O_8{\cdot}2H_2O$ at 700 °C demonstrated an initial discharge capacity of 936.9 mAh/g (at 0.5 C, within 0.02–2.5 V range) over 600 cycles. After 500 cycles, the C-700 electrode maintained a capacity of 246.4 mAh/g, corresponding to a retention of 26.3% [194].

In a similar test, the CN-700 electrode synthesized from Vitamin B5 ($C_9H_{17}NO_5$) exhibited a reversible capacity of 1528 mAh/g after 50 cycles in a Li^+ half-cell at a current density of 100 mA/g. Its rate performance showed 1300 mAh/g at 0.2 A/g and still retained 200 mAh/g at 30 A/g, indicating superior high-rate capability [195].

For KIBs, the NPC electrode derived from sodium citrate and urea achieved a capacity retention of 342.8 mAh/g after 500 cycles, corresponding to 90.9% of the capacity observed at the 21st cycle. It also delivered 419.7 mAh/g at 0.05 A/g and 185.0 mAh/g at 10.0 A/g, demonstrating excellent rate performance [209].

Moreover, N-doped graphene sheets used as LIB anode exhibited a specific lithium storage capacity of 608 mAh/g at 0.5 A/g, surpassing the 445 mAh/g capacity of pristine graphene [210]. In a separate investigation, HN-CNTs demonstrated a reversible specific capacity of 494 mAh/g, nearly double that of non-doped CNTs [199,200].

These studies show that nitrogen doping significantly increases the electrochemical performance of carbon-based anode in MIBs.

8.3. SC as Anode

Heteroatom doping is a prevalent method for modifying the physicochemical characteristics of carbon materials. Nitrogen doping is acknowledged as an excellent method to enhance the ion storage capabilities of carbon. In comparison to NC, SC seems to be more appropriate as an anode for SIBs due to greater covalent radius of sulfur (102 pm) relative to carbon (77 pm) and nitrogen (75 pm). The incorporation of sulfur atoms may substantially increase the interlayer spacing in carbon, hence enhancing the insertion and extraction of Na^+ ions and other larger metal ions [26].

To evaluate the electrochemical properties of the synthesized SC, the working electrode was prepared by coating a slurry composed of the SC mixture, polyvinylidene fluoride (PVDF) as a binder, and super P or conductive carbon. The cast electrode was dried in a vacuum oven at 70–80 °C [211].

SC anode have been investigated mainly in SIBs and KIBs as alternatives to LIBs to mitigate the challenges associated with lithium scarcity. Since Na^+ and K^+ ions are larger than Li^+ ions, SC anode are particularly advantageous because the larger sulfur atoms help expand the interlayer distance, facilitating ion transport.

The long-term cycling performance of SC was evaluated at a current density of 0.5 A/g in SIBs. The reversible capacity reached 384.5 mAh/g in the initial cycle and stabilized at 322.0 mAh/g by the 10th cycle. Even after 700 cycles, a capacity of 303.2 mAh/g was maintained, corresponding to a capacity retention of 94.2%. These results confirm that the increased interlayer spacing in SC ensures structural stability during repeated Na^+ adsorption/desorption processes [26]. Table 1 shows the performance of SIBs with different precursors when utilized for the fabrication of SC as anode.

Table 1. Results obtained from using different precursors for SC.

Precursor for SC	Discharge Capacity	Charge Capacity	Cyclic Efficiency (%)	Columbic Efficiency (%)	References
PEDOT	655.0 mAh/g at 0.1 A/g	482.1 mAh/g at 0.1 A/g	73.6	100.0 after 700 cycles	[26]
Polythiophene	714.0 mAh/g at 0.05 A/g	491.0 mAh/g at 0.05 A/g	69.0	100.0 after 500 cycles	[203]

8.4. Boron-Doped Carbon (BC) as an Anode

The incorporation of Boron into carbonaceous materials is garnering interest for various MIBs such as K, Na, and Ca, including Li as well. Boron doping has been explored to mitigate the restricted specific capacity of 372 mAh/g and subpar rate performance of graphite [212,213].

In addition to graphite, reduced GO and amorphous carbon exhibit inadequate cycling stability and constrained specific capacity, primarily due to the significant volume change induced by substantial metal-ion intercalation. To address this issue, loosely packed nanosheets with increased interlayer spacing are designed to effectively manage significant volume changes and mitigate the structural instability faced by bulk materials [214] as discussed earlier. Furthermore, the introduction of point defects, edges, grain boundaries, and doping is essential for improving the electrochemical performance of graphene. Controllable doping of graphene has been accomplished with atomic precision, demonstrating favorable thermodynamic properties [215]. Owing to the comparable atomic sizes of boron and carbon along with the electron-deficiency characteristics of boron doping, both 3D BC structures [216–218] and B-doped graphene [215,219,220] have been extensively investigated as anode materials for LIBs and SIBs, demonstrating significant capacity and excellent cycling stability. The utilization of the potential of B-doped graphene as an anode material for KIBs is currently under investigation [212].

For LIBs, the galvanostatic charge and discharge profiles of the plasma carbon and BC electrodes during the first and second cycles, within a voltage range of 0.0–2.0 V (vs. Li/Li^+) at a current density of 100 mA/g, were studied to evaluate the effect of boron doping.

In the initial discharge process of the plasma carbon electrode, two distinct voltage plateaus were observed at approximately 0.8 V and 0.2 V. The initial plateau signifies the breakdown of the electrolyte at the anode surface, resulting in the formation of a SEI layer [221,222]. The second plateau, noted at approximately 0.2 V, is ascribed to the intercalation of lithium within the plasma carbon.

Whereas the BC electrodes exhibit an additional plateau near ~1.6 V, likely resulting from the formation of boron–carbon bonds induced by the boron doping process. Furthermore, the increase in charge capacity was observed using BC electrodes, which is attributed to the improved Li-ion intercalation within the graphitic layer due to the presence of boron [212].

8.5. PC as an Anode

As stated before, N doping and S doping are the most extensively studied heteroatoms to develop doped carbon as anode, since they facilitate the adsorption of metal ion and provide many active sites for salt storage, resulting in improved capacity [201,223]. However, the elevated discharge voltage associated with the high average oxidation voltage of nitrogen-related functional groups or reactive sulfur dopants is concerning [224,225]. To address this limitation of N and S doping, P doping has been investigated. P doping enhanced adsorption capacity, and it aids to achieve low discharge voltage (<1.0 V) [226–228]. The stated average potential of P in LIBs is quite high ($\approx$0.8 V against Li/Li^+), and its substantial specific capacity (2590 mA h/g) may offset this deficiency. In SIBs, phosphorus has the greatest theoretical specific capacity (2590 mA h/g) for all anode materials and a very low working potential ($\approx$0.3 V against Na/Na^+), highlighting its superiority [229,230]. Moreover, P has abundant reserves, which makes it economical, with minor pollution potential [231,232]. Consequently, P-based anode materials have significant promise in AIBs and have garnered considerable interest from researchers [229].

Phosphorus doping has been employed to enhance interlayer spacing and active site density. The electrochemical performance of graphene-based phosphorus-doped carbon (GPC) was evaluated as an anode for SIBs as presented in Figure 8 [197].

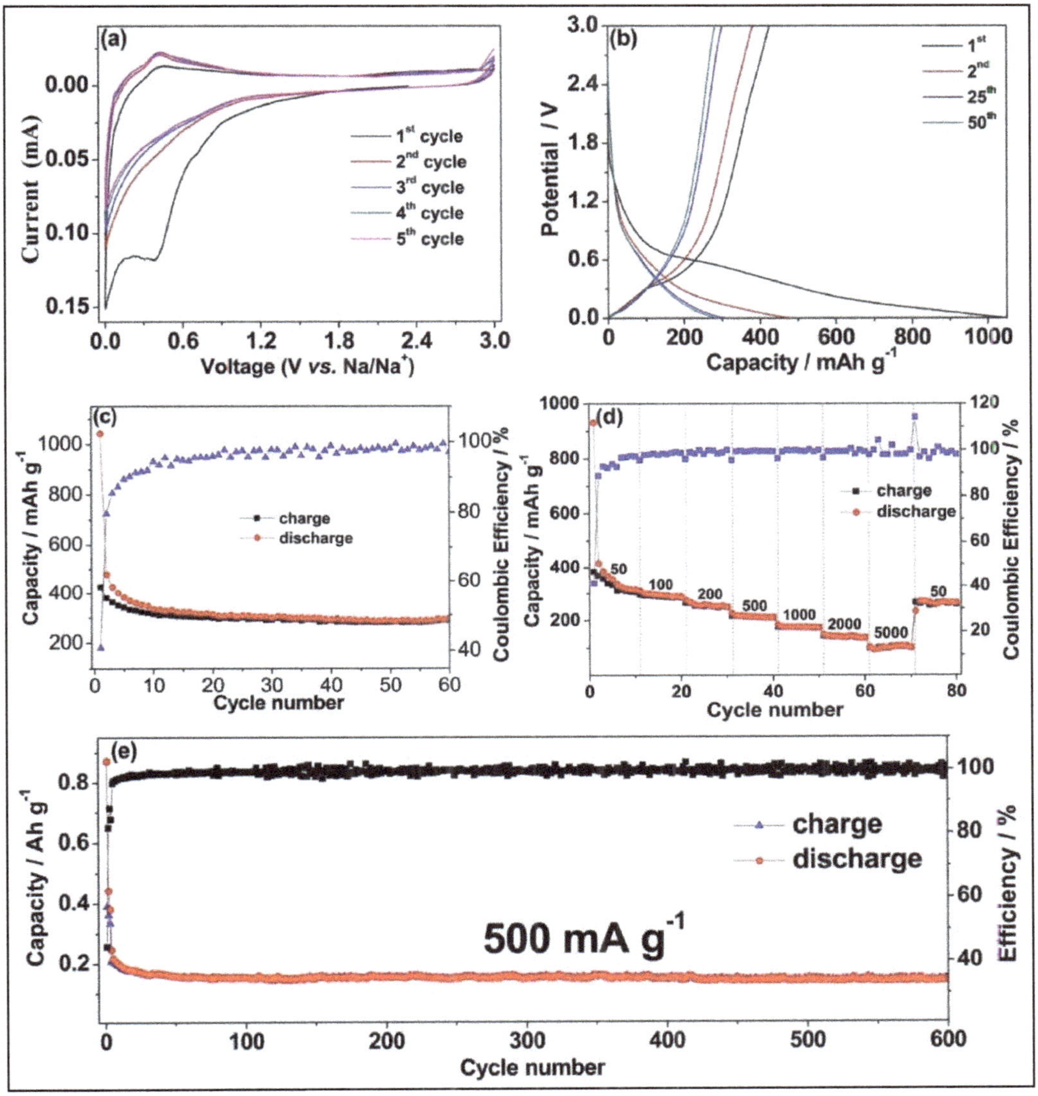

Figure 8. (**a**) CV of GPC electrode between 0.01 and 3.0 V at a scan rate of 0.2 mVs^{-1} for first five cycles. (**b**) The 1st, 2nd, 25th and 50th discharge–charge curves of GPC electrode at a current density of 50 mAg^{-1}. (**c**) Cycle performance of GPC electrode at a current density of 50 mAg^{-1}. (**d**) The rate capability of the GPC electrode. (**e**) Cycle performance of GPC electrodes at a high current density of 500 mAg^{-1} [197].

In contrast, the ultrahigh phosphorus-doped carbon (UPC) anode, tested for SIBs, demonstrated superior electrochemical characteristics as presented in Figure 9 [201]. The introduction of high phosphorus content and the formation of P–(C_3) bonds significantly increased the interlayer spacing of the carbon lattice. This structural modification enhanced Na^+ and K^+ adsorption energy, created more active sites for ion storage, and improved both capacity and rate capability. Consequently, the UPC anode provided an exceptionally high reversible capacity, excellent long-cycle performance, and remarkable rate potential, outperforming GPC and P-doped CNF electrodes [197,201]. These findings make UPC one of the most promising phosphorus-based anode materials for multivalent ion batteries since they demonstrate the critical role that substantial phosphorus doping plays in enhancing ion intercalation kinetics and electrode stability.

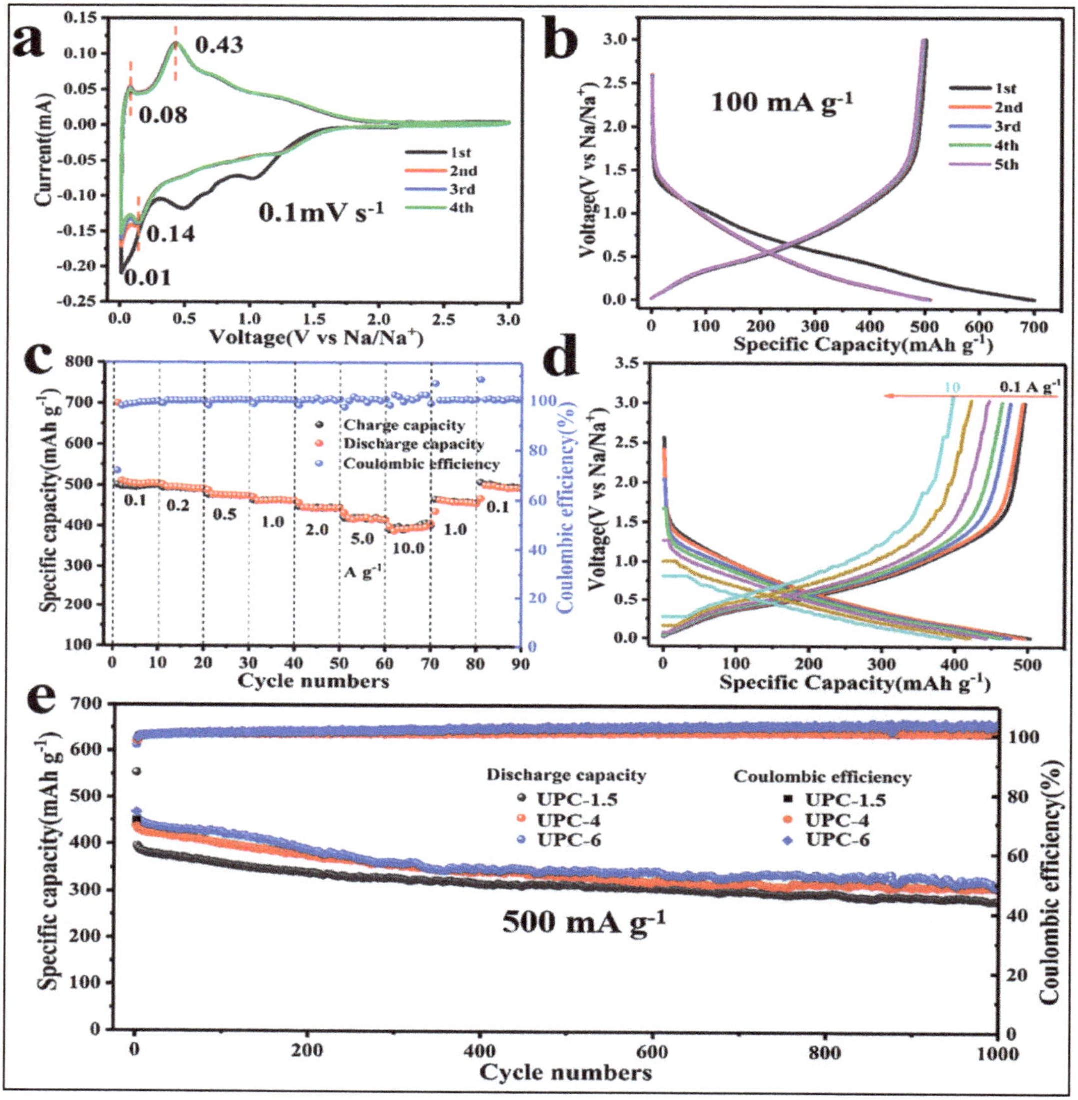

Figure 9. Electrochemical performance of UPCs electrodes (active material mass loading of 1.1–1.4 mg cm^{-2}) in ether electrolyte. (**a**) CV curves of UPC-6 at a scan rate of 0.1 mVs^{-1}, (**b**) charge–discharge curves of UPC-6 at 100 mAg^{-1} for the first five cycles, (**c**) rate capability and (**d**) the corresponding voltage curves of UPC-6, and (**e**) long cycle performance of UPCs at 500 mAg^{-1} [201].

9. Carbon-Based Composite Materials as an Anode for MIBs

Carbon-based composite materials have gained significant attention as anode candidates for MIBs owing to their ability to integrate the structural advantages of carbon with the functional properties of secondary components such as metals, metal oxides, sulfides, and, phosphides [233–236]. By combining carbon's high electrical conductivity, mechanical resilience, and flexible architectures with the high theoretical capacities of active guest materials, these composites effectively address limitations such as sluggish ion diffusion, poor cycling stability, and large volume expansion commonly observed in MIB anode [237,238]. Moreover, carbon matrices including graphene, CNTs, amorphous carbon, and porous carbon frameworks serve as conductive networks that enhance charge transport and buffer structural strain during repeated cycling. Tailoring the composition, morphology, and interfacial chemistry of these composites enables improved ion accessibility, faster charge-transfer kinetics, and enhanced electrode integrity across various MIB (e.g., Li, Na, and Ca-ion batteries). As a result, carbon-based composites represent a versatile and highly tunable class of anode materials with strong potential for advancing high-performance next-generation MIB technologies.

9.1. *Synthesis of Carbon-Based Composite Materials*

Carbon-based composite materials can be synthesized through a range of established methods, including (i) CVD, (ii) hydrothermal techniques, (iii) ball milling, and (iv) arc-discharge approaches, as shown in the schematic below (Figure 10). These diverse methods enable precise control over composition, morphology, and interfacial structures, allowing the resulting composites to be tailored for high-performance MIB applications.

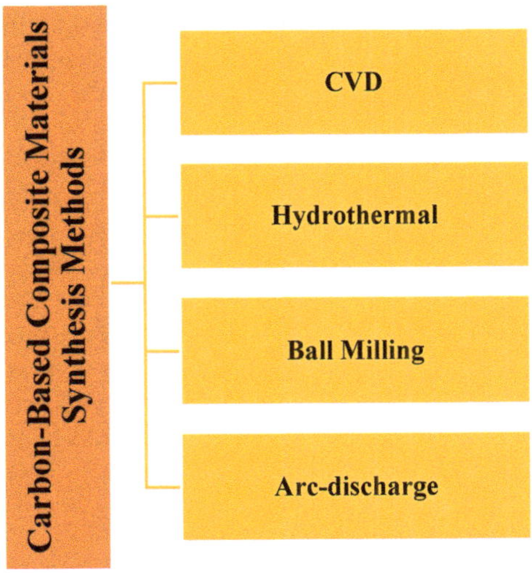

Figure 10. Schematic illustration of the major synthesis methods employed for carbon-based composite materials.

9.1.1. CVD Method

Researchers have extensively explored CVD synthesis for carbon-based composite materials such as silicon-carbon. In CVD process, gaseous precursors are introduced into a tube furnace, where they thermally decompose to generate reactive atomic or molecular species that subsequently deposit onto a substrate surface. This technique is widely used to synthesize graphite-like carbon coatings, carbon-coated silicon, and carbon nanotubes, owing to its precise control over film thickness, uniformity, and composition. CVD can also be employed to deposit silicon using precursors such as SiH_4 or $SiHCl_3$, enabling the fabrication of composite structures tailored for high-performance MIB anode [239]. Jin et al. [240] prepared core–shell structure of Si/C composite via CVD, and showed outstanding electrochemical performance; the composition delivers a high initial capacity (1971 mAh/g), good coulombic efficiency (67.1%), and retains 1441 mAh/g after 100 cycles.

The Si/C composite provides a flexible, conductive network that buffers silicon's volume changes and reduces side reactions, resulting in improved cycling stability and rate capability. Ko and co-workers [241] produced silicon–graphite–carbon (SGC) hybrid anode are engineered with ultrathin Si nanolayers uniformly deposited within the carbon matrix using CVD process, enabling controlled lithiation despite the large volume expansion of Si. In situ TEM studies reveal that these nanolayers lithiate with linear kinetics both in open voids and between graphite sheets, where the lower compressive strength of graphite allows Si to expand with minimal mechanical constraint. This structural synergy maintains electrical integrity, suppresses cracking, and ensures efficient Li transport. As a result, CVD-derived SGC hybrids achieve high volumetric energy density (~1043 Wh/L), rapid stabilization of Coulombic efficiency, and excellent cycling stability highlighting their strong potential for next-generation high-energy LIBs.

9.1.2. Hydrothermal Method

Hydrothermal synthesis has been widely employed to fabricate diverse carbon-based composite materials, including carbon–metal oxide, carbon–metalloid, silicon–carbon, etc. This method has emerged as a highly effective route for producing carbon-based composite materials due to its simplicity, scalability, and ability to precisely tailor structural and chemical features by controlling temperature, pressure, and reaction duration. Using water as a high-temperature, high-pressure solvent, this technique enables the formation of diverse composites for applications in energy storage. It has been successfully applied to fabricate nanostructures such as carbon nanotube/iron oxide hybrids, highlighting its versatility in engineering advanced carbon-based composites [242].

Huyan et al. [243] synthesized carboxyl modification tubular carbon nanofibers and MnO_2 composites (CMTCFs@MNS) through acidification and hydrothermal methods, showed excellent LIB anode performance. Vertically grown δ-MnO_2 nanosheets enhanced ion transport and accommodated volume changes, while CMTCFs ensured electron conductivity and structural stability. Irregular λ-MnO_2 nanoparticles further improved electron transport and lithium storage, making CMTCFs@MNS a promising high-performance electrode material. Similarly, Cao et al. [244] reported dually coated C@SnO_2@C hollow nanospheres, featuring a double carbon shell that effectively buffers SnO_2 volume expansion and enhances conductivity. As an LIB anode, they exhibit excellent cycling stability (78.7% retention over 300 cycles) and high-rate performance, with >400 mAh g^{-1} after 10,000 cycles, highlighting this design as a promising strategy for electrodes prone to large volume changes.

9.1.3. Ball Milling Method

Ball milling is a versatile top-down technique for producing nanomaterials, including metallic, alloyed, multi-metallic, and ceramic composites. It provides a simple, cost-effective, and high-throughput approach, making it efficient for large-scale nanocomposite fabrication. Various types of ball mills exist, including those classified by mechanical energy, operating mode, or wettability, as well as advanced techniques such as plasma-assisted milling (P-milling), microwave-assisted ball milling, electrical discharge milling, magnetic field-induced ball milling, and ultrasonic milling [245–251]. Zhang et al. [252] developed SnSe/C nanocomposites using a simple, low-cost high-energy ball milling method using Sn powder, Se powder, and carbon black as precursors. The process successfully produced SnSe uniformly mixed with carbon black at the nanoscale (50–80 nm), with amorphous carbon serving as a buffering matrix. When employed as anode materials for LIBs and SIBs, the SnSe/C nanocomposite exhibited enhanced electrochemical performance, including high capacity, long-term cycling stability, and good rate capability. In SIBs, an initial capacity of 748.5 mAh/g was obtained, with 324.9 mAh/g maintained at a high current density of 500 mA/g after 200 cycles, corresponding to 72.5% retention of the second-cycle capacity (447.7 mAh/g). In LIBs, the nanocomposite delivered a high initial capacity of ~1097.6 mAh/g, which decreased to 633.1 mAh/g after 100 cycles at 500 mA/g. The improved performance is attributed to the uniform dispersion of active SnSe nanoparticles within the carbon matrix, which buffers volume changes and enhances electron/ion transport, highlighting the potential of SnSe/C composites as high-performance anode for both sodium and lithium storage.

9.1.4. Arc-Discharge Method

The arc discharge method enables the synthesis of carbon-based nanocomposites by generating a high-temperature electric arc between graphite electrodes in an inert or vacuum environment. This vaporizes the electrodes, forming nanostructures such as carbon nanotubes with embedded metal nanoparticles. While the method produces high-quality composites, it often generates mixed carbon forms that require post-synthesis purification. Charinpanitkul et al. [253] reported a single-step arc discharge synthesis of copper–carbon nanocomposites in liquid nitrogen, producing multi-shelled carbon nanocapsules (70–150 nm) with uniformly embedded copper nanoparticles. TEM and spectroscopic analyses confirmed the nanostructure, and partial oxidation resulted in cuprite formation. BET measurements indicated high specific surface areas, highlighting the effectiveness of this approach for metal–carbon nanocomposite fabrication. Similarly, Rivani et al. [254] synthesized Fe_3O_4/C and TiO_2/Fe_3O_4/C nanocomposites via arc discharge in 50% ethanol. Vibrating Sample Magnetometer (VSM) analysis showed that TiO_2/Fe_3O_4/C exhibited higher saturation magnetization than Fe_3O_4/C, likely due to stronger interactions of carbon with the TiO_2 lattice, while both composites retained superparamagnetic behavior.

9.2. Carbon-Based Composite Anode for LIBs

Graphite, despite its widespread use and stable cycling performance, offers a limited theoretical capacity of 372 mAh/g, which is insufficient to meet the growing demand for high-energy-density LIBs. In contrast, high-capacity anode materials such as metal oxides, metals/metalloids, and metal phosphides/sulfides/nitrides (P, S, N) provide significantly greater lithium storage, which are 2–5 times larger than the graphite anode but suffer from severe volume expansion during cycling, leading to rapid capacity fading and structural degradation [234]. To overcome these limitations, extensive research has focused on integrating these high-capacity materials with carbon. Carbon-based composites effectively buffer volume changes, enhance electrical conductivity, and improve structural stability. As a result, these hybrid anode exhibit improved cycling stability, higher reversible capacity, and better rate capability, making them strong candidates for next-generation LIB anode [255–257]. Among different composite designs, embedding Si within conductive and mechanically robust matrices has proven particularly effective. Carbon materials are especially attractive in this regard, owing to their excellent electrical conductivity, mechanical strength, and minimal volume change during cycling [256]. A wide range of carbon forms, including graphite [258], expanded graphite [259], graphene [260], CNTs [261], and amorphous carbon [262], have been employed to create Si/carbon (Si/C) composites with diverse architectures such as core–shell, yolk–shell, honeycomb, sandwich structures, as well as morphologies like spheres, fibers, films, and 3D frameworks [263–266]. Notably, natural graphite is frequently utilized as a matrix in graphite/Si composites due to its superior conductivity, high ICE, and limited volume expansion. These elements improve the electrochemical and mechanical performance of Si anode. To date, common fabrication methods for graphite/Si composites include high-energy mechanical milling, liquid solidification, spray drying, and CVD [267,268]. Li et al. [269] developed a mesoporous Si@amorphous carbon/graphite (Si@C/G) composite through a combination of chemical etching, ball milling, and subsequent annealing process. The mesoporous Si obtained from acid-etched Al–Si alloy buffered volume expansion and promoted Li^+ transport, while the combination with graphite and amorphous carbon improved electronic conductivity. The resulting anode delivered 907.9 mAh/g capacity with a coulombic efficiency of 78.4% and stable cycling over 150 cycles. Although mechanical milling is scalable and low-cost, it often damages particle surfaces, leading to side reactions. Furthermore, carbon-encapsulated structures have emerged as highly effective anode materials because the carbon shell provides a robust barrier against particle aggregation, while the inner nanostructure often created through acid etching accommodates volume changes during cycling. For example, SnO_2/C composites demonstrate excellent lithium-storage capability, delivering a discharge capacity of 745.7 mAh/g at 0.1 A/g after 150 cycles, along with strong cycling stability. These results highlight the suitability of carbon-coated metal oxides and sulfides as high-performance LIB anode [270].

Similarly, encapsulating Fe_3O_4 within tubular mesoporous carbon produces a composite with enhanced conductivity and cycling durability. The carbon matrix not only buffers volume expansion elastically but also provides continuous, short pathways for rapid electron and ion transport while maintaining intimate contact with Fe_3O_4 nanoparticles. This optimized structure leads to outstanding electrochemical performance, including a reversible capacity of about 800 mAh/g at 2 A/g after 1000 cycles. Together, these examples underscore the effectiveness of nanostructured, carbon-coated metal oxide composites for next-generation LIB anode [271].

9.3. Carbon-Based Composite Anode for SIBs

Efforts to advance carbon-based anode for SIBs have increasingly centered on developing carbon-based composites, where carbon is combined with other functional materials to overcome the inherent limits of single-component electrodes [272,273]. This approach capitalizes on carbon's excellent electrical conductivity, mechanical durability, thermal stability, and low cost, while addressing challenges such as limited theoretical capacity and sluggish ion transport [274]. By integrating materials with higher sodium-storage capability, these composites improve ion diffusion, enhance electronic pathways, and maintain structural integrity during cycling, resulting in higher capacity and better long-term stability [275].

Because of their low cost, ease of fabrication, and strong electrochemical performance, carbon-based composites have emerged as a key direction for developing practical SIB anode, especially for large-scale energy storage. Tailoring their composition, porosity, and nanoscale architecture can significantly boost energy density, extend cycle life, and improve safety [276,277]. As a result, the most effective SIB anode today are typically built on composite systems such as carbon–carbon, carbon–alloy, and carbon conversion types (including carbon–metal oxides and carbon–metal sulfides) which collectively offer promising routes toward high-performance SIBs [278]. Among them carbon-alloying is one of the important class of carbon-based composite which integrate with group IV or V elements with carbon matrices [279]. While alloy-type anode generally suffer from severe volume expansion upon Na^+ insertion in SIBs, leading to electrode pulverization and poor cycling stability, coupling them with carbon significantly mitigates this issue [280]. For example, Zhu et al. [281] fabricated a carbon-alloy composites by electrodepositing a tin (Sn) thin film on a conductive wood fiber (Sn@WF), where the Sn@WF electrode maintained structural integrity after cycling. The soft wood fibers accommodate volume expansion during sodiation, while their mesoporous network acts as an electrolyte reservoir, enabling efficient ion transport along both inner and outer surfaces. These advantages are supported by experimental and computational results. The Sn@WF anode delivers stable cycling for 400 cycles with an initial capacity of 339 mAh/g, outperforming many reported Sn nanostructures. This demonstrates that low-cost wood fibers provide a promising platform for advanced SIB anode. Similarly, Nithya and Gopukumar [282] developed Sb–rGO nanosheet composites using GO (prepared via a modified Hummers method) and $SbCl_3$ precursors, with $NaBH_4$ as the reducing agent. Compared to pure Sb and rGO electrodes, the Sb–rGO composite exhibited enhanced electrochemical performance, achieving an ICE of 88% and a reversible capacity of 598 mAh/g at 131 mA/g after 50 cycles. The improved stability was attributed to the rGO matrix, which effectively buffered the volume changes of Sb nanoparticles during cycling. These results highlight the potential of carbon matrices as robust hosts to alleviate volume expansion in metal–carbon alloy composites, thereby improving their electrochemical performance.

9.4. Carbon-Based Composite Anode for KIBs

While carbon remains the most widely used anode material, theoretical studies indicate that several Group IVA and VA elements and their compounds could store significantly larger amounts of K^+-ion storage. Their elevated specific capacities highlight their potential as next-generation anode materials for KIBs [283–288]. Although elements such as P and Si offer very high theoretical storage capacities, their inherently low electrical conductivity severely restricts their practical performance in KIBs [289,290]. As a result, bringing alloy-type anode into real-world KIBs remains challenging. One of the most persistent issues is their rapid loss of capacity during cycling. This degradation is largely driven by slow reaction kinetics caused by the disappearance of short diffusion pathways and by structural pulverization of the active material due to dramatic volume changes [291]. Because K^+- is considerably larger than Li^+ or Na^+, the mechanical strain generated during repeated potassiation and depotassiation is much more intense. This makes it difficult for the electrode to maintain structural integrity and electrical continuity, leading to accelerated capacity decay and poor long-term stability. During cycling, fresh alloy surfaces continuously emerge and react with the electrolyte, leading to repeated SEI formation. This ongoing SEI regeneration restricts charge transfer and consumes active ions, ultimately accelerating performance degradation. As a result, the inherently low ICE caused by the instability and constant reconstruction of the SEI remains a major challenge for KIB alloy anode. To overcome these fundamental limitations, substantial research efforts have focused on redesigning alloy-based electrodes, and notable improvements have been realized. A widely adopted approach involves integrating alloy materials into highly conductive carbon scaffolds such as graphite, porous carbons, or hard carbon. These carbon hosts provide mechanical cushioning for large volume changes, shorten ion-diffusion paths, and improve electronic conductivity, collectively mitigating rapid capacity loss. By coupling high-capacity alloy materials with resilient carbon matrices, researchers have developed carbon–alloy composite anode that show significantly enhanced structural stability and electrochemical performance in KIBs. For example, Sultana and colleagues [291] were the first to develop a Sn–C composite anode via ball milling 70 wt% Sn powder with 30 wt% graphite. The resulting material features Sn nanoparticles uniformly dispersed within an amorphous carbon matrix. Electrochemical evaluation demonstrated that the Sn–C composite functions efficiently as a negative electrode for KIBs, operating within a voltage range of 2.00–0.01 V vs. K/K^+ and achieving a reversible capacity of around 150 mAh/g. XRD measurements showed that crystalline phases form during potassiation and partially disappear during depotassiation, indicating that alloying and dealloying reactions occur within the Sn component. These results broaden the scope of materials suitable for KIB anode and suggest that further investigation of potassium-alloying electrode materials is warranted. Similar to Sn, antimony (Sb) is well known for its alloying behavior with Li and Na, and it can form K_3Sb when alloyed with potassium in KIBs. However, Sb undergoes a large volume expansion of approximately 407%, leading to rapid capacity decay. To address these issues, Han et al. [292] developed Sb nanoparticles confined within a honeycomb-like 3D carbon framework (3D SbNPs@C) using a straightforward freeze-drying method followed by carbothermic reduction, representing its first use as a KIB anode. The robust confinement of Sb nanoparticles within the carbon network stabilizes the electrode during repeated potassiation and depotassiation cycles. Moreover, the conductive carbon matrix improves electron transport and accommodates volume expansion, maintaining the structural integrity of the electrode. In comparison to commercial Sb, the 3D SbNPs@C anode delivers a high reversible capacity of approximately 478 mAh/g at 200 mA/g and excellent rate performance of ~288 mAh/g at 1000 mA/g. Mechanistic investigations show a two-step alloying process, forming an intermediate K_xSb phase before converting to the final K_3Sb phase during cycling. This strategy demonstrates an effective approach for designing alloy-carbon composite anode, offering potential for high-performance KIBs.

10. Summary and Future Prospects

Carbon-based materials have emerged as versatile and high-performance anode for MIBs, offering an exceptional balance of electrical conductivity, structural tunability, cost-effectiveness, and electrochemical stability. Their diverse allotropes, ranging from graphite and graphene to nanotubes, nanofibers, and porous frameworks, allow tailored ion storage and transport properties that address the limitations of other anode materials, such as Si, alloy-type, and transition-metal oxide electrodes.

Furthermore, chemical modifications, including heteroatom doping and hybridization, expand the functionality of carbon materials, enabling enhanced capacity, faster ion diffusion, and broader compatibility with various metal-ion chemistries, including Li, Na, K, and Ca. Despite these advantages, challenges remain in achieving higher specific capacities, optimizing ion transport in complex carbon structures, and developing scalable, cost-effective synthesis routes. Looking forward, the integration of artificial intelligence (AI), machine learning (ML), and advanced computational simulations with experimental strategies promises to accelerate the rational design of carbon anode (Figure 11).

These approaches can predict optimal structures, guide targeted doping, and improve understanding of ion–carbon interactions at multiple scales. Combined with sustainable synthesis techniques, such as biomass-derived carbons and 3D printing, these strategies will enable the production of environmentally friendly, high-performance anode tailored for next-generation energy storage applications.

In summary, carbon-based anode are poised to remain at the forefront of MIB research and development. Their inherent versatility, coupled with advances in computational design and green fabrication, offers a promising path toward high-capacity, long-lasting, and commercially viable batteries, paving the way for sustainable energy storage solutions that can meet the growing global demand.

Despite substantial progress in developing carbon-based anode for MIBs, several challenges remain, such as limited specific capacities compared to alloy or conversion-type materials, sluggish ion diffusion in certain carbon structures, and issues with large-scale, cost-effective synthesis. To overcome these limitations, future research should focus on rational design strategies, advanced fabrication techniques, and predictive modeling approaches.

One of the most promising directions lies in the integration of AI and ML with computational simulations. Traditional trial-and-error experimental methods are time-consuming and resource-intensive. By contrast, AI-driven algorithms can screen vast chemical and structural spaces to identify optimal carbon architectures (e.g., biomass-derived, hard carbon, or heteroatom-doped frameworks) with desirable electrochemical properties. For instance, ML models trained on large datasets can rapidly predict parameters such as ion adsorption energies, diffusion barriers, and electronic conductivity, enabling accelerated material discovery.

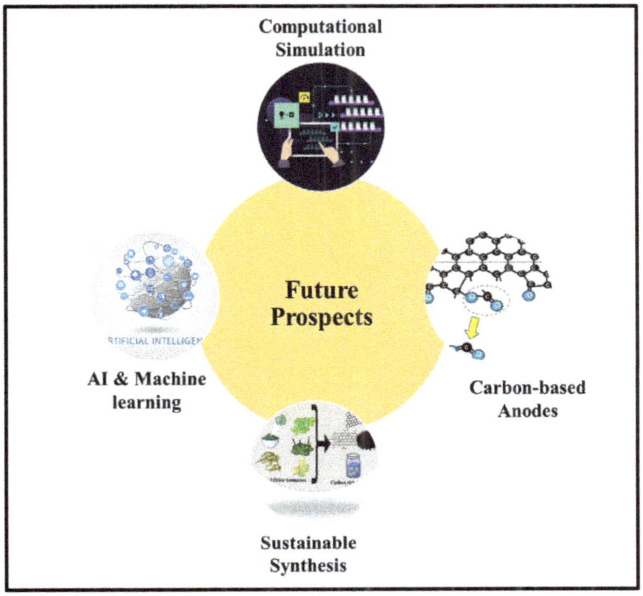

Figure 11. Roadmap for advancing carbon-based anode in MIBs through AI-driven design, computational modeling and green synthesis strategies.

Additionally, first-principles simulations (e.g., density functional theory) and molecular dynamics (MD) studies provide atomistic insights into ion–carbon interactions, charge transfer processes, and structural stability during cycling. Coupling these simulations with AI-based optimization can help design carbon matrices that minimize volume expansion, enhance ion mobility, and improve long-term durability. Emerging methods like high-throughput computational screening and automated materials informatics will likely guide the synthesis of novel doped or hybridized carbon systems tailored for specific metal ions (Li^+, Na^+, K^+, and Ca^{2+}).

Moreover, multi-scale modeling approaches bridging atomic, mesoscopic, and device levels will play a critical role in linking fundamental material properties with practical battery performance metrics such as energy density, rate capability, and cycle life. Integration of such computational insights with experimental validation will accelerate the development of carbon-based anode optimized for diverse applications, from portable electronics to large-scale grid storage.

In parallel, scalable and sustainable synthesis routes, such as biomass-derived carbons, 3D printing, and plasma-assisted methods, are expected to gain momentum. Combining these eco-friendly fabrication methods with computationally guided design will pave the way toward low-cost, high-performance, and environmentally benign anode.

In conclusion, the future of carbon-based anode for MIBs lies in a synergistic approach: leveraging advances in AI, computational modeling, and green synthesis to design next-generation anode with superior capacity, stability, and scalability. Such efforts will be pivotal in addressing the growing global energy demand and advancing the commercialization of sustainable energy storage technologies.

Funding: Author acknowledge the funding by ASGC NASA grant (Award No.: 80NSSC20M0106).

Data Availability Statement: No new data were created or analyzed in this study.

Conflicts of Interest: The authors declare no conflict of interest.

References

1. Amiri, N.; Yacoubi, M.; Messouli, M. Population projections, food consumption, and agricultural production are used to optimize agriculture under climatic constraints. In *Intelligent Solutions for Optimizing Agriculture and Tackling Climate Change: Current and Future Dimensions*; IGI Global: Hershey, PA, USA, 2023; pp. 169–192.
2. Shafiee, S.; Topal, E. When will fossil fuel reserves be diminished? *Energy Policy* **2009**, *37*, 181–189. [CrossRef]
3. Archer, D.; Eby, M.; Brovkin, V.; Ridgwell, A.; Cao, L.; Mikolajewicz, U.; Caldeira, K.; Matsumoto, K.; Munhoven, G.; Montenegro, A.; et al. Atmospheric Lifetime of Fossil Fuel Carbon Dioxide. *Annu. Rev. Earth Planet. Sci.* **2009**, *37*, 117–134. [CrossRef]
4. Lamb, W.F.; Wiedmann, T.; Pongratz, J.; Andrew, R.; Crippa, M.; Olivier, J.G.J.; Wiedenhofer, D.; Mattioli, G.; Al Khourdajie, A.; House, J.; et al. A review of trends and drivers of greenhouse gas emissions by sector from 1990 to 2018. *Environ. Res. Lett.* **2021**, *16*, 073005. [CrossRef]
5. Hu, C.G.; Xiao, Y.; Zou, Y.Q.; Dai, L.M. Carbon-Based Metal-Free Electrocatalysis for Energy Conversion, Energy Storage, and Environmental Protection. *Electrochem. Energy Rev.* **2018**, *1*, 84–112; Correction in *Electrochem. Energy Rev.* **2018**, *1*, 238. [CrossRef]
6. Ellabban, O.; Abu-Rub, H.; Blaabjerg, F. Renewable energy resources: Current status, future prospects and their enabling technology. *Renew. Sustain. Energy Rev.* **2014**, *39*, 748–764. [CrossRef]
7. Winter, M.; Brodd, R.J. What are batteries, fuel cells, and supercapacitors? *Chem. Rev.* **2004**, *104*, 4245–4270. [CrossRef]
8. Zhao, G.W.; Li, W.C.; Ju, M.H.; Liu, Y.; Rao, Q.Q.; Zhou, K.; Ai, F.R. Nitrogen-Rich Multilayered Porous Carbon for an Efficient and Stable Anode. *J. Electron. Mater.* **2021**, *50*, 1002–1009. [CrossRef]
9. Yang, L.; Xi, G. Preparation and electrochemical performance of $LiNi_{1/3}Co_{1/3}Mn_{1/3}O_2$ cathode materials for lithium-ion batteries from spent mixed alkaline batteries. *J. Electron. Mater.* **2016**, *45*, 301–306. [CrossRef]
10. Zhang, J.; Zheng, C.; Li, L.; Xia, Y.; Huang, H.; Gan, Y.; Liang, C.; He, X.; Tao, X.; Zhang, W. Unraveling the intra and intercycle interfacial evolution of Li6PS5Cl-based all-solid-state lithium batteries. *Adv. Energy Mater.* **2020**, *10*, 1903311. [CrossRef]
11. Jin, Y.; Zhu, B.; Lu, Z.; Liu, N.; Zhu, J. Challenges and recent progress in the development of Si anode for lithium-ion battery. *Adv. Energy Mater.* **2017**, *7*, 1700715. [CrossRef]
12. Gopinadh, S.V.; Phanendra, P.V.R.L.; Anoopkumar, V.; John, B.; Mercy, T.D. Progress, challenges, and perspectives on alloy-based anode materials for lithium ion battery: A mini-review. *Energy Fuels* **2024**, *38*, 17253–17277. [CrossRef]
13. Zhao, Y.; Li, X.; Yan, B.; Xiong, D.; Li, D.; Lawes, S.; Sun, X. Recent developments and understanding of novel mixed transition-metal oxides as anode in lithium ion batteries. *Adv. Energy Mater.* **2016**, *6*, 1502175. [CrossRef]
14. Xiao, J.; Han, J.; Zhang, C.; Ling, G.; Kang, F.; Yang, Q.H. Dimensionality, function and performance of carbon materials in energy storage devices. *Adv. Energy Mater.* **2022**, *12*, 2100775. [CrossRef]
15. Huang, H.; Shi, H.; Das, P.; Qin, J.; Li, Y.; Wang, X.; Su, F.; Wen, P.; Li, S.; Lu, P. The chemistry and promising applications of graphene and porous graphene materials. *Adv. Funct. Mater.* **2020**, *30*, 1909035. [CrossRef]
16. Sun, Z.; Fang, S.; Hu, Y.H. 3D graphene materials: From understanding to design and synthesis control. *Chem. Rev.* **2020**, *120*, 10336–10453. [CrossRef] [PubMed]
17. Sarkar, S.; Roy, S.; Hou, Y.; Sun, S.; Zhang, J.; Zhao, Y. Recent progress in amorphous carbon-based materials for anode of sodium-ion batteries: Synthesis strategies, mechanisms, and performance. *ChemSusChem* **2021**, *14*, 3693–3723. [CrossRef] [PubMed]
18. Wang, H.; Shao, Y.; Mei, S.; Lu, Y.; Zhang, M.; Sun, J.-k.; Matyjaszewski, K.; Antonietti, M.; Yuan, J. Polymer-derived heteroatom-doped porous carbon materials. *Chem. Rev.* **2020**, *120*, 9363–9419. [CrossRef] [PubMed]
19. Hu, C.; Dai, L. Doping of carbon materials for metal-free electrocatalysis. *Adv. Mater.* **2019**, *31*, 1804672. [CrossRef]
20. Zhu, Y.Y.; Wang, Y.H.; Wang, Y.T.; Xu, T.J.; Chang, P. Research progress on carbon materials as negative electrodes in sodium-and potassium-ion batteries. *Carbon Energy* **2022**, *4*, 1182–1213. [CrossRef]
21. Abramova, E.N.; Bobyleva, Z.V.; Drozhzhin, O.A.; Abakumov, A.M.; Antipov, E.V. Hard carbon as anode material for metal-ion batteries. *Russ. Chem. Rev.* **2024**, *93*, RCR5100. [CrossRef]
22. Gong, Y.; Xue, Y.-H. Carbon nanomaterials for stabilizing zinc anode in zinc-ion batteries. *New Carbon Mater.* **2023**, *38*, 438–454. [CrossRef]
23. Zou, G.; Hou, H.; Foster, C.W.; Banks, C.E.; Guo, T.; Jiang, Y.; Zhang, Y.; Ji, X. Advanced Hierarchical Vesicular Carbon Co-Doped with S, P, N for High-Rate Sodium Storage. *Adv. Sci.* **2018**, *5*, 1800241. [CrossRef]
24. Stevens, D.; Dahn, J. High capacity anode materials for rechargeable sodium-ion batteries. *J. Electrochem. Soc.* **2000**, *147*, 1271. [CrossRef]

25. Wang, G.; Yu, M.; Feng, X. Carbon materials for ion-intercalation involved rechargeable battery technologies. *Chem. Soc. Rev.* **2021**, *50*, 2388–2443. [CrossRef]
26. Qie, L.; Chen, W.; Xiong, X.; Hu, C.; Zou, F.; Hu, P.; Huang, Y. Sulfur-doped carbon with enlarged interlayer distance as a high-performance anode material for sodium-ion batteries. *Adv. Sci.* **2015**, *2*, 1500195. [CrossRef]
27. Wen, Y.; He, K.; Zhu, Y.; Han, F.; Xu, Y.; Matsuda, I.; Ishii, Y.; Cumings, J.; Wang, C. Expanded graphite as superior anode for sodium-ion batteries. *Nat. Commun.* **2014**, *5*, 4033. [CrossRef]
28. Yoshino, A.; Sanechika, K.; Nakajima, T. Secondary Battery. U.S. Patent 4,668,595, 26 May 1987.
29. Li, Y.; Lu, Y.; Adelhelm, P.; Titirici, M.-M.; Hu, Y.-S. Intercalation chemistry of graphite: Alkali metal ions and beyond. *Chem. Soc. Rev.* **2019**, *48*, 4655–4687. [CrossRef] [PubMed]
30. Aurbach, D.; Markovsky, B.; Shechter, A.; Ein-Eli, Y.; Cohen, H. A comparative study of synthetic graphite and Li electrodes in electrolyte solutions based on ethylene carbonate-dimethyl carbonate mixtures. *J. Electrochem. Soc.* **1996**, *143*, 3809. [CrossRef]
31. Bhattacharyya, S.; Subramanyam, S. Properties, and Applications. In *Electrical and Optical Polymer Systems: Fundamentals: Methods, and Application*; CRC Press: Boca Raton, FL, USA, 1998; p. 201.
32. Tian, L.; Zhuang, Q.; Li, J.; Shi, Y.; Chen, J.; Lu, F.; Sun, S. Mechanism of intercalation and deintercalation of lithium ions in graphene nanosheets. *Chin. Sci. Bull.* **2011**, *56*, 3204–3212. [CrossRef]
33. Chen, S.; Qiu, L.; Cheng, H.-M. Carbon-based fibers for advanced electrochemical energy storage devices. *Chem. Rev.* **2020**, *120*, 2811–2878. [CrossRef]
34. Kothandam, G.; Singh, G.; Guan, X.; Lee, J.M.; Ramadass, K.; Joseph, S.; Benzigar, M.; Karakoti, A.; Yi, J.; Kumar, P. Recent advances in carbon-based electrodes for energy storage and conversion. *Adv. Sci.* **2023**, *10*, 2301045. [CrossRef]
35. Xu, J.; Dou, Y.; Wei, Z.; Ma, J.; Deng, Y.; Li, Y.; Liu, H.; Dou, S. Recent progress in graphite intercalation compounds for rechargeable metal (Li, Na, K, Al)-ion batteries. *Adv. Sci.* **2017**, *4*, 1700146. [CrossRef] [PubMed]
36. Luo, P.; Zheng, C.; He, J.; Tu, X.; Sun, W.; Pan, H.; Zhou, Y.; Rui, X.; Zhang, B.; Huang, K. Structural engineering in graphite-based metal-ion batteries. *Adv. Funct. Mater.* **2022**, *32*, 2107277. [CrossRef]
37. Deng, T.; Zhou, X. The preparation of porous graphite and its application in lithium ion batteries as anode material. *J. Solid State Electrochem.* **2016**, *20*, 2613–2618. [CrossRef]
38. Li, Z.; Peng, K.; Ji, N.; Zhang, W.; Tian, W.; Gao, Z. Advanced mechanisms and applications of microwave-assisted synthesis of carbon-based materials: A brief review. *Nanoscale Adv.* **2025**, *7*, 419–432. [CrossRef] [PubMed]
39. Amini, A.; Latifi, M.; Chaouki, J. Electrification of materials processing via microwave irradiation: A review of mechanism and applications. *Appl. Therm. Eng.* **2021**, *193*, 117003. [CrossRef]
40. Krawczyk, P.; Gurzęda, B.; Bachar, A.; Buchwald, T. Formation of a N_2O_5—Graphite intercalation compound by ozone treatment of natural graphite. *Green Chem.* **2020**, *22*, 5463–5469. [CrossRef]
41. Gopalakrishnan, V.; Sundararajan, A.; Omprakash, P.; Panemangalore, D.B. Energy Storage through Graphite Intercalation Compounds. *J. Electrochem. Soc.* **2021**, *168*, 040541. [CrossRef]
42. Kumar, R.; Sahoo, S.; Joanni, E.; Singh, R.K.; Kar, K.K. Microwave as a tool for synthesis of carbon-based electrodes for energy storage. *ACS Appl. Mater. Interfaces* **2021**, *14*, 20306–20325. [CrossRef]
43. Wang, Z.; Yu, C.; Huang, H.; Guo, W.; Zhao, C.; Ren, W.; Xie, Y.; Qiu, J. Energy accumulation enabling fast synthesis of intercalated graphite and operando decoupling for lithium storage. *Adv. Funct. Mater.* **2021**, *31*, 2009801. [CrossRef]
44. Priecel, P.; Lopez-Sanchez, J.A. Advantages and limitations of microwave reactors: From chemical synthesis to the catalytic valorization of biobased chemicals. *ACS Sustain. Chem. Eng.* **2018**, *7*, 3–21. [CrossRef]
45. Sumdani, M.; Islam, M.; Yahaya, A.; Safie, S. Recent advances of the graphite exfoliation processes and structural modification of graphene: A review. *J. Nanopart. Res.* **2021**, *23*, 253. [CrossRef]
46. Sengupta, R.; Bhattacharya, M.; Bandyopadhyay, S.; Bhowmick, A.K. A review on the mechanical and electrical properties of graphite and modified graphite reinforced polymer composites. *Prog. Polym. Sci.* **2011**, *36*, 638–670. [CrossRef]
47. Viculis, L.M.; Mack, J.J.; Mayer, O.M.; Hahn, H.T.; Kaner, R.B. Intercalation and exfoliation routes to graphite nanoplatelets. *J. Mater. Chem.* **2005**, *15*, 974–978. [CrossRef]
48. Xu, M.; Xing, L.; Zhang, Q.; Pu, J. Ultrasonic-assisted method of graphite preparation from wheat straw. *Bioresources* **2017**, *12*, 6405–6417. [CrossRef]
49. Mendoza-Duarte, J.; Robles-Hernández, F.C.; Gomez-Esparza, C.D.; Miranda-Hernández, J.G.; Garay-Reyes, C.G.; Estrada-Guel, I.; Martínez-Sánchez, R. Exfoliated graphite preparation based on an eco-friendly mechanochemical route. *J. Environ. Chem. Eng.* **2020**, *8*, 104370. [CrossRef]
50. Ahmed, M.; Ahmed, H.; Adebayo, P.; Carter, R.; Khanal, L. Mechanical properties and application of graphite and graphite-based nanocomposite: A review. *Chem. Mater. Res.* **2023**, *15*, 10–32.
51. Cravotto, G.; Cintas, P. Sonication-assisted fabrication and post-synthetic modifications of graphene-like materials. *Chem. A Eur. J.* **2010**, *16*, 5246–5259. [CrossRef]

52. Zhao, N.; Wang, J.; Ding, Y.; Li, Y. Energy consumption calculation and energy-saving measures of substation based on Multi-objective artificial bee colony algorithm. *Int. J. Emerg. Electr. Power Syst.* **2024**, *25*, 25–34. [CrossRef]
53. Perevalov, Y.; Kozulina, T.; Yermekova, M.; Demidovich, V. Digital Shadow Induction Furnace for Heating Carbon Fibers. In Proceedings of the 2021 IEEE Conference of Russian Young Researchers in Electrical and Electronic Engineering (ElConRus), St. Petersburg, Russia, 26–29 January 2021; pp. 1027–1031.
54. Shen, H.; Yao, Z.; Shi, Y.; Hu, J. Study on temperature field induced in high frequency induction heating. *Acta Metall. Sin. (Engl. Lett.)* **2006**, *19*, 190–196. [CrossRef]
55. Mueller, A.; Ziegler, K.; Amsharov, K.Y.; Jansen, M. *In Situ Synthesis of Chlorinated Fullerenes by the High-Frequency Furnace Method*; Wiley Online Library: Hoboken, NJ, USA, 2011.
56. Li, Z.; Zhao, B.; Liu, P.; Zhao, B.; Chen, D.; Zhang, Y. Synthesis of high-quality single-walled carbon nanotubes by high-frequency-induction heating. *Phys. E Low-Dimens. Syst. Nanostruct.* **2008**, *40*, 452–456. [CrossRef]
57. Zahid, M.; Abuzairi, T. Sustainable graphene production: Flash joule heating utilizing pencil graphite precursors. *Nanomaterials* **2024**, *14*, 1289. [CrossRef]
58. Liu, Y.; Liu, Q.; Gu, J.; Kang, D.; Zhou, F.; Zhang, W.; Wu, Y.; Zhang, D. Highly porous graphitic materials prepared by catalytic graphitization. *Carbon* **2013**, *64*, 132–140. [CrossRef]
59. Li, R.; Zhang, Y.; Chu, X.; Gan, L.; Li, J.; Li, B.; Du, H. Design and Numerical Study of Induction-Heating Graphitization Furnace Based on Graphene Coils. *Appl. Sci.* **2024**, *14*, 2528. [CrossRef]
60. Adham, K.; Bowes, G. Natural graphite purification through chlorination in fluidized bed reactor. In Proceedings of the Extraction 2018: First Global Conference on Extractive Metallurgy, Ottawa, ON, Canada, 26–29 August 2018; Springer: Berlin/Heidelberg, Germany, 2018.
61. Chen, D.; Feng, H.; Li, J. Graphene oxide: Preparation, functionalization, and electrochemical applications. *Chem. Rev.* **2012**, *112*, 6027–6053. [CrossRef] [PubMed]
62. Chen, X.; Qu, Z.; Liu, Z.; Ren, G. Mechanism of oxidization of graphite to graphene oxide by the hummers method. *ACS Omega* **2022**, *7*, 23503–23510. [CrossRef] [PubMed]
63. Silva, W.C.H.; Zafar, M.A.; Allende, S.; Jacob, M.V.; Tuladhar, R. Sustainable synthesis of graphene oxide from waste sources: A comprehensive review of methods and applications. *Mater. Circ. Econ.* **2024**, *6*, 23. [CrossRef]
64. Bannov, A.; Manakhov, A.; Shibaev, A.; Ukhina, A.; Polčák, J.; Maksimovskii, E. Synthesis dynamics of graphite oxide. *Thermochim. Acta* **2018**, *663*, 165–175. [CrossRef]
65. Inagaki, M.; Iwashita, N.; Kouno, E. Potential change with intercalation of sulfuric acid into graphite by chemical oxidation. *Carbon* **1990**, *28*, 49–55. [CrossRef]
66. Lavin-Lopez, M.d.P.; Romero, A.; Garrido, J.; Sanchez-Silva, L.; Valverde, J.L. Influence of different improved hummers method modifications on the characteristics of graphite oxide in order to make a more easily scalable method. *Ind. Eng. Chem. Res.* **2016**, *55*, 12836–12847. [CrossRef]
67. Kuanyshbekov, T.; Akatan, K.; Kabdrakhmanova, S.; Nemkaeva, R.; Aitzhanov, M.; Imasheva, A.; Kairatuly, E. Synthesis of Graphene Oxide from Graphite by the Hummers Method. *Oxid. Commun.* **2021**, *44*, 356.
68. Yadav, N.; Lochab, B. A comparative study of graphene oxide: Hummers, intermediate and improved method. *FlatChem* **2019**, *13*, 40–49. [CrossRef]
69. Bai, L.-Z.; Zhao, D.-L.; Zhang, T.-M.; Xie, W.-G.; Zhang, J.-M.; Shen, Z.-M. A comparative study of electrochemical performance of graphene sheets, expanded graphite and natural graphite as anode materials for lithium-ion batteries. *Electrochim. Acta* **2013**, *107*, 555–561. [CrossRef]
70. Wang, F.; Yi, J.; Wang, Y.; Wang, C.; Wang, J.; Xia, Y. Graphite Intercalation Compounds (GICs): A New Type of Promising Anode Material for Lithium-Ion Batteries. *Adv. Energy Mater.* **2014**, *4*, 1300600. [CrossRef]
71. Liu, Y.; Shi, H.; Wu, Z.-S. Recent status, key strategies and challenging perspectives of fast-charging graphite anode for lithium-ion batteries. *Energy Environ. Sci.* **2023**, *16*, 4834–4871. [CrossRef]
72. Yang, C.; Zhang, X.; Li, J.; Ma, J.; Xu, L.; Yang, J.; Liu, S.; Fang, S.; Li, Y.; Sun, X. Holey graphite: A promising anode material with ultrahigh storage for lithium-ion battery. *Electrochim. Acta* **2020**, *346*, 136244. [CrossRef]
73. Wang, Z.; Li, Y.; Zhou, Q.; Li, Q.; Zhao, R.; Qiu, Z.; Zhang, R.; Sun, Y.; Wu, F.; Wu, C. Multi-ion strategies toward advanced rechargeable batteries: Materials, properties, and prospects. *Energy Mater. Adv.* **2024**, *5*, 0109. [CrossRef]
74. Yuan, F.; Wu, Z.; Li, Z.; Sun, Q.; Wang, Q.; Li, R.; Wang, W.; Zhang, D.; Wang, B. Decoupling KOH Activation Path to Construct Graphitic Porous Carbon Anode for Enhanced Potassium Ion Storage. *Small* **2025**, *21*, 2505910. [CrossRef]
75. Tai, Z.; Zhang, Q.; Liu, Y.; Liu, H.; Dou, S. Activated carbon from the graphite with increased rate capability for the potassium ion battery. *Carbon* **2017**, *123*, 54–61. [CrossRef]
76. Shim, J.-H.; Lee, S. Characterization of graphite etched with potassium hydroxide and its application in fast-rechargeable lithium ion batteries. *J. Power Sources* **2016**, *324*, 475–483. [CrossRef]

77. Mendiboure, A.; Delmas, C.; Hagenmuller, P. Electrochemical intercalation and deintercalation of NaxMnO$_2$ bronzes. *J. Solid State Chem.* **1985**, *57*, 323–331. [CrossRef]
78. Ellis, B.L.; Nazar, L.F. Sodium and sodium-ion energy storage batteries. *Curr. Opin. Solid State Mater. Sci.* **2012**, *16*, 168–177. [CrossRef]
79. Sun, Y.; Wu, Q.; Liang, X.; Xiang, H. Recent developments in carbon-based materials as high-rate anode for sodium ion batteries. *Mater. Chem. Front.* **2021**, *5*, 4089–4106. [CrossRef]
80. Chayambuka, K.; Mulder, G.; Danilov, D.L.; Notten, P.H. From li-ion batteries toward Na-ion chemistries: Challenges and opportunities. *Adv. Energy Mater.* **2020**, *10*, 2001310. [CrossRef]
81. Chen, J.; Chua, D.H.; Lee, P.S. The advances of metal sulfides and in situ characterization methods beyond Li ion batteries: Sodium, potassium, and aluminum ion batteries. *Small Methods* **2020**, *4*, 1900648. [CrossRef]
82. Olsson, E.; Yu, J.; Zhang, H.; Cheng, H.M.; Cai, Q. Atomic-scale design of anode materials for alkali metal (Li/Na/K)-ion batteries: Progress and perspectives. *Adv. Energy Mater.* **2022**, *12*, 2200662. [CrossRef]
83. Xu, J.; Dou, S.; Wang, Y.; Yuan, Q.; Deng, Y.; Chen, Y. Development of metal and metal-based composites anode materials for potassium-ion batteries. *Trans. Tianjin Univ.* **2021**, *27*, 248–268. [CrossRef]
84. Nason, C.A.; Vijaya Kumar Saroja, A.P.; Lu, Y.; Wei, R.; Han, Y.; Xu, Y. Layered potassium titanium niobate/reduced graphene oxide nanocomposite as a potassium-ion battery anode. *Nano-Micro Lett.* **2024**, *16*, 1. [CrossRef]
85. Zhou, J.; Liu, Y.; Zhang, S.; Zhou, T.; Guo, Z. Metal chalcogenides for potassium storage. *InfoMat* **2020**, *2*, 437–465. [CrossRef]
86. Jin, J.; Schwingenschlögl, U. Exploration of the two-dimensional transition metal phosphide MoP$_2$ as anode for Na/K ion batteries. *npj 2D Mater. Appl.* **2024**, *8*, 31. [CrossRef]
87. Wu, X.; Chen, Y.; Xing, Z.; Lam, C.W.K.; Pang, S.S.; Zhang, W.; Ju, Z. Advanced carbon-based anode for potassium-ion batteries. *Adv. Energy Mater.* **2019**, *9*, 1900343. [CrossRef]
88. Yu, J.; Jiang, M.; Zhang, W.; Li, G.; Soomro, R.A.; Sun, N.; Xu, B. Advancements and prospects of graphite anode for potassium-ion batteries. *Small Methods* **2023**, *7*, 2300708. [CrossRef]
89. Zhang, L.; Wang, W.; Lu, S.; Xiang, Y. Carbon anode materials: A detailed comparison between Na-ion and K-ion batteries. *Adv. Energy Mater.* **2021**, *11*, 2003640. [CrossRef]
90. Raghavan, P.; Das, A.; Jabeen Fatima, M.J. *Advanced Technologies for Rechargeable Batteries: Metal Ion, Hybrid, and Metal-Air Batteries*; CRC Press: Boca Raton, FL, USA, 2024.
91. Cerrato, J.M.; Barrows, C.J.; Blue, L.Y.; Lezama-Pacheco, J.S.; Bargar, J.R.; Giammar, D.E. Effect of Ca^{2+} and Zn^{2+} on UO_2 dissolution rates. *Environ. Sci. Technol.* **2012**, *46*, 2731–2737. [CrossRef]
92. Aurbach, D.; Skaletsky, R.; Gofer, Y. The electrochemical behavior of calcium electrodes in a few organic electrolytes. *J. Electrochem. Soc.* **1991**, *138*, 3536. [CrossRef]
93. Ponrouch, A.; Frontera, C.; Bardé, F.; Palacín, M.R. Towards a calcium-based rechargeable battery. *Nat. Mater.* **2016**, *15*, 169–172. [CrossRef] [PubMed]
94. Wang, D.; Gao, X.; Chen, Y.; Jin, L.; Kuss, C.; Bruce, P.G. Plating and stripping calcium in an organic electrolyte. *Nat. Mater.* **2018**, *17*, 16–20. [CrossRef] [PubMed]
95. Richard Prabakar, S.; Ikhe, A.B.; Park, W.B.; Chung, K.C.; Park, H.; Kim, K.J.; Ahn, D.; Kwak, J.S.; Sohn, K.S.; Pyo, M. Graphite as a long-life Ca^{2+}-intercalation anode and its implementation for rocking-chair type calcium-ion batteries. *Adv. Sci.* **2019**, *6*, 1902129. [CrossRef]
96. Iijima, S. Helical microtubules of graphitic carbon. *Nature* **1991**, *354*, 56–58. [CrossRef]
97. Qiu, Z.; Cao, F.; Pan, G.X.; Lid, C.; Chen, M.H.; Zhang, Y.Q.; He, X.P.; Xia, Y.; Xia, X.H.; Zhang, W.K. Carbon materials for metal-ion batteries. *Chemphysmater* **2023**, *2*, 267–281. [CrossRef]
98. Chen, Y.Q.; Kang, Y.Q.; Zhao, Y.; Wang, L.; Liu, J.L.; Li, Y.X.; Liang, Z.; He, X.M.; Li, X.; Tavajohi, N.; et al. A review of lithium-ion battery safety concerns: The issues, strategies, and testing standards. *J. Energy Chem.* **2021**, *59*, 83–99. [CrossRef]
99. Tong, Z.; Lv, C.; Bai, G.D.; Yin, Z.W.; Zhou, Y.; Li, J.T. A review on applications and challenges of carbon nanotubes in lithium-ion battery. *Carbon Energy* **2025**, *7*, e643. [CrossRef]
100. Landi, B.J.; Ganter, M.J.; Cress, C.D.; DiLeo, R.A.; Raffaelle, R.P. Carbon nanotubes for lithium ion batteries. *Energy Environ. Sci.* **2009**, *2*, 638–654. [CrossRef]
101. Xie, X.; Kretschmer, K.; Zhang, J.; Sun, B.; Su, D.; Wang, G. Sn@ CNT nanopillars grown perpendicularly on carbon paper: A novel free-standing anode for sodium ion batteries. *Nano Energy* **2015**, *13*, 208–217. [CrossRef]
102. Zeng, Y.; Zhang, X.; Qin, R.; Liu, X.; Fang, P.; Zheng, D.; Tong, Y.; Lu, X. Dendrite-free zinc deposition induced by multifunctional CNT frameworks for stable flexible Zn-ion batteries. *Adv. Mater.* **2019**, *31*, 1903675. [CrossRef] [PubMed]
103. Khan, N.; Han, G.; Mazari, S.A. Carbon nanotubes-based anode materials for potassium ion batteries: A review. *J. Electroanal. Chem.* **2022**, *907*, 116051. [CrossRef]
104. Szabó, A.; Perri, C.; Csató, A.; Giordano, G.; Vuono, D.; Nagy, J.B. Synthesis methods of carbon nanotubes and related materials. *Materials* **2010**, *3*, 3092–3140. [CrossRef]

105. Journet, C.; Maser, W.K.; Bernier, P.; Loiseau, A.; de La Chapelle, M.L.; Lefrant, D.S.; Deniard, P.; Lee, R.; Fischer, J.E. Large-scale production of single-walled carbon nanotubes by the electric-arc technique. *Nature* **1997**, *388*, 756–758. [CrossRef]
106. Guo, T.; Nikolaev, P.; Rinzler, A.G.; Tomanek, D.; Colbert, D.T.; Smalley, R.E. Self-assembly of tubular fullerenes. *J. Phys. Chem.* **1995**, *99*, 10694–10697. [CrossRef]
107. Hafner, J.H.; Bronikowski, M.J.; Azamian, B.R.; Nikolaev, P.; Rinzler, A.G.; Colbert, D.T.; Smith, K.A.; Smalley, R.E. Catalytic growth of single-wall carbon nanotubes from metal particles. *Chem. Phys. Lett.* **1998**, *296*, 195–202. [CrossRef]
108. Sinnott, S.B.; Andrews, R. Carbon nanotubes: Synthesis, properties, and applications. *Crit. Rev. Solid State Mater. Sci.* **2001**, *26*, 145–249. [CrossRef]
109. José-Yacamán, M.; Miki-Yoshida, M.; Rendón, L.; Santiesteban, J.G. Catalytic growth of carbon microtubules with fullerene structure. *Appl. Phys. Lett.* **1993**, *62*, 657–659. [CrossRef]
110. Teo, K.B.; Singh, C.; Chhowalla, M.; Milne, W.I. Catalytic synthesis of carbon nanotubes and nanofibers. *Encycl. Nanosci. Nanotechnol.* **2003**, *10*, 1–22.
111. Journet, C.; Bernier, P. Production of carbon nanotubes. *Appl. Phys. A Mater. Sci. Process.* **1998**, *67*, 1–9. [CrossRef]
112. Breczko, J.; Wysocka-Żołopa, M.; Grądzka, E.; Winkler, K. Zero-Dimensional carbon nanomaterials for electrochemical energy storage. *ChemElectroChem* **2024**, *11*, e202300752. [CrossRef]
113. Brun, N.; Sakaushi, K.; Yu, L.; Giebeler, L.; Eckert, J.; Titirici, M.M. Hydrothermal carbon-based nanostructured hollow spheres as electrode materials for high-power lithium–sulfur batteries. *Phys. Chem. Chem. Phys.* **2013**, *15*, 6080–6087. [CrossRef] [PubMed]
114. Gogotsi, Y.; Libera, J.A.; Yoshimura, M. Hydrothermal synthesis of multiwall carbon nanotubes. *J. Mater. Res.* **2000**, *15*, 2591–2594. [CrossRef]
115. Gogotsi, Y.; Naguib, N.; Libera, J. In situ chemical experiments in carbon nanotubes. *Chem. Phys. Lett.* **2002**, *365*, 354–360. [CrossRef]
116. Yahyazadeh, A.; Nanda, S.; Dalai, A.K. Carbon nanotubes: A review of synthesis methods and applications. *Reactions* **2024**, *5*, 429–451. [CrossRef]
117. Baig, N.; Kammakakam, I.; Falath, W. Nanomaterials: A review of synthesis methods, properties, recent progress, and challenges. *Mater. Adv.* **2021**, *2*, 1821–1871. [CrossRef]
118. Manafi, S.; Nadali, H.; Irani, H. Low temperature synthesis of multi-walled carbon nanotubes via a sonochemical/hydrothermal method. *Mater. Lett.* **2008**, *62*, 4175–4176. [CrossRef]
119. Moreno, J.M.C.; Swamy, S.S.; Fujino, T.; Yoshimura, M. Carbon nanocells and nanotubes grown in hydrothermal fluids. *Chem. Phys. Lett.* **2000**, *329*, 317–322. [CrossRef]
120. Stevens, M.G.; Subramoney, S.; Foley, H.C. Spontaneous formation of carbon nanotubes and polyhedra from cesium and amorphous carbon. *Chem. Phys. Lett.* **1998**, *292*, 352. [CrossRef]
121. Hsu, W.; Hare, J.; Terrones, M.; Kroto, H.; Walton, D.; Harris, P. Condensed-phase nanotubes. *Nature* **1995**, *377*, 687. [CrossRef]
122. Bai, J.; Hamon, A.-L.; Marraud, A.; Jouffrey, B.; Zymla, V. Synthesis of SWNTs and MWNTs by a molten salt (NaCl) method. *Chem. Phys. Lett.* **2002**, *365*, 184–188. [CrossRef]
123. Zeng, Q.; Li, Z.; Zhou, Y. Synthesis and application of carbon nanotubes. *J. Nat. Gas Chem.* **2006**, *15*, 235–246. [CrossRef]
124. Grobert, N. Carbon nanotubes–becoming clean. *Mater. Today* **2007**, *10*, 28–35. [CrossRef]
125. Novoselova, I.; Oliinyk, N.; Volkov, S.; Konchits, A.; Yanchuk, I.; Yefanov, V.; Kolesnik, S.; Karpets, M. Electrolytic synthesis of carbon nanotubes from carbon dioxide in molten salts and their characterization. *Phys. E Low-Dimens. Syst. Nanostruct.* **2008**, *40*, 2231–2237. [CrossRef]
126. Rusman, E.; Nulu, A.; Sohn, K.Y. N-doped CNT assisted GeO-Ge nanoparticles as a high-capacity and durable anode material for lithium-ion batteries. *RSC Adv.* **2025**, *15*, 28841–28852. [CrossRef]
127. Liu, B.; Sun, X.L.; Liao, Z.Q.; Lu, X.Y.; Zhang, L.; Hao, G.P. Nitrogen and boron doped carbon layer coated multiwall carbon nanotubes as high performance anode materials for lithium ion batteries. *Sci. Rep.* **2021**, *11*, 5633. [CrossRef]
128. Al-Samet, M.A.M.M.; Burgaz, E. Improving the lithium-ion diffusion and electrical conductivity of $LiFePO_4$ cathode material by doping magnesium and multi-walled carbon nanotubes. *J. Alloy Compd.* **2023**, *947*, 169680. [CrossRef]
129. Hoseini, S.A.; Mohajerzadeh, S.; Sanaee, Z. Flaky sputtered silicon MWCNTs core-shell structure as a freestanding binder-free electrode for lithium-ion battery. *Sci. Rep.* **2025**, *15*, 3733. [CrossRef]
130. Zhong, Y.; Deng, K.; Zheng, J.; Zhang, T.T.; Liu, P.; Lv, X.B.; Tian, W.; Ji, J.Y. One-step growth of the interconnected carbon nanotubes/graphene hybrids from cuttlebone-derived bi-functional catalyst for lithium-ion batteries. *J. Mater. Sci. Technol.* **2023**, *149*, 205–213. [CrossRef]
131. Doñoro, A.; Muñoz-Mauricio, A.; Etacheri, V. High-Performance Lithium Sulfur Batteries Based on Multidimensional Graphene-CNT-Nanosulfur Hybrid Cathodes. *Batteries* **2021**, *7*, 26. [CrossRef]
132. Wang, C.; Wu, Y.; Gao, J.; Sun, X.L.; Zhao, Q.; Si, W.Y.; Zhang, Y.; Wang, K.; Zhao, F.H.; Ohsaka, T.; et al. Synergistic Defect Engineering and Interface Stability of Activated Carbon Nanotubes Enabling Ultralong Lifespan All-Solid-State Lithium-Sulfur Batteries. *ACS Appl. Mater. Interfaces* **2023**, *15*, 40496–40507. [CrossRef]

133. Cai, Y.Z.; Chen, M.X.; Cheng, L.F.; Hu, Z.Y.; Guo, S.Y.; Yuan, Y.B.; Ren, S.X.; Yu, Z.X.; Chai, Y.L.; Huang, X. Multidimensional, Superflexible, and Binder-free CNT-rGO/Si Buckypaper as Anode for Lithium-Ion Batteries and Electrochemical Performance. *ACS Appl. Energy Mater.* **2024**, *7*, 9194–9206. [CrossRef]
134. Chen, H.; Wang, C.; Fan, Z.; Hao, L.; Pan, L.J. Facile fabrication of binder-free carbon nanotube-carbon nanocoil hybrid films for anode of lithium-ion batteries. *J. Solid State Electrochem.* **2024**, *28*, 3325–3335. [CrossRef]
135. Kim, J.S.; Baek, I.G.; Nyamaa, O.; Goo, K.M.; Uyanga, N.; Kim, K.S.; Nam, T.H.; Yang, J.H.; Noh, J.P. Flexible, binder-free, freestanding silicon/oxidized carbon nanotubes composite anode for lithium-ion batteries with enhanced electrochemical performance through chemical reduction. *Mater. Sci. Eng. B* **2025**, *313*, 117971. [CrossRef]
136. Kim, J.H.; Kim, S.; Han, J.H.; Seo, S.B.; Choi, Y.R.; Lim, J.; Kim, Y.A. Perspective on carbon nanotubes as conducting agent in lithium-ion batteries: The status and future challenges. *Carbon Lett.* **2023**, *33*, 325–333. [CrossRef]
137. Zhu, S.; Sheng, J.; Chen, Y.; Ni, J.F.; Li, Y. Carbon nanotubes for flexible batteries: Recent progress and future perspective. *Natl. Sci. Rev.* **2021**, *8*, nwaa261. [CrossRef]
138. Bai, X.; Wu, N.N.; Yu, G.C.; Li, T. Recent Advances in Anode Materials for Sodium-Ion Batteries. *Inorganics* **2023**, *11*, 289. [CrossRef]
139. Fan, H.; Huang, Z.; Zhang, S.Y.; Li, Z.Y.; Zhang, D.D.; Jiang, G.D.; Xiong, J.; Yuan, S.D. Co-oxidation GN/CNT 3D network enhances the cathode performance of NVPF@O-GN/CNT sodium-ion battery. *Ionics* **2025**, *31*, 10449–10460. [CrossRef]
140. He, Y.; Chen, T.R.; Zettsu, N. Conductive 3D SW-/MW-CNTs hybrid frameworks for ultra-high-content Prussian white cathodes in sodium-ion batteries. *Mater. Adv.* **2025**, *6*, 6931–6943. [CrossRef]
141. Huang, J.R.; Zhang, Z.H.; Chen, D.Q.; Yu, H.S.; Wu, Y.; Chen, Y.F. Spray-Drying Synthesis of $Na_4Fe_3(PO4)_2P_2O_7$@CNT Cathode for Ultra-Stable and High-Rate Sodium-Ion Batteries. *Molecules* **2025**, *30*, 753. [CrossRef] [PubMed]
142. Li, L.; Li, H.; Liu, L.X.; Yan, X.C.; Long, Y.Z.; Han, W.P. Amorphous FeO Anchored on N-Doped Graphene with Internal Micro-Channels as an Active and Durable Anode for Sodium-Ion Batteries. *Nanomaterials* **2024**, *14*, 937. [CrossRef]
143. Liu, Z.G.; Lu, Z.Y.; Guo, S.H.; Yang, Q.H.; Zhou, H.S. Toward High Performance Anode for Sodium-Ion Batteries: From Hard Carbons to Anode-Free Systems. *ACS Cent. Sci.* **2023**, *9*, 1076–1087. [CrossRef]
144. Jia, Q.X.; Li, Z.Y.; Ruan, H.L.; Luo, D.W.; Wang, J.J.; Ding, Z.Y.; Chen, L.N. A Review of Carbon Anode Materials for Sodium-Ion Batteries: Key Materials, Sodium-Storage Mechanisms, Applications, and Large-Scale Design Principles. *Molecules* **2024**, *29*, 4331. [CrossRef]
145. Rehman, A.U.; Saleem, S.; Ali, S.; Abbas, S.M.; Choi, M.; Choi, W. Recent advances in alloying anode materials for sodium-ion batteries: Material design and prospects. *Energy Mater.* **2024**, *4*, 400068. [CrossRef]
146. Jia, X.X.; Liu, C.F.; Neale, Z.G.; Yang, J.H.; Cao, G.Z. Active Materials for Aqueous Zinc Ion Batteries: Synthesis, Crystal Structure, Morphology, and Electrochemistry. *Chem. Rev.* **2020**, *120*, 7795–7866. [CrossRef] [PubMed]
147. Wei, C.; Tao, Y.; Fei, H.; An, Y.; Tian, Y.; Feng, J.; Qian, Y. Recent advances and perspectives in stable and dendrite-free potassium metal anode. *Energy Storage Mater.* **2020**, *30*, 206–227. [CrossRef]
148. Chen, D.M.; Huang, Z.Q.; Sun, S.Q.; Zhang, H.Y.; Wang, W.J.; Yu, G.X.; Chen, J. A Flexible Multi-Channel Hollow CNT/Carbon Nanofiber Composites with S/N Co-Doping for Sodium/Potassium Ion Energy Storage. *ACS Appl. Mater. Interfaces* **2021**, *13*, 44369–44378. [CrossRef]
149. Yao, Y.; Qi, E.; Sun, M.Z.; Wei, Z.X.; Jiang, H.; Du, F. Unlocking the potential of potassium-ion batteries: Anode material mechanisms, challenges, and future directions. *Nanoscale* **2025**, *17*, 19021–19054. [CrossRef] [PubMed]
150. Zhang, C.L.; Zhao, H.P.; Lei, Y. Recent Research Progress of Anode Materials for Potassium-ion Batteries. *Energy Environ. Mater.* **2020**, *3*, 105–120. [CrossRef]
151. Yuan, F.; Li, Y.A.; Zhang, D.; Li, Z.J.; Wang, H.; Wang, B.; Wu, Y.S.; Wu, Y.M.A. A comprehensive review of carbon anode materials for potassium-ion batteries based on specific optimization strategies. *Inorg. Chem. Front.* **2023**, *10*, 2547–2573. [CrossRef]
152. Wang, J.L.; Wang, H.W.; Zang, X.B.; Zhai, D.Y.; Kang, F.Y. Recent Advances in Stability of Carbon-Based Anode for Potassium-Ion Batteries. *Batter. Supercaps* **2021**, *4*, 554–570. [CrossRef]
153. Thakur, A.K.; Ahmed, M.S.; Park, J.; Prabakaran, R.; Sidney, S.; Sathyamurthy, R.; Kim, S.C.; Periasamy, S.; Kim, J.; Hwang, J.Y. A review on carbon nanomaterials for K-ion battery anode: Progress and perspectives. *Int. J. Energy Res.* **2022**, *46*, 4033–4070. [CrossRef]
154. Tang, Z.; Zhou, S.; Huang, Y.; Wang, H.; Zhang, R.; Wang, Q.; Sun, D.; Tang, Y.; Wang, H. Improving the initial coulombic efficiency of carbonaceous materials for Li/Na-ion batteries: Origins, solutions, and perspectives. *Electrochem. Energy Rev.* **2023**, *6*, 8. [CrossRef]
155. Ma, X.; Ji, C.; Li, X.; Liu, Y.; Xiong, X. Red@ Black phosphorus core–shell heterostructure with superior air stability for high-rate and durable sodium-ion battery. *Mater. Today* **2022**, *59*, 36–45. [CrossRef]
156. Tan, S.; Yang, H.; Zhang, Z.; Xu, X.; Xu, Y.; Zhou, J.; Zhou, X.; Pan, Z.; Rao, X.; Gu, Y. The progress of hard carbon as an anode material in sodium-ion batteries. *Molecules* **2023**, *28*, 3134. [CrossRef]

157. Wang, K.; Jin, Y.; Sun, S.; Huang, Y.; Peng, J.; Luo, J.; Zhang, Q.; Qiu, Y.; Fang, C.; Han, J. Low-cost and high-performance hard carbon anode materials for sodium-ion batteries. *ACS Omega* **2017**, *2*, 1687–1695. [CrossRef]
158. Wang, S.; Dai, G.; Yang, H.; Luo, Z. Lignocellulosic biomass pyrolysis mechanism: A state-of-the-art review. *Progress Energy Combust. Sci.* **2017**, *62*, 33–86. [CrossRef]
159. Hu, B.; Zhang, B.; Xie, W.-l.; Jiang, X.-y.; Liu, J.; Lu, Q. Recent progress in quantum chemistry modeling on the pyrolysis mechanisms of lignocellulosic biomass. *Energy Fuels* **2020**, *34*, 10384–10440. [CrossRef]
160. Dahbi, M.; Kiso, M.; Kubota, K.; Horiba, T.; Chafik, T.; Hida, K.; Matsuyama, T.; Komaba, S. Synthesis of hard carbon from argan shells for Na-ion batteries. *J. Mater. Chem. A* **2017**, *5*, 9917–9928. [CrossRef]
161. Xiao, L.; Lu, H.; Fang, Y.; Sushko, M.L.; Cao, Y.; Ai, X.; Yang, H.; Liu, J. Low-defect and low-porosity hard carbon with high coulombic efficiency and high capacity for practical sodium ion battery anode. *Adv. Energy Mater.* **2018**, *8*, 1703238. [CrossRef]
162. Zhen, Y.; Chen, Y.; Li, F.; Guo, Z.; Hong, Z.; Titirici, M.-M. Ultrafast synthesis of hard carbon anode for sodium-ion batteries. *Proc. Natl. Acad. Sci. USA* **2021**, *118*, e2111119118. [CrossRef] [PubMed]
163. Guo, S.; Chen, Y.; Tong, L.; Cao, Y.; Jiao, H.; Qiu, X. Biomass hard carbon of high initial coulombic efficiency for sodium-ion batteries: Preparation and application. *Electrochim. Acta* **2022**, *410*, 140017. [CrossRef]
164. Kitchamsetti, N.; Kim, K.-H.; Han, H.; Mhin, S. Biomass-Derived Hard Carbon Anode for Sodium-Ion Batteries: Recent Advances in Synthesis Strategies. *Nanomaterials* **2025**, *15*, 1554. [CrossRef] [PubMed]
165. Qin, D.; Liu, Z.; Zhao, Y.; Xu, G.; Zhang, F.; Zhang, X. A sustainable route from corn stalks to N, P-dual doping carbon sheets toward high performance sodium-ion batteries anode. *Carbon* **2018**, *130*, 664–671. [CrossRef]
166. Zhang, B.; Ghimbeu, C.M.; Laberty, C.; Vix-Guterl, C.; Tarascon, J.M. Correlation between microstructure and Na storage behavior in hard carbon. *Adv. Energy Mater.* **2016**, *6*, 1501588. [CrossRef]
167. Zhang, X.; Dong, X.; Qiu, X.; Cao, Y.; Wang, C.; Wang, Y.; Xia, Y. Extended low-voltage plateau capacity of hard carbon spheres anode for sodium ion batteries. *J. Power Sources* **2020**, *476*, 228550. [CrossRef]
168. Fan, C.; Zhang, R.; Luo, X.; Hu, Z.; Zhou, W.; Zhang, W.; Liu, J.; Liu, J. Epoxy phenol novolac resin: A novel precursor to construct high performance hard carbon anode toward enhanced sodium-ion batteries. *Carbon* **2023**, *205*, 353–364. [CrossRef]
169. Lu, Y.; Zhao, C.; Qi, X.; Qi, Y.; Li, H.; Huang, X.; Chen, L.; Hu, Y.S. Pre-oxidation-tuned microstructures of carbon anode derived from pitch for enhancing Na storage performance. *Adv. Energy Mater.* **2018**, *8*, 1800108. [CrossRef]
170. Chen, X.; Tian, J.; Li, P.; Fang, Y.; Fang, Y.; Liang, X.; Feng, J.; Dong, J.; Ai, X.; Yang, H. An overall understanding of sodium storage behaviors in hard carbons by an "adsorption-intercalation/filling" hybrid mechanism. *Adv. Energy Mater.* **2022**, *12*, 2200886. [CrossRef]
171. Sun, H.; Zhang, Q.; Ma, Y.; Li, Z.; Zhang, D.; Sun, Q.; Wang, Q.; Liu, D.; Wang, B. Unraveling the mechanism of sodium storage in low potential region of hard carbons with different microstructures. *Energy Storage Mater.* **2024**, *67*, 103269. [CrossRef]
172. Zhang, S.-W.; Lv, W.; Luo, C.; You, C.-H.; Zhang, J.; Pan, Z.-Z.; Kang, F.-Y.; Yang, Q.-H. Commercial carbon molecular sieves as a high performance anode for sodium-ion batteries. *Energy Storage Mater.* **2016**, *3*, 18–23. [CrossRef]
173. Kamiyama, A.; Kubota, K.; Igarashi, D.; Youn, Y.; Tateyama, Y.; Ando, H.; Gotoh, K.; Komaba, S. MgO-template synthesis of extremely high capacity hard carbon for Na-ion battery. *Angew. Chem. Int. Ed.* **2021**, *60*, 5114–5120. [CrossRef]
174. Yin, X.; Lu, Z.; Wang, J.; Feng, X.; Roy, S.; Liu, X.; Yang, Y.; Zhao, Y.; Zhang, J. Enabling fast Na+ transfer kinetics in the whole-voltage-region of hard-carbon anode for ultrahigh-rate sodium storage. *Adv. Mater.* **2022**, *34*, 2109282. [CrossRef] [PubMed]
175. Cheng, D.; Li, Z.; Zhang, M.; Duan, Z.; Wang, J.; Wang, C. Engineering ultrathin carbon layer on porous hard carbon boosts sodium storage with high initial coulombic efficiency. *ACS Nano* **2023**, *17*, 19063–19075. [CrossRef]
176. Yin, T.; Guo, Y.; Huang, X.; Yang, X.; Qin, L.; Ning, T.; Tan, L.; Li, L.; Zou, K. Heteroatom doping strategy of advanced carbon for alkali Metal-Ion capacitors. *Batteries* **2025**, *11*, 69. [CrossRef]
177. Wang, H.; Liu, S.; Lei, C.; Qiu, H.; Jiang, W.; Sun, X.; Zhang, Y.; He, W. P-doped hard carbon material for anode of sodium ion battery was prepared by using polyphosphoric acid modified petroleum asphalt as precursor. *Electrochim. Acta* **2024**, *477*, 143812. [CrossRef]
178. Xie, L.; Tang, C.; Bi, Z.; Song, M.; Fan, Y.; Yan, C.; Li, X.; Su, F.; Zhang, Q.; Chen, C. Hard carbon anode for next-generation Li-ion batteries: Review and perspective. *Adv. Energy Mater.* **2021**, *11*, 2101650. [CrossRef]
179. Cao, W.; Zheng, J.; Adams, D.; Zheng, J.P. Comparative study of the power performance for advanced Li-ion capacitors with various carbon anode. *ECS Trans.* **2014**, *61*, 37. [CrossRef]
180. Liu, L.; Tian, Y.; Abdussalam, A.; Gilani, M.R.H.S.; Zhang, W.; Xu, G. Hard carbons as anode in sodium-ion batteries: Sodium storage mechanism and optimization strategies. *Molecules* **2022**, *27*, 6516. [CrossRef] [PubMed]
181. Bashir, T.; Zhou, S.; Yang, S.; Ismail, S.A.; Ali, T.; Wang, H.; Zhao, J.; Gao, L. Progress in 3D-MXene electrodes for lithium/sodium/potassium/magnesium/zinc/aluminum-ion batteries. *Electrochem. Energy Rev.* **2023**, *6*, 5. [CrossRef]
182. Ge, P.; Fouletier, M. Electrochemical intercalation of sodium in graphite. *Solid State Ion.* **1988**, *28*, 1172–1175. [CrossRef]

183. Vaalma, C.; Buchholz, D.; Weil, M.; Passerini, S. A cost and resource analysis of sodium-ion batteries. *Nat. Rev. Mater.* **2018**, *3*, 18013. [CrossRef]
184. Chen, D.; Zhang, W.; Luo, K.; Song, Y.; Zhong, Y.; Liu, Y.; Wang, G.; Zhong, B.; Wu, Z.; Guo, X. Hard carbon for sodium storage: Mechanism and optimization strategies toward commercialization. *Energy Environ. Sci.* **2021**, *14*, 2244–2262. [CrossRef]
185. Li, W.; Li, J.; Biney, B.W.; Yan, Y.; Lu, X.; Li, H.; Liu, H.; Xia, W.; Liu, D.; Chen, K. Innovative synthesis and sodium storage enhancement of closed-pore hard carbon for sodium-ion batteries. *Energy Storage Mater.* **2025**, *74*, 103867. [CrossRef]
186. Lin, X.; Liu, Y.; Tan, H.; Zhang, B. Advanced lignin-derived hard carbon for Na-ion batteries and a comparison with Li and K ion storage. *Carbon* **2020**, *157*, 316–323. [CrossRef]
187. Qiu, P.; Chen, H.; Zhang, H.; Wang, H.; Wang, L.; Guo, Y.; Qi, J.; Yi, Y.; Zhang, G. Hard Carbon as Anode for Potassium-Ion Batteries: Developments and Prospects. *Inorganics* **2024**, *12*, 302. [CrossRef]
188. Kim, H.; Hyun, J.C.; Jung, J.I.; Lee, J.B.; Choi, J.; Cho, S.Y.; Jin, H.-J.; Yun, Y.S. Potassium-ion storage behavior of microstructure-engineered hard carbons. *J. Mater. Chem. A* **2022**, *10*, 2055–2063. [CrossRef]
189. Zhong, L.; Zhang, W.; Sun, S.; Zhao, L.; Jian, W.; He, X.; Xing, Z.; Shi, Z.; Chen, Y.; Alshareef, H.N. Engineering of the crystalline lattice of hard carbon anode toward practical potassium-ion batteries. *Adv. Funct. Mater.* **2023**, *33*, 2211872. [CrossRef]
190. Lei, H.; Li, J.; Zhang, X.; Ma, L.; Ji, Z.; Wang, Z.; Pan, L.; Tan, S.; Mai, W. A review of hard carbon anode: Rational design and advanced characterization in potassium ion batteries. *InfoMat* **2022**, *4*, e12272. [CrossRef]
191. Wu, F.; Dong, R.Q.; Bai, Y.; Li, Y.; Chen, G.H.; Wang, Z.H.; Wu, C. Phosphorus-Doped Hard Carbon Nanofibers Prepared by Electrospinning as an Anode in Sodium-Ion Batteries. *ACS Appl. Mater. Interfaces* **2018**, *10*, 21335–21342. [CrossRef]
192. Li, Y.; Chen, M.H.; Liu, B.; Zhang, Y.; Liang, X.Q.; Xia, X.H. Heteroatom Doping: An Effective Way to Boost Sodium Ion Storage. *Adv. Energy Mater.* **2020**, *10*, 2000927. [CrossRef]
193. Chen, W.M.; Wan, M.; Liu, Q.; Xiong, X.Q.; Yu, F.Q.; Huang, Y.H. Heteroatom-Doped Carbon Materials: Synthesis, Mechanism, and Application for Sodium-Ion Batteries. *Small Methods* **2019**, *3*, 1800323. [CrossRef]
194. Guo, W.; Li, X.; Xu, J.T.; Liu, H.K.; Ma, J.M.; Dou, S.X. Growth of Highly Nitrogen-Doped Amorphous Carbon for Lithium-ion Battery Anode. *Electrochim. Acta* **2016**, *188*, 414–420. [CrossRef]
195. Wang, L.; Yang, C.L.; Dou, S.; Wang, S.Y.; Zhang, J.T.; Gao, X.; Ma, J.M.; Yu, Y. Nitrogen-doped hierarchically porous carbon networks: Synthesis and applications in lithium-ion battery, sodium-ion battery and zinc-air battery. *Electrochim. Acta* **2016**, *219*, 592–603. [CrossRef]
196. Tanaka, U.; Sogabe, T.; Sakagoshi, H.; Ito, M.; Tojo, T. Anode property of boron-doped graphite materials for rechargeable lithium-ion batteries. *Carbon* **2001**, *39*, 931–936. [CrossRef]
197. Ma, G.Y.; Xiang, Z.H.; Huang, K.S.; Ju, Z.C.; Zhuang, Q.C.; Cui, Y.H. Graphene-Based Phosphorus-Doped Carbon as Anode Material for High-Performance Sodium-Ion Batteries. *Part. Part. Syst. Charact.* **2017**, *34*, 1600315. [CrossRef]
198. Xie, J.M.; Zhuang, R.; Du, Y.X.; Pei, Y.W.; Tan, D.M.; Xu, F. Advances in sulfur-doped carbon materials for use as anode in sodium-ion batteries. *New Carbon Mater.* **2023**, *38*, 305–316. [CrossRef]
199. Li, X.F.; Liu, J.; Zhang, Y.; Li, Y.L.; Liu, H.; Meng, X.B.; Yang, J.L.; Geng, D.S.; Wang, D.N.; Li, R.Y.; et al. High concentration nitrogen doped carbon nanotube anode with superior Li storage performance for lithium rechargeable battery application. *J. Power Sources* **2012**, *197*, 238–245. [CrossRef]
200. Shaker, M.; Ghazvini, A.A.S.; Shahalizade, T.; Gaho, M.A.; Mumtaz, A.; Javanmardi, S.; Riahifar, R.; Meng, X.M.; Jin, Z.; Ge, Q. A review of nitrogen-doped carbon materials for lithium-ion battery anode. *New Carbon Mater.* **2023**, *38*, 247–278. [CrossRef]
201. Yan, J.; Li, H.M.; Wang, K.L.; Jin, Q.Z.; Lai, C.L.; Wang, R.X.; Cao, S.L.; Han, J.; Zhang, Z.C.; Su, J.Z.; et al. Ultrahigh Phosphorus Doping of Carbon for High-Rate Sodium Ion Batteries Anode. *Adv. Energy Mater.* **2021**, *11*, 2003911. [CrossRef]
202. Bommier, C.; Mitlin, D.; Ji, X.L. Internal structure—Na storage mechanisms—Electrochemical performance relations in carbons. *Prog. Mater. Sci.* **2018**, *97*, 170–203. [CrossRef]
203. Neff, T.; Hessdörfer, J.; Bilican, A.; Kolb, L.; Reinert, F.; Krueger, A. Superior sulfur-doped carbon anode for sodium-ion batteries through incorporation of onion-like carbon. *Electrochim. Acta* **2025**, *537*, 146912. [CrossRef]
204. Wu, J.X.; Pan, Z.Y.; Zhang, Y.; Wang, B.J.; Peng, H.S. The recent progress of nitrogen-doped carbon nanomaterials for electrochemical batteries. *J. Mater. Chem. A* **2018**, *6*, 12932–12944. [CrossRef]
205. Jeon, I.Y.; Noh, H.J.; Baek, J.B. Nitrogen-Doped Carbon Nanomaterials: Synthesis, Characteristics and Applications. *Chem-Asian J.* **2020**, *15*, 2282–2293. [CrossRef] [PubMed]
206. Wei, D.C.; Liu, Y.Q.; Wang, Y.; Zhang, H.L.; Huang, L.P.; Yu, G. Synthesis of N-Doped Graphene by Chemical Vapor Deposition and Its Electrical Properties. *Nano Lett.* **2009**, *9*, 1752–1758. [CrossRef] [PubMed]
207. Sun, Y.; Huang, F.Z.; Li, S.K.; Shen, Y.H.; Xie, A.J. Novel porous starfish-like Co O @nitrogen-doped carbon as an advanced anode for lithium-ion batteries. *Nano Res.* **2017**, *10*, 3457–3467. [CrossRef]
208. Gomez-Martin, A.; Martinez-Fernandez, J.; Ruttert, M.; Winter, M.; Placke, T.; Ramirez-Rico, J. An electrochemical evaluation of nitrogen-doped carbons as anode for lithium ion batteries. *Carbon* **2020**, *164*, 261–271. [CrossRef]

209. Li, D.P.; Ren, X.H.; Ai, Q.; Sun, Q.; Zhu, L.; Liu, Y.; Liang, Z.; Peng, R.Q.; Si, P.C.; Lou, J.; et al. Facile Fabrication of Nitrogen-Doped Porous Carbon as Superior Anode Material for Potassium-Ion Batteries. *Adv. Energy Mater.* **2018**, *8*, 1802386. [CrossRef]
210. Cai, D.D.; Wang, S.Q.; Lian, P.C.; Zhu, X.F.; Li, D.D.; Yang, W.S.; Wang, H.H. Superhigh capacity and rate capability of high-level nitrogen-doped graphene sheets as anode materials for lithium-ion batteries. *Electrochim. Acta* **2013**, *90*, 492–497. [CrossRef]
211. Ding, J.; Wang, H.; Li, Z.; Kohandehghan, A.; Cui, K.; Xu, Z.; Zahiri, B.; Tan, X.; Lotfabad, E.M.; Olsen, B.C. Carbon nanosheet frameworks derived from peat moss as high performance sodium ion battery anode. *ACS Nano* **2013**, *7*, 11004–11015. [CrossRef] [PubMed]
212. Chae, Y.J.; Kim, S.O.; Lee, J.K. Employment of boron-doped carbon materials for the anode materials of lithium ion batteries. *J. Alloy Compd.* **2014**, *582*, 420–427. [CrossRef]
213. Yoshio, M.; Wang, H.Y.; Fukuda, K.; Hara, Y.; Adachi, Y. Effect of carbon coating on electrochemical performance of treated natural graphite as lithium-ion battery anode material. *J. Electrochem. Soc.* **2000**, *147*, 1245–1250. [CrossRef]
214. Su, J.C.; Pei, Y.; Yang, Z.H.; Wang, X.Y. Ab initio study of graphene-like monolayer molybdenum disulfide as a promising anode material for rechargeable sodium ion batteries. *RSC Adv.* **2014**, *4*, 43183–43188. [CrossRef]
215. Zhou, L.J.; Hou, Z.F.; Gao, B.; Frauenheim, T. Doped graphenes as anode with large capacity for lithium-ion batteries. *J. Mater. Chem. A* **2016**, *4*, 13407–13413. [CrossRef]
216. Stadie, N.P.; Billeter, E.; Piveteau, L.; Kravchyk, K.V.; Döbeli, M.; Kovalenko, M.V. Direct Synthesis of Bulk Boron-Doped Graphitic Carbon. *Chem. Mater.* **2017**, *29*, 3211–3218. [CrossRef]
217. Joshi, R.P.; Ozdemir, B.; Barone, V.; Peralta, J.E. Hexagonal BC: A Robust Electrode Material for Li, Na, and K Ion Batteries. *J. Phys. Chem. Lett.* **2015**, *6*, 2728–2732. [CrossRef] [PubMed]
218. Kim, C.; Fujino, T.; Hayashi, T.; Endo, M.; Dresselhaus, M.S. Structural and electrochemical properties of pristine and B-doped materials for the anode of Li-ion secondary batteries. *J. Electrochem. Soc.* **2000**, *147*, 1265–1270. [CrossRef]
219. Ling, C.; Mizuno, F. Boron-doped graphene as a promising anode for Na-ion batteries. *Phys. Chem. Chem. Phys.* **2014**, *16*, 10419–10424. [CrossRef] [PubMed]
220. Wan, W.; Wang, H.D. First-Principles Investigation of Adsorption and Diffusion of Ions on Pristine, Defective and B-doped Graphene. *Materials* **2015**, *8*, 6163–6178. [CrossRef]
221. Buqa, H.; Würsig, A.; Vetter, J.; Spahr, M.E.; Krumeich, F.; Novák, P. SEI film formation on highly crystalline graphitic materials in lithium-ion batteries. *J. Power Sources* **2006**, *153*, 385–390. [CrossRef]
222. Ma, X.; Wang, E.G.; Tilley, R.D.; Jefferson, D.A.; Zhou, W. Size-controlled short nanobells: Growth and formation mechanism. *Appl. Phys. Lett.* **2000**, *77*, 4136–4138. [CrossRef]
223. Saurel, D.; Orayech, B.; Xiao, B.W.; Carriazo, D.; Li, X.L.; Rojo, T. From Charge Storage Mechanism to Performance: A Roadmap toward High Specific Energy Sodium-Ion Batteries through Carbon Anode Optimization. *Adv. Energy Mater.* **2018**, *8*, 1703268. [CrossRef]
224. Hong, Z.S.; Zhen, Y.C.; Ruan, Y.R.; Kang, M.L.; Zhou, K.Q.; Zhang, J.M.; Huang, Z.G.; Wei, M.D. Rational Design and General Synthesis of S-Doped Hard Carbon with Tunable Doping Sites toward Excellent Na-Ion Storage Performance. *Adv. Mater.* **2018**, *30*, 1802035. [CrossRef]
225. Li, J.H.; El-Demellawi, J.K.; Sheng, G.; Bjoerk, J.; Zeng, F.S.; Zhou, J.; Liao, X.X.; Wu, J.W.; Rosen, J.; Liu, X.J.; et al. Pseudocapacitive Heteroatom-Doped Carbon Cathode for Aluminum-Ion Batteries with Ultrahigh Reversible Stability. *Energy Environ. Mater.* **2024**, *7*, e12733. [CrossRef]
226. Li, S.N.; Zhao, J.M.; Li, L.L.; Dong, W. Sodium Adsorption and Intercalation in Bilayer Graphene Doped with B, N, Si and P: A First-Principles Study. *J. Electron. Mater.* **2020**, *49*, 6336–6347. [CrossRef]
227. Yan, J.; Li, W.; Feng, P.Y.; Wang, R.X.; Jiang, M.; Han, J.; Cao, S.L.; Wang, K.L.; Jiang, K. Enhanced Na pseudocapacitance in a P, S co-doped carbon anode arising from the surface modification by sulfur and phosphorus with C-S-P coupling. *J. Mater. Chem. A* **2020**, *8*, 422–432. [CrossRef]
228. Liu, Z.G.; Zhao, J.H.; Yao, H.; He, X.X.; Zhang, H.; Qiao, Y.; Wu, X.Q.; Li, L.; Chou, S.L. P-doped spherical hard carbon with high initial coulombic efficiency and enhanced capacity for sodium ion batteries. *Chem. Sci.* **2024**, *15*, 8478–8487. [CrossRef]
229. Zhou, J.H.; Shi, Q.T.; Ullah, S.; Yang, X.Q.; Bachmatiuk, A.; Yang, R.Z.; Rummeli, M.H. Phosphorus-Based Composites as Anode Materials for Advanced Alkali Metal Ion Batteries. *Adv. Funct. Mater.* **2020**, *30*, 2004648. [CrossRef]
230. Xia, Q.B.; Li, W.J.; Miao, Z.C.; Chou, S.L.; Liu, H.K. Phosphorus and phosphide nanomaterials for sodium-ion batteries. *Nano Res.* **2017**, *10*, 4055–4081. [CrossRef]
231. Fu, Y.Q.; Wei, Q.L.; Zhang, G.X.; Wang, X.M.; Zhang, J.H.; Hu, Y.F.; Wang, D.N.; Zuin, L.C.; Zhou, T.; Wu, Y.C.; et al. High-Performance Reversible Aqueous Zn-Ion Battery Based on Porous MnO Nanorods Coated by MOF-Derived N-Doped Carbon. *Adv. Energy Mater.* **2018**, *8*, 1801445. [CrossRef]
232. Qin, X.Y.; Yan, B.Y.; Yu, J.; Jin, J.; Tao, Y.; Mu, C.; Wang, S.C.; Xue, H.G.; Pang, H. Phosphorus-based materials for high-performance rechargeable batteries. *Inorg. Chem. Front.* **2017**, *4*, 1424–1444. [CrossRef]

233. Li, M.-Y.; Wang, Y.; Liu, C.-L.; Zhang, C.; Dong, W.-S. Synthesis of carbon/tin composite anode materials for lithium-ion batteries. *J. Electrochem. Soc.* **2011**, *159*, A91. [CrossRef]
234. Zhao, Y.; Wang, L.P.; Sougrati, M.T.; Feng, Z.; Leconte, Y.; Fisher, A.; Srinivasan, M.; Xu, Z. A review on design strategies for carbon based metal oxides and sulfides nanocomposites for high performance Li and Na ion battery anode. *Adv. Energy Mater.* **2017**, *7*, 1601424. [CrossRef]
235. Yue, L.; Liang, J.; Wu, Z.; Zhong, B.; Luo, Y.; Liu, Q.; Li, T.; Kong, Q.; Liu, Y.; Asiri, A.M. Progress and perspective of metal phosphide/carbon heterostructure anode for rechargeable ion batteries. *J. Mater. Chem. A* **2021**, *9*, 11879–11907. [CrossRef]
236. Shi, H.; Zhang, W.; Wang, J.; Wang, D.; Wang, C.; Xiong, Z.; Wu, J.; Bai, Z.; Yan, X. Scalable synthesis of a porous structure silicon/carbon composite decorated with copper as an anode for lithium ion batteries. *Appl. Surf. Sci.* **2023**, *620*, 156843. [CrossRef]
237. Wu, S.; Ge, R.; Lu, M.; Xu, R.; Zhang, Z. Graphene-based nano-materials for lithium–sulfur battery and sodium-ion battery. *Nano Energy* **2015**, *15*, 379–405. [CrossRef]
238. Wu, S.; Xu, R.; Lu, M.; Ge, R.; Iocozzia, J.; Han, C.; Jiang, B.; Lin, Z. Graphene-containing nanomaterials for lithium-ion batteries. *Adv. Energy Mater.* **2015**, *5*, 1500400. [CrossRef]
239. Shi, Q.; Zhou, J.; Ullah, S.; Yang, X.; Tokarska, K.; Trzebicka, B.; Ta, H.Q.; Ruemmeli, M.H. A review of recent developments in Si/C composite materials for Li-ion batteries. *Energy Storage Mater.* **2021**, *34*, 735–754. [CrossRef]
240. Jin, H.-c.; Sun, Q.; Wang, J.-t.; Ma, C.; Ling, L.-c.; Qiao, W.-m. Preparation and electrochemical properties of novel silicon-carbon composite anode materials with a core-shell structure. *New Carbon Mater.* **2021**, *36*, 390–400. [CrossRef]
241. Ko, M.; Chae, S.; Ma, J.; Kim, N.; Lee, H.-W.; Cui, Y.; Cho, J. Scalable synthesis of silicon-nanolayer-embedded graphite for high-energy lithium-ion batteries. *Nat. Energy* **2016**, *1*, 16113. [CrossRef]
242. Hsiao, C.; Lee, C.; Tai, N. High retention supercapacitors using carbon nanomaterials/iron oxide/nickel-iron layered double hydroxides as electrodes. *J. Energy Storage* **2022**, *46*, 103805. [CrossRef]
243. Huyan, Y.; Chen, J.; Yang, K.; Zhang, Q.; Zhang, B. Tailoring carboxyl tubular carbon nanofibers/MnO_2 composites for high-performance lithium-ion battery anode. *J. Am. Ceram. Soc.* **2021**, *104*, 1402–1414. [CrossRef]
244. Cao, B.; Liu, Z.; Xu, C.; Huang, J.; Fang, H.; Chen, Y. High-rate-induced capacity evolution of mesoporous C@SnO_2@C hollow nanospheres for ultra-long cycle lithium-ion batteries. *J. Power Sources* **2019**, *414*, 233–241. [CrossRef]
245. Mordyuk, B.; Prokopenko, G. Mechanical alloying of powder materials by ultrasonic milling. *Ultrasonics* **2004**, *42*, 43–46. [CrossRef]
246. Dapeng, W.; Xiaodong, L.; Minggang, Z.; Wei, L.; Min, Q. Anisotropic Sm_2Co_{17} nano-flakes produced by surfactant and magnetic field assisted high energy ball milling. *J. Rare Earths* **2013**, *31*, 366–369.
247. Gorrasi, G.; Sorrentino, A. Mechanical milling as a technology to produce structural and functional bio-nanocomposites. *Green Chem.* **2015**, *17*, 2610–2625. [CrossRef]
248. Zhao, Z. Microwave-assisted synthesis of vanadium and chromium carbides nanocomposite and its effect on properties of WC-8Co cemented carbides. *Scr. Mater.* **2016**, *120*, 103–106. [CrossRef]
249. Wei, X.; Wang, X.; Gao, B.; Zou, W.; Dong, L. Facile ball-milling synthesis of CuO/biochar nanocomposites for efficient removal of reactive red 120. *ACS Omega* **2020**, *5*, 5748–5755. [CrossRef] [PubMed]
250. Bor, A.; Jargalsaikhan, B.; Lee, J.; Choi, H. Effect of different milling media for surface coating on the copper powder using two kinds of ball mills with discrete element method simulation. *Coatings* **2020**, *10*, 898. [CrossRef]
251. Joy, J.; Krishnamoorthy, A.; Tanna, A.; Kamathe, V.; Nagar, R.; Srinivasan, S. Recent developments on the synthesis of nanocomposite materials via ball milling approach for energy storage applications. *Appl. Sci.* **2022**, *12*, 9312. [CrossRef]
252. Zhang, Z.; Zhao, X.; Li, J. SnSe/carbon nanocomposite synthesized by high energy ball milling as an anode material for sodium-ion and lithium-ion batteries. *Electrochim. Acta* **2015**, *176*, 1296–1301. [CrossRef]
253. Charinpanitkul, T.; Soottitantawat, A.; Tonanon, N.; Tanthapanichakoon, W. Single-step synthesis of nanocomposite of copper and carbon nanoparticles using arc discharge in liquid nitrogen. *Mater. Chem. Phys.* **2009**, *116*, 125–128. [CrossRef]
254. Rivani, D.A.; Retnosari, I.; Saraswati, T.E. Influence of TiO_2 addition on the magnetic properties of carbon-based iron oxide nanocomposites synthesized using submerged arc-discharge. In *IOP Conference Series: Materials Science and Engineering*; IOP Publishing: Bristol, UK, 2019.
255. Zhang, J.; Yu, A. Nanostructured transition metal oxides as advanced anode for lithium-ion batteries. *Sci. Bull.* **2015**, *60*, 823–838. [CrossRef]
256. Dou, F.; Shi, L.; Chen, G.; Zhang, D. Silicon/carbon composite anode materials for lithium-ion batteries. *Electrochem. Energy Rev.* **2019**, *2*, 149–198. [CrossRef]
257. Liang, S.; Cheng, Y.J.; Zhu, J.; Xia, Y.; Müller-Buschbaum, P. A chronicle review of nonsilicon (Sn, Sb, Ge)-based lithium/sodium-ion battery alloying anode. *Small Methods* **2020**, *4*, 2000218. [CrossRef]
258. Fuchsbichler, B.; Stangl, C.; Kren, H.; Uhlig, F.; Koller, S. High capacity graphite–silicon composite anode material for lithium-ion batteries. *J. Power Sources* **2011**, *196*, 2889–2892. [CrossRef]

259. Xu, L.; Quan, Z.; Wang, F.; Lu, A.; Zhao, Q.; Zhang, W.; Tang, Z.; Dang, D.; Liu, Q.; Zhang, C. Scalable synthesis of nano silicon-embedded graphite for high-energy and low-expansion lithium-ion batteries. *J. Power Sources* **2025**, *656*, 238022. [CrossRef]
260. Ren, J.G.; Wu, Q.H.; Hong, G.; Zhang, W.J.; Wu, H.; Amine, K.; Yang, J.; Lee, S.T. Silicon–Graphene Composite Anode for High-Energy Lithium Batteries. *Energy Technol.* **2013**, *1*, 77–84.
261. Zhang, H.; Zhang, X.; Jin, H.; Zong, P.; Bai, Y.; Lian, K.; Xu, H.; Ma, F. A robust hierarchical 3D Si/CNTs composite with void and carbon shell as Li-ion battery anode. *Chem. Eng. J.* **2019**, *360*, 974–981.
262. Datta, M.K.; Maranchi, J.; Chung, S.J.; Epur, R.; Kadakia, K.; Jampani, P.; Kumta, P.N. Amorphous silicon–carbon based nano-scale thin film anode materials for lithium ion batteries. *Electrochim. Acta* **2011**, *56*, 4717–4723.
263. Demirkan, M.T.; Yurukcu, M.; Dursun, B.; Demir-Cakan, R.; Karabacak, T. Evaluation of double-layer density modulated Si thin films as Li-ion battery anode. *Mater. Res. Express* **2017**, *4*, 106405.
264. Li, X.; Zhang, M.; Yuan, S.; Lu, C. Research progress of silicon/carbon anode materials for lithium-ion batteries: Structure design and synthesis method. *ChemElectroChem* **2020**, *7*, 4289–4302.
265. You, S.; Tan, H.; Wei, L.; Tan, W.; Chao Li, C. Design strategies of Si/C composite anode for lithium-ion batteries. *Chem.-A Eur. J.* **2021**, *27*, 12237–12256. [CrossRef]
266. Cong, R.; Choi, J.-Y.; Song, J.-B.; Jo, M.; Lee, H.; Lee, C.-S. Characteristics and electrochemical performances of silicon/carbon nanofiber/graphene composite films as anode materials for binder-free lithium-ion batteries. *Sci. Rep.* **2021**, *11*, 1283. [CrossRef]
267. Nzabahimana, J.; Liu, Z.; Guo, S.; Wang, L.; Hu, X. Top-down synthesis of silicon/carbon composite anode materials for lithium-ion batteries: Mechanical milling and etching. *ChemSusChem* **2020**, *13*, 1923–1946. [CrossRef]
268. Park, I.; Lee, H.; Chae, O.B. Synthesis Methods of Si/C Composite Materials for Lithium-Ion Batteries. *Batteries* **2024**, *10*, 381. [CrossRef]
269. Li, X.; Yang, D.; Hou, X.; Shi, J.; Peng, Y.; Yang, H. Scalable preparation of mesoporous Silicon@ C/graphite hybrid as stable anode for lithium-ion batteries. *J. Alloy Compd.* **2017**, *728*, 1–9. [CrossRef]
270. Zhang, Y.; Wang, C. Environment-friendly synthesis of carbon-encapsulated SnO_2 core-shell nanocubes as high-performance anode materials for lithium ion batteries. *Mater. Today Energy* **2020**, *16*, 100406. [CrossRef]
271. Cao, Z.; Ma, X. Encapsulated Fe_3O_4 into tubular mesoporous carbon as a superior performance anode material for lithium-ion batteries. *J. Alloy Compd.* **2020**, *815*, 152542. [CrossRef]
272. Bommier, C.; Luo, W.; Gao, W.-Y.; Greaney, A.; Ma, S.; Ji, X. Predicting capacity of hard carbon anode in sodium-ion batteries using porosity measurements. *Carbon* **2014**, *76*, 165–174. [CrossRef]
273. Luo, W.; Jian, Z.; Xing, Z.; Wang, W.; Bommier, C.; Lerner, M.M.; Ji, X. Electrochemically expandable soft carbon as anode for Na-ion batteries. *ACS Cent. Sci.* **2015**, *1*, 516–522. [CrossRef]
274. Qiu, S.; Xiao, L.; Sushko, M.L.; Han, K.S.; Shao, Y.; Yan, M.; Liang, X.; Mai, L.; Feng, J.; Cao, Y. Manipulating adsorption–insertion mechanisms in nanostructured carbon materials for high-efficiency sodium ion storage. *Adv. Energy Mater.* **2017**, *7*, 1700403. [CrossRef]
275. Song, M.; Song, Q.; Zhang, T.; Huo, X.; Lin, Z.; Hu, Z.; Dong, L.; Jin, T.; Shen, C.; Xie, K. Growing curly graphene layer boosts hard carbon with superior sodium-ion storage. *Nano Res.* **2023**, *16*, 9299–9309. [CrossRef]
276. Wang, K.; Sun, F.; Wang, H.; Wu, D.; Chao, Y.; Gao, J.; Zhao, G. Altering thermal transformation pathway to create closed pores in coal-derived hard carbon and boosting of Na+ plateau storage for high-performance sodium-ion battery and sodium-ion capacitor. *Adv. Funct. Mater.* **2022**, *32*, 2203725. [CrossRef]
277. Xu, W.; Li, H.; Zhang, X.; Chen, T.Y.; Yang, H.; Min, H.; Shen, X.; Chen, H.Y.; Wang, J. Regulating Graphitic Microcrystalline and Single-Atom Chemistry in Hard Carbon Enables High-Performance Potassium Storage. *Adv. Funct. Mater.* **2024**, *34*, 2309509. [CrossRef]
278. David, L.; Singh, G. Reduced graphene oxide paper electrode: Opposing effect of thermal annealing on Li and Na cyclability. *J. Phys. Chem. C* **2014**, *118*, 28401–28408. [CrossRef]
279. Yang, G.; Ilango, P.R.; Wang, S.; Nasir, M.S.; Li, L.; Ji, D.; Hu, Y.; Ramakrishna, S.; Yan, W.; Peng, S. Carbon-based alloy-type composite anode materials toward sodium-ion batteries. *Small* **2019**, *15*, 1900628. [CrossRef] [PubMed]
280. Zhang, H.; Hasa, I.; Passerini, S. Beyond insertion for Na-ion batteries: Nanostructured alloying and conversion anode materials. *Adv. Energy Mater.* **2018**, *8*, 1702582. [CrossRef]
281. Zhu, H.; Jia, Z.; Chen, Y.; Weadock, N.; Wan, J.; Vaaland, O.; Han, X.; Li, T.; Hu, L. Tin anode for sodium-ion batteries using natural wood fiber as a mechanical buffer and electrolyte reservoir. *Nano Lett.* **2013**, *13*, 3093–3100. [CrossRef]
282. Nithya, C.; Gopukumar, S. rGO/nano Sb composite: A high performance anode material for Na+ ion batteries and evidence for the formation of nanoribbons from the nano rGO sheet during galvanostatic cycling. *J. Mater. Chem. A* **2014**, *2*, 10516–10525. [CrossRef]
283. Tran, T.T.; Obrovac, M. Alloy negative electrodes for high energy density metal-ion cells. *J. Electrochem. Soc.* **2011**, *158*, A1411. [CrossRef]

284. Liu, Y.; Fan, F.; Wang, J.; Liu, Y.; Chen, H.; Jungjohann, K.L.; Xu, Y.; Zhu, Y., Bigio, D.; Zhu, T. In situ transmission electron microscopy study of electrochemical sodiation and potassiation of carbon nanofibers. *Nano Lett.* **2014**, *14*, 3445–3452. [CrossRef]
285. Wang, Q.; Zhao, X.; Ni, C.; Tian, H.; Li, J.; Zhang, Z.; Mao, S.X.; Wang, J.; Xu, Y. Reaction and capacity-fading mechanisms of tin nanoparticles in potassium-ion batteries. *J. Phys. Chem. C* **2017**, *121*, 12652–12657. [CrossRef]
286. Zhang, W.; Mao, J.; Li, S.; Chen, Z.; Guo, Z. Phosphorus-based alloy materials for advanced potassium-ion battery anode. *J. Am. Chem. Soc.* **2017**, *139*, 3316–3319. [CrossRef]
287. Huang, J.; Lin, X.; Tan, H.; Zhang, B. Bismuth microparticles as advanced anode for potassium-ion battery. *Adv. Energy Mater.* **2018**, *8*, 1703496. [CrossRef]
288. Zhang, Q.; Mao, J.; Pang, W.K.; Zheng, T.; Sencadas, V.; Chen, Y.; Liu, Y.; Guo, Z. Boosting the potassium storage performance of alloy-based anode materials via electrolyte salt chemistry. *Adv. Energy Mater.* **2018**, *8*, 1703288. [CrossRef]
289. Wu, Y.; Huang, H.B.; Feng, Y.; Wu, Z.S.; Yu, Y. The promise and challenge of phosphorus-based composites as anode materials for potassium-ion batteries. *Adv. Mater.* **2019**, *31*, 1901414. [CrossRef] [PubMed]
290. Loaiza, L.C.; Monconduit, L.; Seznec, V. Si and Ge-based anode materials for Li-, Na-, and K-ion batteries: A perspective from structure to electrochemical mechanism. *Small* **2020**, *16*, 1905260. [CrossRef] [PubMed]
291. Sultana, I.; Ramireddy, T.; Rahman, M.M.; Chen, Y.; Glushenkov, A.M. Tin-based composite anode for potassium-ion batteries. *Chem. Commun.* **2016**, *52*, 9279–9282. [CrossRef] [PubMed]
292. Han, C.; Han, K.; Wang, X.; Wang, C.; Li, Q.; Meng, J.; Xu, X.; He, Q.; Luo, W.; Wu, L. Three-dimensional carbon network confined antimony nanoparticle anode for high-capacity K-ion batteries. *Nanoscale* **2018**, *10*, 6820–6826. [CrossRef]

MDPI AG
Grosspeteranlage 5
4052 Basel
Switzerland
Tel.: +41 61 683 77 34

Batteries Editorial Office
E-mail: batteries@mdpi.com
www.mdpi.com/journal/batteries